P9-AQF-641

OPTICAL INFORMATION PROCESSING

Volume 2

OPTICAL INFORMATION PROCESSING

Volume 2

Edited by

Euval S. Barrekette
International Business Machines Corporation
Yorktown Heights, New York

George W. Stroke
State University of New York
Stony Brook, New York

Yu. E. Nesterikhin
Institute of Automation
Academy of Sciences of the USSR, Siberian Branch
Novosibirsk, USSR

and

Winston E. Kock
University of Cincinnati
Cincinnati, Ohio

PLENUM PRESS · NEW YORK AND LONDON

Library of Congress Cataloging in Publication Data

US-USSR Science Cooperative Seminar on Optical Information Processing, 2d, Novosibirsk, Russian S.F.S.R., 1976.
Optical information processing, volume 2.

Sponsored by the National Science Foundation.
Includes index.
1. Optical data processing–Congresses. I. Barrekette, Euval Salomon, 1931-
II. United States. National Science Foundation. III. Title. TA1630.U54 1976
ISBN 0-306-34472-6 621.38'0414 77-17579

Proceedings of the US–USSR Science Cooperation Seminar on Optical Information Processing, sponsored by the National Science Foundation and by the USSR Academy of Sciences-Siberian Branch, held in Novosibirsk, USSR, July 10–16, 1976

A Division of Plenum Publishing Corporation
227 West 17th Street, New York, N.Y. 10011

Printed in the United States of America

Foreword

This is the second volume on "Optical Information Processing" within the scope of the US-USSR Science Cooperation Program co-sponsored by the US National Science Foundation and the USSR Academy of Sciences Siberian Branch.

Volume I was published in 1976, also by Plenum Press, and contained the papers presented by a group of US and USSR scientists at the First US-USSR Science Cooperation Seminar "Optical Information Processing" held at the US National Academy of Sciences in Washington, D.C. from 16 to 20 June 1975. The seminar was followed by a series of visits to US scientific research laboratories and universities, to which the visiting USSR scientists were escorted by Dr. W. E. Kock and Dr. G. W. Stroke. The visits included Bell Laboratories, IBM Thomas J. Watson Research Laboratory and M.I.T., as reported in detail in the FOREWORD of Volume I.

Volume II now presents the papers presented by another group of US and USSR scientists, some having participated in the first seminar: this series of papers was presented at the Second US-USSR Science Cooperation Seminar on "Optical Information Processing" held at the USSR Academy of Sciences Siberian Branch Institute of Automation and Electrometry in the famous "science city" of Akademgorodok, near Novosibirsk in Siberia, USSR from 10 to 16 July 1976. Like its US counterpart, the Siberian seminar was followed by a series of visits to USSR scientific research laboratories and universities, to which the US scientists were escorted by Dr. Yu. E. Nesterikhin and Dr. Voldemar P. Koronkevich with colleagues. The visits included - in addition to the Institute of Automation and Electrometry and its remarkable opto-digital image processing facilities and work, as directed by Academician Yu. E. Nesterikhin - the famous Akademgorodok Institute of Nuclear Physics, where the visit was hosted by Academician A. N. Skrinsky, a participant in our seminar; the A. F. Ioffe Physico-Technical Institute laboratories in Leningrad, directed by Dr. S. B. Gurevich Dr. Yu. I. Ostrovsky and

Dr. M. P. Petrov, Deputy Director of the Institute; the Moscow Kinofoto Institute, where the visit was hosted by Prof. V. Komar and included a remarkable display of white-light reflection holograms; the USSR Academy of Sciences Institute of Radio Engineers and Electronics in Moscow, where the visit was hosted by its Deputy Director, Dr. Yurii V. Gulayev; and the visits were topped by visits to both of the Soviet laser-fusion installations at the Lebedev Institute in Moscow, directed respectively by Nobel Laureates Academicians N. Basov and A. M. Prokhorov.

The US-USSR science cooperation program is under the direction, in the USA, of Dr. John R. Thomas, Program Director of the US-USSR program at the National Science Foundation who also participated in the July 1976 seminar and post-conference site-visits in the USSR. The U.S. organizing chairman for both the 1975 and 1976 US-USSR seminars was Prof. George W. Stroke, with Dr. Winston E. Kock, co-chairman, while the USSR organizing chairman for both the seminars was Prof. Yuri E. Nesterikhin, who co-directs the US-USSR science cooperation program in this field of "optical information processing" for the USSR with Academician A. M. Prokhorov, Nobel Laureate.

The U.S. papers and authors, as scheduled at the July 1976 Akademgorodok seminar were, in their order: G. W. Stroke, SUNY, Stony Brook, "A New Assessment of Optical and Digital Image Processing for Real-World Applications;" W. E. Kock, University of Cincinnati, "Optical Processing in Radar and Sonar;" E. S. Barrekette, IBM Thomas J. Watson Research Laboratory, "Real-Time Optical Processing;" W. K. Pratt, Image Processing Institute, University of Southern California, "Fast Sequential SVD Pseudo-inverse Image Restoration;" M. P. Zampino, Citibank, Operating Group, "Electronic Image Filing Systems;" H. Kogelnik, Bell Laboratories, Holmdel, N.J., "Integrated Optics;" S. H. Lee, University of California, San Diego, "Non-Linear Optical Processing;" A. Kozma, Environmental Research Institute of Michigan (ERIM), Ann Arbor, "Laser Speckle;" and J. C. Urbach, Xerox, Palo Alto, "Thermoplastic and Elastometric Optical Storage Media." The U.S. papers were introduced by an opening address on the NSF US-USSR program by Dr. John R. Thomas, its director.

The USSR papers, and authors, as scheduled at the July 1976 Akademgorodok seminar, were in their order: Yu. E. Nesterikhin, Institute of Automation and Electrometry in Novosibirsk, "Optical Information Storage and Processing;" A. N. Skrinsky, Institute of Nuclear Physics in Novosibirsk, "Synchrotron Holography," S. B. Gurevich, "Informational Characteristics of Space-Time Light Modulators;" A. L. Mikaeliane, "Holographic Memories with Information Recording in Crystals;" V. A. Zverev, Radiotechnical Institute in Gorky, "Optical Processing in Radiophysical Problems;" V. N. Sintsov, S. I. Vavilov Institute in Leningrad, "Optical Aperture Synthesis;"

V. M. Morozov, Lebedev Institute of Physics in Moscow, and I. N. Kompanets, of the same USSR Academy of Sciences Institute, "Optical Data Transformation and Coding in Electro-Optical Processors;" V. K. Bykhovsky, Institute of Control Sciences in Moscow, "Optical Interpolative Memory and Optical Processor Architecture;" P. E. Tverdokhleb, Institute of Automation in Novosibirsk, with I. S. Gibin, Yu. V. Chugui and others, "Multi-Channel Information Retrieval in Non-Coherent Optical Storage Systems" and "Investigation of a Version of a Holographic Character Memory Device;" V. P. Koronkevich, of the same Institute, with a further report on his pioneering work in "Kinoforms" in thin films of chalcogenide vitreous semi-conductors; V. M. Efimov, R. D. Baghlay and E. S. Nezhevenko, all also of the same Institute; and Dr. V. K. Sokolov of the A. F. Ioffe Physico-Technical Institute in Leningrad.

All these US and USSR papers, some with additions and modified titles, and some additional papers are to be found in the present volume. With one exception, all the Soviet papers are printed in the form submitted by their authors without any editing of the language, for the reasons explained in the "FOREWORD" of Volume I, notably also in order to indicate to English-speaking readers the high level of competence in English of the Russian-speaking authors.

The editors thank the contributors for preparing their papers in "camera-ready" form in a timely fashion and hope that the quick presentation of this work will help the readers with an introduction to the rapidly developing ramifications of this field.

E. S. Barrekette

G. W. Stroke

Yu. E. Nesterikhin

W. E. Kock

Contents

X-RAY HOLOGRAPHY OF MICROOBJECTS

A. M. Kondratenko and A. N. Skrinsky

Institute of Nuclear Physics

Novosibirsk 90, USSR

Possibilities of x-ray holography with the use of synchrotron radiation of modern electron storage rings are investigated. The required exposure time of a hologram and also that required of detected quanta are found, both are determined by quantum noises and allowing to choose an optimum reference wave under experiment conditions. The most interesting case of low-contrast objects is specially considered.

1. INTRODUCTION

In recent years a new type of X-ray radiation has appeared. These are the modern storage rings of high energy electrons whose spectrum brightness of synchrotron radiation is about two-three orders higher compared with that in the characteristic lines of the best sharp-focus X-ray tubes with a rotating anode. There is no, technical difficulty to increase a brightness of synchrotron radiation additionally by two-three orders by using the magnetic "snakes" and decreasing the beam size at a point of radiation.

The availability of such bright sources makes us to look in a new fashion at the possibilities of different versions of X-ray microscopy, such as the contrast, projection and scanning microscopies, the diffraction analysis of periodical structures, the X-ray holography.

Our study is devoted to the analysis of X-ray holography. This method of research is attractive due to the one-to-one reconstruction of a wave came from the object under study, including its phase, and has no principal restrictions on the minimum size of the resolved elements of the object up to a wavelength of the radiation used (as well as the diffraction structure analysis). A possibility is important to work with low-contrast objects. After passing through such objects the variations in both an amplitude and phase of the X-ray wave are very small.

The new prospects of X-ray holography is also related with the availability of the coordinate-sensitive detectors of X-ray

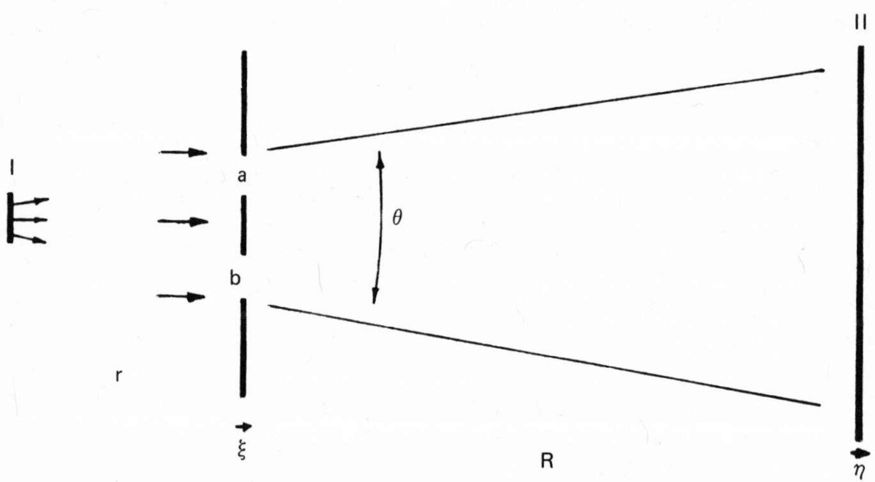

Figure 1. The scheme of a lensless Fourier-transform holography. I - Radiation source, II - Recording plane.

The object, used as an illuminator and generating the known reference radiation, is placed into a diaphragm a. The object under study is into diaphragm b.

radiation with unit-order recording efficiency which are designed on the basis of wire proportional spark chambers (7).

A recording efficiency of such single-quantum detectors is by some orders higher in comparison with that of conventional photodetectors, they have a good spatial resolution (parts of millimetre). The information is introduced directly into a computer which processes and transforms it in the form most convenient to an experimenter.

The Young experiment scheme is the most natural one in X-ray holography. Such a scheme of lensless Fourier-transform holography was proposed by Stroke in 1965 (2). The purpose of this work is to analyze the Young-Stroke scheme, to examine its possibilities and the problems which arise.

2. DESCRIPTION OF THE HOLOGRAPHY SCHEME

In the scheme of a lensless Fourier-transform holography the illuminator is placed closely to the object under study and an angle between their radiation directions towards the points of a remote detector appears to be small (see Fig. 1).

According to an interference picture which is readily detected in a far zone (where the distances between interference maxima can as much as one likes exceed the radiation wavelength λ) the distribution of amplitudes and phases of wave came from the object under study may be reconstructed. The resulting hologram with the angular dimensions $\Theta_1 \times \Theta_2$ makes it possible to distinguish the object components with the following transverse dimensions (in a plane $\vec{\xi}$ perpendicular to the wavefront propagation direction) (3,4):

$$\delta^2 = \delta_1 \cdot \delta_2 = \frac{\lambda}{\Theta_1} \cdot \frac{\lambda}{\Theta_2} .$$

A possible number of the resolved elements is about L^2/δ^2 ($L^2 = L^2 = L_1 = L_2$ are the object sizes in a plane $\vec{\xi}$). If the object is not two-dimensional, then for the same number of resolved elements a longitudinal location of elements along the wavefront propagation can also be specified from a hologram; in a low-angular holography a longitudinal resolution δ_{11} turns out to be considerably worse than the transverse (3,4):

$$\delta_{11} \approx \frac{\lambda}{\Theta^2} .$$

To record an interference picture a detector with a larger number of resolved elements than that of objects in each of directions is needed. A distance to the detector R must be sufficient that its coordinate resolution $h_1 \times h_2$ be enough for resolving the very near maxima:

$$R > \frac{\ell_1 h_2}{\lambda}, \frac{\ell_2 h_2}{\lambda}.$$

In addition, if the following condition is satisfied

$$R > \frac{\ell_1^2}{\lambda}, \frac{\ell_2^2}{\lambda}, \tag{1}$$

The image reconstruction of the objects from the hologram reduces to the Fourier-transform for which the very efficient algorithms have been developed allowing a fast computer processing (see, for example, (5)).

To obtain, in the recording plane, a contrast interference picture allowing to reconstruct an image of the object under study with a given resolution the radiation should have a sufficient degree of monochromaticity and spatial coherence. In case the radiation is not monochromatic (as synchrotron radiation) it is possible spatially separate the wave with length λ, for example, by a crystal-monochromator. The necessary degree of monochromaticity is determined by a possibility to distinguish the nearest interference maxima at a periphery of the hologram. For the objects with sizes $\ell_1 \times \ell_2 \times \ell_{11}$ this means that the allowable relative range of the wavelength $\Delta\lambda/\lambda$ does not exceed the value

$$\Delta\lambda/\lambda < \min \frac{\delta_2}{\ell_1}, \frac{\delta_2}{\ell_2}, \frac{\delta_{11}}{\ell_{11}}, \frac{\delta^2}{\ell_{11}\lambda}.$$

To form an interference picture the object sizes are required to be not larger than the size of a spatial coherence region. The size of a coherence area $a_1 \times a_2$ is determined by the source size $\sigma_1 \times \sigma_2$ and by the distance from it to the objects τ (6):

$$a_1 \approx \lambda \frac{\tau}{\sigma_1}, \quad a_2 \approx \lambda \frac{\tau}{\sigma_2}.$$

If the object sizes ℓ_1, ℓ_2 are more than those in the coherence region a_1, a_2 it is then useful to increase the size of the latter when setting the diaphragm near the source (or increasing a distance τ). Note that, as a rule, the allowable longitudinal object sizes are determined by the time incoherence and are at least Θ^{-1} times larger than the allowable transverse size.

The main source characteristic allowing to evaluate quantitatively its application possibilities in holography is the mutually-coherent radiation power P (7). The power P is equal to a part of the source radiation within the spatial coherence angle $\Delta\Omega \lesssim a_1 a_2/\tau^2$. The value P is proportional to the source brightness and equals the total number of quanta, useful for holography, per second. When diaphragming the total number of mutually-coherent quanta is maintained the same (provided that, of course, the diaphragm size is considerably larger than λ/α_λ, where α_λ is the characteristic angle intervals in which the radiation falling on the diaphragm is concentrated). For a full use of the mutually-coherent quanta the diaphragm size should be choiced in such a way that the transverse sizes of objects be the same as the coherence area size.

3. INTERFERENCE PICTURE AND ITS ANALYSIS

Let us denote the complex amplitudes of the wave of the illuminator and the object under study in a plane $\vec{\xi}$ by $_\tau(\vec{\xi})$ and $(\vec{\xi})$, respectively. The illuminator and the object are placed in the radiation coherence region (see Fig. 1). The value $(| \;|^2 + |\;_2|^2)d\vec{\xi}$ is equal to the number of quanta coming from the objects per time unit. In a far zone with condition (1) fulfilled, the field amplitude in the recording plane is obtained by the Fourier-transform for amplitudes $_2$ and . The formula for the number of quanta arriving, per time unit, at a point $\vec{\eta}$ of the detector placed in the far zone is written as follows

$$dI = |U_\tau(\vec{K}) + U(\vec{K})|^2 d\vec{K} = [|U_2|^2 + |U|^2 + U^*U_\tau + UU^*_\tau]d\vec{K} , \quad (2)$$

where $\vec{K}$ is a wave vector $\vec{K} = 2\pi(\vec{R} + \vec{\eta})/\lambda|\vec{R} + \vec{\eta}|$, $d\vec{K} = K^2 d\Omega \approx$ $K^2\, d\eta_1 d\eta_2/R^2$, the amplitudes $U_\tau = 1/2\pi \int \mathcal{U}_\tau\, e^{-i\vec{K}\vec{\xi}}\, d\vec{\xi}$ and $U = 1/2\pi \int \mathcal{U} e^{-i\vec{K}\vec{\xi}}\, d\vec{\xi}$ are the spatial Fourier-harmonics of images of both the illuminator and object under study. It is clear that an unknown amplitude $\mathcal{U}$ can be reconstructed (if the reference wave amplitude is known) from interference terms, e.g., from the term proportional to UU^*_τ.

As the illustration, let us consider the simplest case when a small hole, whose area is S_τ, is used as an illuminator. Let a value U_τ be zero throughout except the hole region where it is constant. In principle, the hole size limits the resolution over the object under study and, therefore, should be smaller than that of the resolved details of the object.*

---

* If the dependence $\mathcal{U}_\tau(\vec{\xi})$ has been determined previously then, of course, it is possible to reconstruct the amplitude $\mathcal{U}$ whose the hole size is larger than δ. In this example a simpler version is suggested wherein the hole size is much smaller than δ.

In this example the value U^*_τ is

$$U^*_\tau = \frac{1}{2\pi} U^*_\tau S_\tau e^{-i\vec{K}\vec{d}} \tag{3}$$

where $\vec{d}$ is the radius-vector plotted in a plane $\vec{\xi}$ from the center of the object to the hole.

To find the amplitude $\mathcal{U}$ it is necessary to perform the Fourier-transform for the expression $|U_\tau + U|^2$ (known from the measurements of the number of quanta falling, per time unit, on the detector elements) divided by U^*_τ. As a result of this transform, the images corresponding to the other terms of formula (2) appear in addition to the amplitude $\mathcal{U}$ obtained from the interference term UU^*_τ. To spatially separate the image of the object under study it is necessary that the distance d should be longer than 3/2 ℓ_1 (ℓ_1 is the object size in the direction of the vector $\vec{d}$) (3,4):

$$d > 3/2\ \ell_1 \ .$$

The possibility to reconstruct the amplitude $\mathcal{U}$ with the arbitrary reference object wave (of the illuminator) was shown in (8). In optical holography, when reconstructing with the source of a small size one uses the filters produced by a photomethod (9,10). The direct reconstruction is achieved by a computer performing the Fourier-transform for the expression $dI/d\vec{K}$ divided by U^*_τ. To separate the necessary amplitude the distance d between the centers of both the illuminator and object under study is required to be sufficiently large. In a given case, the amplitude $\mathcal{U}$ in the object region equals

$$= \frac{1}{2\pi} {}_{H} \, e^{i\vec{K}\vec{\xi}} \frac{dI}{U^*_\tau} \tag{4}$$

To obtain the complete amplitude of the field came from the object, integration in this formula must be performed over the whole detector region in which $|U| \neq 0$. However, only certain regions of spatial harmonics of the image having the necessary information on the object are often to be examined. Then integration will apparently be carried out over that detector region which is interesting for an experimenter. In this case, the amplitude $\mathcal{U}$ will be an informative part of the complete amplitude of the field came from the object. The investigation of low-contrast objects which will be considered in section 7, is an important example of that kind. The main part of the radiation power of a low-contrast object evidently falls on the central region of the detector. A small power part responsible for the detailed information on the object is connected with a change of the field passed through the object and falls on the detector at larger angles. Usually, therefore, it is useful to neglect the radiation detection in the central region at all (similar to the conventional diffraction analysis).

When reconstructing a wave with the aid of an arbitrary illuminator it is often required to considerably increase the distance d (especially in the cases when on a hologram there are points $\vec{K}$ in which $U_\tau = 0$) what usually reduces the number of the mutually-coherent quanta used of the radiation source. One can do without increasing this distance, provided that several additional transformations will be performed by a computer.

Let us subtract from the expression (2) the terms $|U|^2 d\vec{K}$ and $|U_\tau|^2 d\vec{K}$ determined by separate exposures. As a result, we get, instead of (2), the expression $(U^*U_\tau + UU^*_\tau)d\vec{K}$ from which the necessary amplitude can be separated with the Fourier-transform. It is only sufficient that

$$d > 1/2\ (\ell_1 + \ell_1^\tau)\ .$$

(ℓ_1^τ is the illuminator size in the direction $\vec{d}$). This condition is satisfied even if the illuminator is near the object during this exposure.

If, for some reasons, it is impossible to determine a value $|U|^2$ with the required accuracy the Fourier-transform can be performed by a computer in the way similar to that proposed in optical holography (11). The Fourier-transform is first performed without dividing by U^*_τ:

$$\int e^{i\vec{K}\vec{\xi}}\, dI = [\mathcal{U}(\vec{\xi} + \vec{\xi}')\,\mathcal{U}^*_\tau(\vec{\xi}') + \ldots + \mathcal{U}_\tau(\vec{\xi} + \vec{\xi}')\,\mathcal{U}^*(\vec{\xi}')]d\vec{\xi}'.$$

Under condition

$$d > 1/2\ (\ell_1 + \ell_1^\tau) + \max\ (\ell_1, \ell_1^\tau) \qquad (5)$$

The autocorrelation of images of both the illuminator and object can be separat4d and the following Fourier-transform can be performed only over it

$$\frac{1}{4\pi^2}\quad e^{-i\vec{K}\vec{\xi}}\ \mathcal{U}(\vec{\xi} + \vec{\xi}')\,\mathcal{U}^*_\tau\ (\vec{\xi}')d\vec{\xi}d\vec{\xi}' = UU^*_\tau\ .$$

It is the Fourier-harmonics of this expression divided by U^*_τ that will be the unknown amplitude, separated under condition (5).

4. EXPOSURE TIME REQUIRED AND NUMBER REQUIRED OF DETECTED QUANTA

In principle, the reconstructed amplitude of the radiation field will be different from the true one. This difference may be connected, for example, with insufficient number of elements of the detector, its parasitic light, insufficient knowledge of the illuminator field, partial radiation incoherence, the finite number of detected quanta, etc.

Let us consider the question on the number of quanta which are required to be detected, and also on the exposure time necessary for the reconstruction of the wave amplitude with a given accuracy. This question is the most actual in the X-ray range where the limits for the source brightness are the most essential and becomes real the single-quantum detection. The formula for the required exposure time and required number of detected quanta makes it possible to estimate a realizability of different versions of X-ray holography under certain conditions.

It is quite obvious that the infinite number of detected quanta is required to faithfully reconstruct the field amplitude. Let us calculate the average-probable value of the reconstructed amplitude $\hat{\mathcal{U}}$ and its average deviation at the finite exposure time T. Let us assume that the detector has "infinite" dynamic detection range and its region is divided by a large number of the (resolved) elements whose coordinates are $\vec{\eta}$. Let N_n quanta be detected in n-cell during time T. Under $T \to \infty$ the ratio N_n/T will tend to the value

$$\frac{N_n}{T} \to \varepsilon I_n$$

where $I_n = |U + U^*_\tau|^2 \Delta\vec{K}_n$ is the power of the radiation falling on n-cell, ε is the recording efficiency (the ratio of detected quanta from those falling on the detector). In the investigated object region the reconstructed amplitude will be calculated with the formula (4) which can be rewritten in the form

$$\hat{\mathcal{U}} = \frac{1}{2\pi\varepsilon T} \int_H e^{i\vec{K}\vec{\xi}} \frac{dN}{U^*_\tau} = \frac{1}{2\pi\varepsilon T} \sum_n \frac{N_n}{U^*_\tau(\vec{K}_n)} e^{i\vec{K}_n\vec{\xi}} . \qquad (6)$$

Under $T \to \infty$ the amplitude tends to .

It is well known that the probability $W(N_n)$ of recording a certain number N_n of quanta in a cell during time T does not depend on the number of detections in the other cells and it is determined by the Poisson distribution* for which

*We neglect "non-quantum" changes in the wave field since time T considerably exceeds the time of their correlations determined by the coherence time and by the times of the radiation power changes (times

$$\overline{N}_n = \underset{N_n}{\int} N_n W(N_n) = \varepsilon I_n T$$

$$\overline{N_n^2} - \overline{N}_n^2 = \varepsilon I_n T$$

Thus, an average-probable value $\overline{\hat{\mathcal{U}}}$ is equal to $\mathcal{U}$, as it should be, and coincides with the amplitude of the field obtained within the limit $T \to \infty$. For the average square deviation $\hat{\mathcal{U}}$ from $\mathcal{U}$ we have

$$D \equiv |\mathcal{U}|^2 - |\mathcal{U}|^2 = \frac{1}{\delta^2 \varepsilon T} \left(1 + \left\langle \left| \frac{U}{U_\tau} \right|^2 \right\rangle\right), \tag{7}$$

where $\delta^2 = 4\pi^2 [d\vec{K}]^{-1}$ is the area of a minimum resolved element, the brackets $\langle \ldots \rangle$ denote the averaging over a hologram:

$$\left\langle \left| \frac{U}{U_\tau} \right|^2 \right\rangle = \int_H \left| \frac{U}{U_\tau} \right|^2 d\vec{K} \cdot \left(\int_H d\vec{K}\right)^{-1}.$$

From formula (7) it is seen that the dispersion $\sqrt{D}$ determining the level of quantum noises has the same value at all the points $\vec{\xi}$ in the investigated object region.

A given level $D\hat{\mathcal{U}}$ determines the exposure time and the number of quanta to be detected. The allowable value $D\hat{\mathcal{U}}$ is, of course, given by the specific purposes of research and the object properties. Without essential restrictions this value may be chosen to be equal to

$$D\hat{\mathcal{U}} = \frac{\partial e^2}{S} \int_H |U|^2 d\vec{K} = \frac{\partial e^2}{\Sigma} \int |\mathcal{U}|^2 d\vec{\xi} = \partial e^2 \langle |\mathcal{U}|^2 \rangle \tag{8}$$

where $S = \ell_1 \ell_2$ is the investigated object area. A small number ∂e is a free parameter giving the accuracy which is necessary to know an informative amplitude $\mathcal{U}$, and showing how many times the average-probable deviation $\hat{\mathcal{U}}$ from $\mathcal{U}$ must be smaller than the average level $\sqrt{\langle |\mathcal{U}|^2 \rangle}$. A value ∂e determines the statistical accuracy of reconstructing the amplitude $\mathcal{U}$ from quantum noises.

By formulas (7) and (8) we can find the required exposure time T and the number of quanta $N = \sum_n N_n$, which is required to be detected for reconstructing the amplitude with an accuracy ∂e:

$$N = \frac{M}{\partial e^2} \left(1 + \frac{I_\tau}{I_o}\right) \left(1 + \left\langle \left| \frac{U}{U_\tau} \right|^2 \right\rangle\right), \tag{9a}$$

$$T = \frac{M}{\varepsilon \partial e^2 I_o} \left(1 + \left\langle \left| \frac{U}{U_\tau} \right|^2 \right\rangle\right), \tag{9b}$$

---

for the power change of the source are of the order of the reversion period of an electron bunch in a storage ring for synchrotron radiation). Therefore, by I_n in all the formulas we mean average values of the radiation power falling on n-cell.

where $M = \frac{S}{\delta^2}$ is the number of resolved elements of the object, $I_\tau = \int_H |U_\tau|^2 d\vec{K}$ and $I_o = \int_H |U|^2 d\vec{K}$ are the numbers of quanta falling on the hologram, per time unit, from both the illuminator and object.

Thus, knowing the value $|U|^2$ (e.g., radiating the detector during a short time by the investigated object radiation) one can determine time T and the number of quanta N. These values become infinitely large if there exist the hologram regions in which $U_\tau = 0$. This is connected with the fact that the values U in these regions are left unknown in the image reconstruction from the interference term UU_τ^*. The illuminator should, therefore, illuminate the whole hologram region needed for the reconstruction. On the other hand, for a full use of the reference radiation it must, as far as possible, be concentrated only on this informative region.

Formulas (9) allow one to find an optimum illuminator, as well as an optimum ratio between the flux of quanta from the illuminator and object. In optimal case, in order to reconstruct an amplitude the minimum detected quanta (and minimum computer time) is required. Investigation of formula (9a) over the extremum shows that the field of an optimum illuminator must satisfy the condition

$$|U_\tau|^2 = \sqrt{<|U|^2>}\,|U|. \tag{10}$$

An optimum ratio of the fluxes of quanta falling on the hologram is expressed by the formula

$$\left(\frac{I_\tau}{I_o}\right)_{opt} = \frac{<|U|>}{\sqrt{<|U|^2>}} . \tag{11}$$

In this case, the number of detected quanta which is necessary for the reconstruction is minimum and equals

$$N_{opt} = \frac{M}{\partial e^2}\left(1 + \frac{<|U|>}{\sqrt{<|U|^2>}}\right)^2 \tag{12a}$$

It is seen from the formula (11) that in optimum case the quantum flux on the hologram from the illuminator does not exceed that from the object:

$$0 < (I_\tau/I_o)_{opt} < 1 .$$

Usually $(I_\tau/I_o)_{opt}$ has the order of unity, however, if almost the whole quantum flux falls on a very small region of the hologram area, then $(I_\tau/I_o)_{opt} \ll 1$.

To fulfill the condition (11) it is necessary that the quantum fluxes from the source should fall on the illuminator and the object under study in a certain ratio. Let us find the quantum fluxes falling on the illuminator F_τ and the object F_o by using the whole possible flux of mutually-coherent quanta:

$$F_\tau + F_o = F = \text{const}$$

Taking into account the condition (11), it is easy to obtain that

$$(F_\tau)_{opt} = \frac{q_o \langle |U| \rangle}{q_o \langle |U| \rangle + q_\tau \sqrt{\langle |U|^2 \rangle}} F \ ,$$

$$(F_o)_{opt} = \frac{q_\tau \sqrt{\langle |U|^2 \rangle}}{q_o \langle |U| \rangle + q_\tau \sqrt{\langle |U|^2 \rangle}} F \ , \qquad (14)$$

where $q_\tau = I_\tau/F_\tau$ and $q_o = I_o/F_o$ are the coefficients of using the radiations, i.e. fractions of quanta both from the illuminator and object, falling on the hologram from the number of those falling on these objects.

With (14) the formula for the required exposure time in an optimum case is written in the form

$$T_{opt} = \frac{M}{\varepsilon \partial e^2 q_o F} \left(1 + \frac{\langle |U| \rangle}{\sqrt{\langle |U|^2 \rangle}}\right) \left(1 + \frac{q_o}{q_\tau} \frac{\langle |U| \rangle}{\sqrt{\langle |U|^2 \rangle}}\right) . \qquad (12b)$$

As seen, if the reference radiation is used completely, (at $q_\tau = 1$), then time T_{opt} is the smallest.

Investigation of formula (9b) over the extremum under condition (13) shows that time T is minimum at the same illuminator structure ($|U_\tau|^2 \sim |U|$) and the ratio I_τ/I_o is equal to

$$\left(\frac{I_\tau}{I_o}\right)_{T=T_{min}} = \sqrt{\frac{q_\tau}{q_o}} \frac{\langle |U| \rangle}{\sqrt{\langle |U|^2 \rangle}} . \qquad (15)$$

In this case,

$$N_{T=t_{min}} = \frac{M}{\partial e^2} \left(1 + \sqrt{\frac{q_\tau}{q_o}} \frac{\langle |U| \rangle}{\sqrt{\langle |U|^2 \rangle}}\right) \left(1 + \sqrt{\frac{q_o}{q_\tau}} \frac{\langle |U| \rangle}{\sqrt{\langle |U|^2 \rangle}}\right) ,$$

$$T_{min} = \frac{M}{\varepsilon \partial e^2 q_\tau q_o F} \left(\sqrt{q_\tau} + \sqrt{q_o} \frac{\langle |U| \rangle}{\sqrt{\langle |U|^2 \rangle}}\right)^2 . \qquad (16)$$

The number of quanta N and the exposure time T are minimum simultaneously only at $q_\tau = q_o$. The value T_{opt} is always of the same order as T_{min} irrespective of the ratio q_τ and q_o:

$$T_{min} < T_{opt} < \left(1 + \frac{\langle |U| \rangle}{\sqrt{\langle |U|^2 \rangle}}\right) T_{min} < T_{min}$$

The number $N_{T=T_{min}}$ of detected quanta required significantly exceeds N_{opt} with essential difference between values q_τ and q_o. For an illuminator with a full use of the falling radiation ($q_\tau = 1$) time T_{min} becomes smaller, and $N_{T=T_{min}}$ may either be increased or decreased.

Production of optimal illuminators is technically difficult and sometimes impossible task. But there are a lot of illuminators when using of which the required exposure time and the required number of detected quanta insignificantly exceed their optimum values. Usually, it is sufficient that the characteristic sizes of the illuminator elements be of the order of the resolution δ, and the quantum fluxes falling on the informative hologram region from both the object and illuminator be of the same order.

Nearly optimum is the case when the reference object (generating the reference wave) is similar to the object under study. Then, from formulas (9), substituting $|U_\tau| = |U|$ we get

$$N = \frac{4M}{\partial e^2}\ , \quad T = \frac{2M}{\varepsilon \partial e^2 I_o}$$

The choice of a reference radiation similar to that under study may be useful for the investigation of the very small variations in objects.

5. SMALL REFERENCE HOLE CASE

Let us consider an example of practical importance when a small hole is used as an illuminator. Such an illuminator is the most appropriative for the examination of various objects. In this case, in the hologram region $|U_\tau|$ = const and from (9) we get the following formulas for the required number of detected quanta and the required exposure time

$$N = \frac{M}{\partial e^2}\frac{(I_\tau + I_o)^2}{I_\tau I_o}\ , \quad T = \frac{M}{\varepsilon \partial e^2}\left(\frac{1}{I_o} + \frac{1}{I_\tau}\right). \tag{17}$$

Only a fraction $I_\tau = \frac{S_\tau}{\partial^2} F_\tau$ of the total radiation power F_τ falling on the hologram region from the small hole is used, therefore, the size of the this hole is desired to be as near as possible to that of the resolved element ($q_\tau = S_\tau/\partial^2 \approx 1$).

The formulas of the previous section for an optimum illuminator can be applied in the case under consideration after substitution

$<|U|>/\sqrt{<|U|^2>} \to 1$. So, for example, the number of quanta is minimum, if $I_\tau = I_o$ (compare with (11)). Then, the formulas for N and T have the form (compare with (12)):

$$N_{opt} = \frac{4M}{\partial e^2}, \quad T_{opt} = \frac{2M}{\varepsilon \partial e^2} \frac{q_\tau + q_o}{q_\tau q_o} . \tag{18}$$

In particular, if the whole detector area is used in the interference picture analysis then in an optimum case for phase-contrast (low-absorbing) objects ($q_o = 1$) one half of the flux of mutually-coherent quanta must pass through the reference hole, and the other half through the object under study.

The object field recording with a small hole has some disadvantages connected with the small reference wave power* (except the cases of the very small values of q_o). If $I_\tau \ll I_o$ then the required number of detected quanta and the required exposure time are significantly exceed optimum values:

$$N = \frac{M}{\partial e^2} \frac{q_o F_o}{q_\tau F_\tau}, \qquad T = \frac{M}{\varepsilon \partial e^2} \frac{1}{q_\tau F_\tau} . \tag{19}$$

In this case, the detector is overloaded with quanta of the object under study. If the reference wave power increase is impossible, it is useful to decrease the power I_o to a value of the order of I_τ, placing the radiation "attenuator," which reduces a value q_o, in front of the object. The exposure time, of course, remains the same.

The reference wave power increase can be achieved with the aid of the radiation focussing on the small hole, i.e., the illuminator. The Fresnel zone plates can serve as an example showing the X-ray focussing possibility when using the diffraction phenomena (as known, the use of the lenses similar to the optical, is impossible because of a very small distinction between the refraction index and the unity in X-ray range). The width of the narrowest (external) ring should be equal to δ-diameter of the illuminator hole, and the plate area is determined by a portion of

---

*Note that it is impossible to decrease the values N and T with the simple increase of the hole size, in spite of that the radiation power increases proportionally to the hole area, since in this case, the illumination of only a small central detector region will be intensified not allowing to obtain the required resolution for the details with the sizes smaller than those of this hole.

the flux of mutually-coherent quanta which must pass through the illuminator.

6. THE OTHER VERSIONS OF ILLUMINATORS

The use of the reference objects of the same sizes as the object under study may become the means of the reference wave power increase. If the illuminator field is unknown previously, it is then possible to know it from the hologram obtained, for example, during a long single exposure with a small reference hole (or with the help of another known illuminator).

An important version of such reference objects for the investigation of arbitrary objects is a diffraction illuminator with the randomly located scattering centers which scatters the radiation uniformly throughout the hologram. Such a "random" diffractor is equivalent to diffusers applied in optical holography. The randomly located holes, absorbing "inclusions," different irregularities of the material surfaces etc. can diffuse the radiation. The characteristic size of the scattering components of an optimum diffractor should be of the same order as that of the required resolution.

In the treatment of an interference picture recorded with a "not-small" reference object the pecularities appear which are connected with the existence of points $\vec{K}$ in which an amplitude U_T becomes zero and in which it is impossible to reconstruct an amplitude U during the finite time of exposure. Small neighborhoods of these points should not be taken into account in the interference picture treatment. Ejected neighborhoods should be chosen in such a way as, on the one hand, to distort the object image to a small degree and, on the other, to increase insignificantly the required exposure time and the required number of detected quanta.

It should be said a few words about one-dimensional holography when it is sufficient to know the dependence of a wave amplitude only on the single coordinate (e.g., in the investigation of the field of linear objects). In such cases, in all formulas for time T and number N of quanta, the value M, instead of S/δ^2, will be evidently equal to ℓ_1/δ_1. In one-dimensional holography a slit may be used similarly to a reference hole. Instead of the two-dimensional Fresnel plates, if it is needed, the one-dimensional plates may be used. Thus, to obtain the spatial resolution over the single degree of freedom which is just the same as that over two degree of freedom and with the same radiation source, the exposure time and the number of detected quanta are required to be significantly smaller.

7. INVESTIGATION OF LOW-CONTRAST OBJECTS

For a large class of objects (e.g., biological) both the amplitude and phase of a X-ray wave suffer small alterations after passing through these objects. The important advantage of holography lies the fact that the contrast of a hologram does not directly depend on that of an object and can be high even for very low-contrast objects.

For a low-contrast weakly-absorbing object almost the total power F_o falls on a central part of the detector region at the angles of the order of λ/ℓ. A small part $F'_o(F'_o << F_o)$, connected with a small amplitude variation of the field came from the object, falls on the detector at the significantly larger angles $\lambda/\delta << \lambda/\ell$.

Let us first consider the case of recording the total radiation of objects and the whole interference picture analysis. Note that $q_o = 1$. It is necessary to choose the dispersion $\sqrt{D\mathfrak{U}}$ sufficiently small so that the level of quantum noises should not exceed the small variations under consideration of an amplitude $\mathfrak{U}$: $\partial e^2 = \frac{F'_o}{F_o} \partial e_o^2 << \partial e_o^2$. In the recording by an optimum illuminator satisfying condition (10) the required number N and the required time T, as seen from (12), equal ($q_\tau = 1$):

$$N_{opt} = \frac{M}{\partial e_o^2} \frac{F_o}{F'_o} , \quad T_{opt} = \frac{M}{\varepsilon \partial e_o^2 F'_o} . \tag{20}$$

When obtaining (20) the fact has been used that a value $\frac{<|U|>}{\sqrt{<|U|^2>}}$ is small since almost the total radiation power falls on the central small hologram region

$$\frac{<|U|>}{\sqrt{<|U|^2>}} \approx \frac{1}{\sqrt{M}} + \sqrt{\frac{F'_o}{F_o}} << 1 .$$

In a given case, the quantum flux falling on the illuminator is equal approximately to (see (14))

$$(F_\tau)_{opt} \approx (\frac{1}{\sqrt{M}} + \sqrt{\frac{F'_o}{F_o}}) F$$

and consists of a small fraction of the total flux falling on these objects.

In the recording with the small hole the optimum version occurs if $I_\tau = I_o = F_o$, and the flux on the illuminator should be not smaller than that on the object under study. In this version, number N and time T are approximately four times larger compared with those when recording with an optimum illuminator.

When recording and analyzing the whole interference picture of objects the hologram turns out to be overloaded with the information on the small spatial frequencies of an image (on the object form) which is often no interest.* In these cases, it is useful to avoid the radiation detection in the central region at all. Then an amplitude , from formula (4), will be equal to only a small informative part of the whole amplitude of the field came from the object. In the recording with a small hole the illumination of the hologram from the hole should be equal, as before, to the illumination from the object, in an optimum variant. But now this means that $q_\tau F_\tau = F'_o$, i.e. through the hole the quantum flux should certainly be $F'_o/q_\tau F_o$ times smaller than that through the object. Number N and time T, as seen from (18), are equal to $(q_o = F'_o/F_o \ll q_\tau = S/\delta^2,\ \partial e = \partial e_o)$:

$$N_{opt} = \frac{4M}{\partial e_o^2}, \quad T_{opt} = \frac{2M}{\varepsilon \partial e_o^2 F'_o}.$$

In the recording by an optimum illuminator the values N_{opt}, T_{opt} and the ratio of fluxes F_τ and F_o will be approximately the same as those given above (in simple cases a value $\langle|U|\rangle/\sqrt{\langle|U|^2\rangle}$ will be of the order of unity when eliminating the central region).

Thus, in the analysis the elimination of the central part of the detector allows to reduce considerably (F_o/F'_o times the necessary number of detected quanta. With optimum illumination the reference wave power is required to be F_o/F'_o times smaller compared with the case of a small hole.

The contrast of interference fringes of the informative region of the hologram becomes totally contrasted. It is desirable to use a diaphragm, into which the object under study is placed, with the most smooth dependence of the X-ray absorbtion index at the edges. Such a diaphragm produces minimum parasitic illumination of a hologram. The limiting resolution over the contrast of an object is determined by the background scattering on the object material.

8. ACCOUNT OF PARTIAL INCOHERENCE OF FALLING RADIATION

Of interest is to examine how a partial radiation incoherence may affect the results. As known, a time incoherence decreases the interference picture contrast towards the periphery of a hologram thereby restricting the limiting resolution over the object which becomes equal approximately to $l\ \Delta\lambda/\lambda$. In order to distinguish

*A form of the object under study (from the central hologram region) should be examined separately during a special (and shorter) exposure.

details of smaller sizes it is necessary to increase sharply the exposure time and the number of detected quanta (inversely proportional to the contrast square of the interference fringes at the periphery of a hologram). A spatial incoherence decreases the interference picture contrast throughout the hologram. This also leads to the necessity to increase time T and number N of quanta for the investigation of the details of the object.

Let us restrict ourselves the quantitative analysis of the simplest case wth a small reference hole for the partially spatially-incoherent radiation. Let $\Gamma(\vec{\xi})$ be a complex degree of the spatial radiation coherence between a point of the object $\vec{\xi}$ and the reference hole.

Now a value $\overline{\mathfrak{u}}$ defined by averaging formula (6) will be equal $\Gamma\mathfrak{u}$. Instead of condition (8) we have

$$|\overline{\hat{\mathfrak{u}}}|^2 - |\Gamma\mathfrak{u}|^2 = \partial e^2 \langle|\Gamma\mathfrak{u}|^2\rangle = \frac{\partial e^2}{S}\int|\Gamma\mathfrak{u}|^2 d\vec{\xi} \quad . \tag{21}$$

Thus, formulas for the exposure time and the needed number of detected quanta, instead of (17), are written as follows

$$T = \frac{M}{\varepsilon\partial e^2}\,\frac{(1 + I_o/I_\tau}{\int|\Gamma\mathfrak{u}|^2 d\vec{\xi}} \quad , \quad N = \frac{M}{\partial e^2}\,\frac{(I_\tau + I_o)^2}{I_\tau\int|\Gamma\mathfrak{u}|^2 d\vec{\xi}} \quad .$$

In case of complete coherence of the falling radiation $|\Gamma| = 1$ these formulas apparently concide with (17). When decreasing the value $|\Gamma|$ time T and number N increase about $|\Gamma|^{-2}$ times.

9. QUANTITATIVE EXAMPLES

To represent the orders of the values which may be under consideration today let us consider the following example. Assume that the storage ring VEPP-3 (INP, Novosibirsk) is a radiation source. Its present-day X-ray mutually-coherent radiation power P equals (7):

$$P = 10^8 \frac{\Delta\lambda}{\lambda}\lambda^2(\text{Å})\ \frac{\text{quanta}}{\text{sec}}$$

Let in the object under study with a contrast of the order of unity $M = 50 \times 50 = 2{,}5.10^3$ details be required to resolved. An interference picture will be recorded by a detector with the recording efficiency close to unity (designed on the basis of wire spark proportional chambers) of 20×20 cm^2 and with spatial resolution $0{,}5 \times 0{,}5$ mm^2. This detector will be placed at 10 meters from objects.

Let us first consider the case when the recording is carried out with a small hole of the size δ. For the radiation with a wavelength $\lambda = 1\mathring{A}$ the resolution over the object is as large as the value $\delta^2 = 50 \times 50\ \mathring{A}^2$, the longitudinal resolution is about 0,25 microns. If to choose $\partial e = 0,1$ then the number of quanta to be detected, as follows from formulas (17), equals $N \approx 10^6$ quanta (if a power I_o is decreased down to approximately I_τ). From formula (19) for the exposure time T we get ($F_\tau \approx P/2M_\tau$, $\Delta\lambda/\lambda \approx 1/50$):

$$T \approx 10 \text{ minutes}$$

For the radiation with a wavelength $\lambda = 10\mathring{A}$ the resolution $\delta^2 = 500 \times 500\ \mathring{A}^2$, the longitudinal resolution is about 2,5 microns, the required number of detected quanta is the same, the needed exposure time is

$$T \approx 6 \text{ sec}$$

If the recording is carried out by a hole with optimum focussing by a zone plate, the number of quanta N is the same, the required exposure time is $T \approx 1$ sec for the radiation with a wavelength 1 $\mathring{A}$, $T \approx 10^{-2}$ sec for the radiation with a wavelength 10 $\mathring{A}$. When studying low-contrast objects the required exposure times will be increased about F_o/F'_o times.

It should be noted that in the near future the mutually-coherent radiation power of the storage-ring VEPP-3 is assumed to be increased about 100 times due to the use of magnetic "snakes" and the beam size decrease at a point of radiation.

10. PROBLEMS OF X-RAY HOLOGRAPHY AND METHODS OF SOLUTION

One of the most important components of the X-ray holography scheme is a reference object. In the simplest version it is a small hole of the size δ. However, the difficulties already appear in mechanical production of the foil holes of micron sizes. To increase the power of a reference wave it is reasonable to use the Fresnel zone plate with the width of the external, narrowest ring δ which focusses the total needed radiation on a hole. The optical quality of both the plate and hole may be low: it affects only the efficiency of using the flux of mutually-coherent quanta rather than the reconstruction accuracy. Acceptable is the use of a "random" diffractor with δ sizes of details as a reference object. In production such optical components the difficulties are just the same as those confronted in production of the size δ.

The production problem of illuminators can be solved, for example, by means of modern microelectronics. So, by using the technology of the modern ultra-violet or X-ray microphotolithography a layer from heavy atoms can be superimposed on a light foil, say, from berillium, not absorbing X-rays. It is possible to "embed" heavy atoms in a light foil matrix with the aid of ion optics of high resolution.

When using wide-band (not monochromatic) radiation the necessity for monochromatization arises. An ideal monochromator must separate from the spectrum an interval $(\Delta\lambda/\lambda)_{max}$ of the order of inverse magnitude of the number of the object's components resolved in one direction without making the spatial radiation coherence worse. These requirements for a crystal-monochromator often turn out to be incompatible (for such monochromators $\Delta\lambda/\lambda$ is usually equal approximately to the angle spread introduced in the flux of quanta: $\Delta\lambda/\lambda \approx \theta_M$). Such a situation occurs in low-angular holography if a monochromator is placed in front of the object under study. To keep the spatial coherence the angle spread $\Delta\theta_M$ should not exceed the angle spread of incidence from a source. But in this case, a value $\Delta\lambda/\lambda$ becomes considerably smaller than the allowable $(\Delta\lambda/\lambda)_{max}$ and the power losses of mutually-coherent quanta are at least $\delta/\lambda \approx \theta^{-1}$ times.

There are two most efficient position of a monochromator (when the flux of the usefulquanta has practically no losses): near a source and in front of a detector. If $(\Delta\lambda/\lambda)_{max}$ is less than the radiation angles of a source α_λ it is reasonable to place the monochromator near a source, namely, at τ_M distance which is not more than $\sigma(\Delta\lambda/\lambda)^{-1}_{max}$ (σ is the source size).

The distance τ_M appears to be at least δ/λ times shorter than the distance to the objects τ. In this case, the introduced angle spread $\Delta\theta_M$ does not exceed the angle spread of the radiation arriving at the monochromator and the spatial coherence degree does not decrease.

In cases when a wider spectrum range $((\Delta\lambda/\lambda)_{max} > \alpha_\lambda)$ is desired to be used or for technical reasons it is impossible to place the monochromator near the source it is then useful to place it not far from the detector, when there is no requirement of keeping the spatially-coherent properties of radiation. It is only necessary that within the distance from the monochromator to the detector the neighboring interference maxima of the monochromator should not intermix. Such a position of the monochromator makes it possible to maintain the flux of the useful quanta, but sharply increases an irradiation of the object under study with parasitic quanta.

Important for the development of X-ray microholography is the problem to design single-quantum coordinate-sensitive detectors of

X-ray radiation on the basis, for example, of wire spark proportional chambers and coordinate-sensitive semi-conductive detectors whose recording efficiency considerably exceeds that of photodetectors. Detectors can apparently be a matrix with the gaps between the recording centers. Already by now one-coordinate single-quantum detectors with 0,1 mm spatial resolution and two-coordinate with 0,7 x 0,7 mm^2 resolution with 30% recording efficiency (on a wavelength of about 1 Å) have been produced on the basis of wire chambers. Such detectors have a high-speed response (load is about 10^6 quanta/sec) and a possibility to operate online with a computer (1).

A principal step will be the development of detectors measuring not only the coordinates of quanta, but simultaneously their energies with a sufficiently high energy resolution. The semi-conductive detectors measuring the energy of quanta with a good resolution (of the order of 1%) have been already now produced. But they record all the quanta within the input diaphragm. One can hope that a complex detector, made with the modern solid-state microelectronics combining the resolution of two kinds mentioned above will be created. This enables us to do without monochromators when using the synchrotron radiation. A possibility will arise to use all the spatially-coherent quanta of a source and also to obtain a complete holographic information on an object over all the wavelength range at once.

The authors acknowledge the useful discussions with G. N. Kulipanov in the course of carrying out this study.

REFERENCES

1. S. Ye. Baru, G. D. Mokul'skaia, M. A. Mokul'sky, V. A. Sidorov, A. G. Habahpashev. Dokl. Acad. Nauk SSSR, 227, 82 (1976).

2. G. W. Stroke, Appl. Phys. Lett. 6, 201 (1965).

3. G. W. Stroke, "An Introduction to Coherent Optics and Holography," Academic Press, New York, 1966.

4. L. M. Soroko, "Principles of Holography and Coherent Optics," "Hauka," Moscow, 1971.

5. "Holography, Methods and Apparatus," Editors: V. M. Ginzburg, B. M. Stepanova, Chapter 6, Moscow, 1974.

6. M. Born, E. Wolf, "Principles of Optics," Pergamon Press, Oxford, 1965.

7. A. M. Kondratenko, A. N. Skrinsky, "Optics and Spectroscopy," (Sov. J.) 42, 378 (1977).

8. G. W. Stroke, R. Restrick, A. Funkhouser, D. Brumm, Phys. Lett. 18, 274 (1965).

9. G. W. Stroke, R. G. Zech, Phys. Lett. A25, 89 (1967).

10. A. W. Lohmann, H. W. Werlich, Phys. Lett. A25, 570 (1967).

11. G. W. Stroke, M. Halioua, Phys. Lett. A33, 3 (1970).

3-D IMAGE RECONSTRUCTION AND DISPLAY IN X-RAY CRYSTALLOGRAPHY AND IN ELECTRON MICROSCOPY USING HOLOGRAPHIC OPTO-DIGITAL COMPUTING

GEORGE W. STROKE and MAURICE HALIOUA

State University of New York at Stony Brook

Stony Brook, N.Y.11794

There is a need in crystallography for displaying 3-D structures directly from crystallographic data(structure-factor amplitudes with determined phases,e.g.as obtained with the aid of computerized diffractometers) without(or prior to) manual model building[1-7](*).This is required notably for large molecules in X-ray crystallography and should be helpful also in electron microscopy of viruses and other biological structures[8-9].

For example,even the most powerful purely digital computation methods have been unable so far to produce 3-D display of actual images of large molecules directly in the sense of optical imaging. Purely optical methods used in the past(the so-called "optical transform methods"[10-13]) have not provided the hoped-for solution of such image display in X-ray crystallography,because,unfortunately, they only provided scrambled projections of the structure images,e.g. onto a plane,since they started from precession photographs(or their computer equivalent),with the phases superimposed in various fashions,i.e. from sections in the reciprocal(i.e.structure-factor or Fourier space) which by optical 2-D Fourier transformation produce,unfortunately,only scrambled projections of the structure,and not the required unscrambled sections.

When a molecular biologist needs to visualize the 3-D structure of a molecule,he has to compute a series of maps of the electron density distribution(rather than optical images) in sections,section by section,throughout the molecule,then transcribe the maps(which look somewhat like

(*)For additional background and references see [15-18].

topographical maps in geography)onto large plastic sheets, locate the atomic peaks or other recognizable features visually(with the aid of prior chemical knowledge,using human pattern recognition rather than automated optical imaging) and finally,with the aid of a two-way beam-splitting mirror(the famous "Richards'box"[4,14])construct a model by hand,fitting wooden parts or balls("atoms")to the recognized parts of the structure or the atomic peaks. Other methods, also starting from computed structures,make use of computerized interactive graphics for displaying projected views(often in the form of stereographic projections of packing diagrams)onto various types of display tubes(storage CRT-cathode ray tube,refresh CRT or,most recently,digital write-only memory plasma display panels[19]). Anaglyph(e.g.two-color) stereo-view printing has also been widely used(see e.g.[4]).Such methods continue to be useful, as far as they go,but they have the evident and well-recognized limitations,as described,most notably their lack of automatic pattern recognition which should result from opto-digital imaging,as we have now already shown to be the case for the small molecule studied as our first test so far[15-18](see FIG.1.and FIG.2).

Our new opto-digital 3-D structure computing-display method produces actual unscrambled images of the atoms in the crystal structure sections,section by section[15-17],in the hoped-for sense of the Bragg "X-ray microscope"(but now with unscrambled images in sections of the structure, in contrast to the scrambled projections of the structure as produced by the so-called "optical transform methods" in the past),and,moreover, a further recent extension[18] now has permitted us also to produce a hologram(3-D superposition hologram) displaying the complete 3-D structure of the molecule,with the atoms radiating diffusely,as if one were looking at the actual molecule visually,with enormous magnifications and highest possible resolution,and this without any manual model building.

FIG.1. shows the principal steps of our "hybrid opto--digital computation & display"method for the computation and display of 3-D structures(on the right) and which permits us,first of all,as shown,to obtain the unscrambled images of sections[15-18] in comparison to methods requiring manual model building(on the left).The basic principles of our method with full mathematical details and references are given in our references [15-17]and further details are being prepared for publication in our reference[18].

FIG.2. shows in S_1 and S_2 two sections (images of the identified atoms Mg,O,C and Br) of a triclinic crystal magnesium bromide tetrahydrofuran complex $MgBr_2 \cdot 4(C_4H_8O) \cdot 2H_2O$ reconstructed according to our method of FIG.1.[15-18]

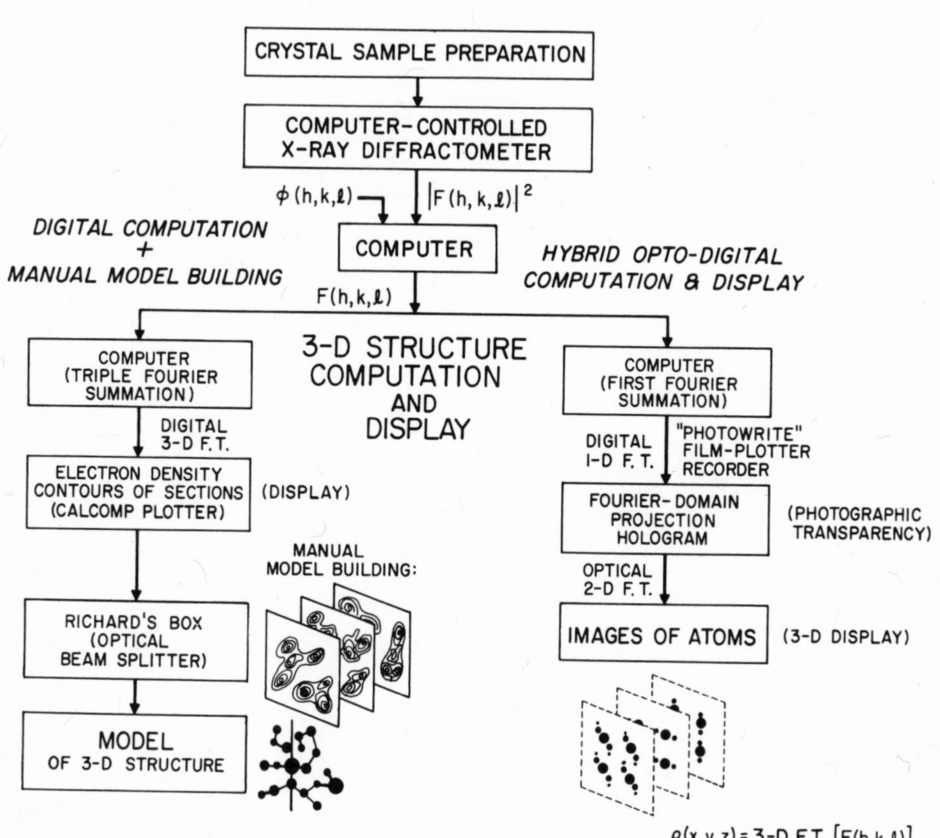

FIG.1. SCHEMATIC DIAGRAM SHOWING NEW OPTO-DIGITAL METHOD FOR 3-D STRUCTURE COMPUTATION AND DISPLAY IN COMPARISON TO PURELY DIGITAL COMPUTATION METHOD(WHICH REQUIRES MANUAL "MODEL BUILDING"TO ACHIEVE COMPARABLE DISPLAY).(From G.W.Stroke,M.Halioua,R.Sarma and V.Srinivasan,IEEE Proc.65,589-591[Apr.1977]).

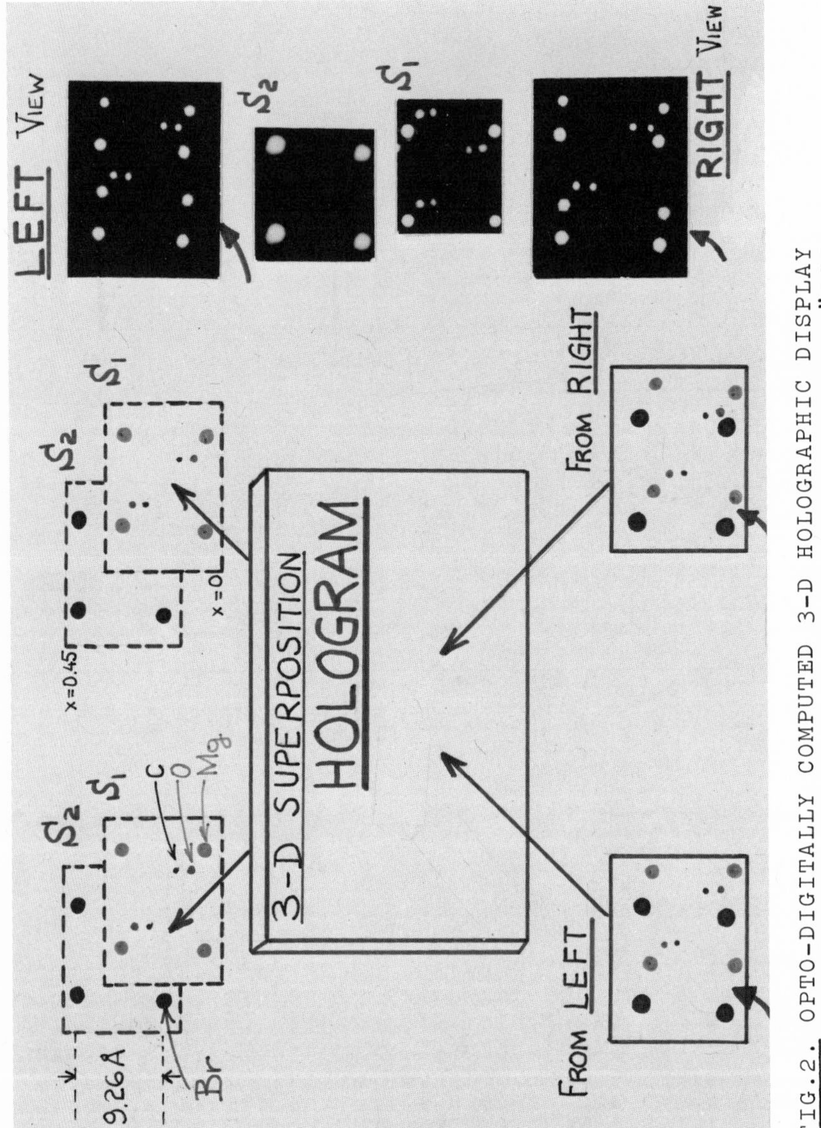

FIG.2. OPTO-DIGITALLY COMPUTED 3-D HOLOGRAPHIC DISPLAY USING "FOURIER-DOMAIN PROJECTION HOLOGRAPHY"ACCORDING TO THE PRINCIPLES DESCRIBED BY G.W.Stroke,M. Halioua,R.Sarma and V.Srinivasan(Ref.15-18)[SEE TEXT].

together with two early tests("LEFT View" and "RIGHT View") demonstrating the possibility of displaying such sections in a full 3-D structure diffusely[18]using the principles of "superposition holography"[20]. Some fundamental relations between X-ray crystallography and holography which are basic to this work were already given by one of us (GWS) in the early 1960's[21].

ACKNOWLEDGMENTS. It is a pleasure to acknowledge collaboration in this work with our colleagues V.SRINIVASAN and R.SARMA and to thank the NATIONAL SCIENCE FOUNDATION,Division of Mathematical and Computer Sciences,for invaluable support and guidance of this work(under NSF Grant MCS--76-11010). One of us(GWS) also further wants to acknowledge with much gratitude the kind interest,encouragement,constructive criticism and many fruitful suggestions from several colleagues and friends,notably from DR.J.R.PASTA, DR.A.KLUG, SIR JOHN KENDREW,Nobel Laureate, PROF.W.HOPPE, DR.J.GASSMANN,DR.JACK B.KINSINGER,DR.C.N.YANG,Nobel Laureate and PROF.MARTIN J.BUERGER,among others.Assistance with some aspects of the experiments in FIG.2. by our graduate student SHELDON CANTOR is also gratefully acknowledged.

REFERENCES

[1] A.McPHERSON JR.,"The Analysis of Biological Structuture with X-Ray Diffraction Techniques",pp.117-240 in Principles and Techniques of Electron Microscopy, Vol.6,Edited by M.A.Hayat(New York,1976:Van Nostrand Reinhold Co.).

[2] J.C.KENDREW,G.BODO,H.M.DINTZIS,R.G.PARRISH,H.WYCKOFF and D.C.PHILLIPS,"A Three-Dimensional Model of the Myoglobin Molecule Obtained by X-Ray Diffraction", Nature,181,662- (1958).

[3] A.C.T.NORTH,"X-Ray Crystallography of Large Molecules of Biological Importance",pp.107-145 in Progress in Biophysics,"Biophysics", (New York,1969:W.A.Benjamin).

[4] H.W.WYCKOFF,D.TSERNOGLOU,A.W.HANSON,J.R.KNOX AND F.M. RICHARDS,"The Three-Dimensional Structure of Ribonuclease-S",J.of Biological Chem.,245,305-328(1970).

[5] R.SARMA and G.ZALOGA"Structure Studies on Styren-treated Immunoglobin Crystals",J.Mol.Biol.98,479-484(1975).

[6] J.W.BECKER,G.N.REEKE JR.,B.A.CUNNINGHAM and G.M.EDELMAN, "New Evidence on the Location of the Saccharide-Binding Site of Concanavalin A",Nature,259,406-409(1976).

[7] R.A.PULLEN,D.G.LINDSAY,S.P.WOOD,I.J.TICKLE & T.L.BLUNDELL, A.WOLLMER,G.KRAIL,D.BRANDENBURG & H.ZAHN,J.GLIEMANN & S.GAMMELTOFT,"Receptor-Binding Region of Insulin", Nature,259,369-373(1976).

[8] D.J.DE ROSIER and A.KLUG,"Reconstruction of Three-Dimensional Structures from Electron Micrographs" Nature,217,130-134(1968).

[9] W.HOPPE,R.LANGER,G.KNESCH and C.H.POPPE,"Protein-Kristalstrukturanalyse mit Elektronenstrahlen",Naturwissenschaften,55,333-336(1968).

[10] W.L.BRAGG,"A New Type of X-Ray Microscope",Nature, 143,678- (1939).

[11] M.J.BUERGER,"Optically Reciprocal Gratings and Their Application to Syntheses of Fourier Series",Proc. Nat.Acad.Sci.,US,27,117-124(1941);"The Photography of Atoms in Crystals",ibid.,36,330-335(1950).

[12] H.LIPSON and C.A.TAYLOR,"X-Ray Crystal Structure Determinations as a Branch of Physical Optics" in Progress in Optics,Vol.5,Edited by E.Wolf,pp.289--348(Amsterdam,1966:North Holland Publishing Co.).

[13] H.S.LIPSON,Editor,Optical Transforms(New York,1972: Academic Press).

[14] F.M.J.RICHARDS,J.Mol.Biol.37,225- (1968).

[15] G.W.STROKE and M.HALIOUA,"Three-Dimensional Reconstructions in X-Ray Crystallography and Electron Microscopy by Reduction to Two-Dimensional Holographic Implementation",Trans.Amer.Crystallographic Assoc. 12,27-41(1976)[First presented by G.W.Stroke on 19 Jan.1976 upon invitation at the Symposium on "Instrumentation for Tomorrow's Crystallography" of the Amer.Crystallographic Assoc.in Clemson,N.C.).

[16] G.W.STROKE,M.HALIOUA,F.THON and D.WILLASCH,"Image Improvement and Three-Dimensional Reconstruction Using Holographic Image Processing"(invited paper), IEEE Proc. 65, 39-62(January 1977).

[17] G.W.STROKE,M.HALIOUA,R.SARMA and V.SRINIVASAN,IEEE Proc.65,589-591(April 1977).

[18] G.W.STROKE,M.HALIOUA,R.SARMA and V.SRINIVASAN(being completed for publication).

[19] M.D.GLICK,T.J.ANDERSON,W.A.BUTLER and E.R.COREY, "Interactive Graphics for Structural Chemistry", Computers & Chemistry,Vol.1,pp.75-78(London,1976: Pergamon Press).[One of us,GWS,wishes to thank Prof. Milton J.Glick and Prof.Frank H.Westervelt of Wayne State University for private communications and several constructive suggestions in April 1977].

[20]a.G.W.STROKE,An Introduction to Coherent Optics and Holography(New York,1969:Academic Press,Second Ed.).

b.D.GABOR and G.W.STROKE et al,"Optical Image Synthesis (Complex Amplitude Addition and Subtraction)by Holographic Fourier Transformation",Physics Lett.18,116-(1965).

[21]a.G.W.STROKE,"Attainment of High Resolutions in Image-Forming X-Ray Microscopy with Lensless Fourier-Transform Holograms and Correlative Source-Effect Compensation",in X-Ray Optics and Microanalysis(Fourth International Congress on X-Ray Optics and Microanalysis,Orsay 1965:invited paper),Edited by R.Castaing,(Paris,1966:Hermann Publ.)pp.30-46.

b.G.W.STROKE,"A Reformulated General Theory of Holography"(see notably Section III."Holographic Image Reconstruction from Holograms Recorded without Separated Coherent Background[X-Ray Crystallography Applications]"):invited paper,presented at "Symposium on Modern Optics"at Polytechnic Institute of Brooklyn,22-24 March 1967(Brooklyn,N.Y.,1967:Polytechnic Press).

c.P.TOLLIN,P.MAIN,M.G.ROSSMANN,G.W.STROKE and R.C.RESTRICK,"Holography and Its Crystallographic Equivalent", Nature.209,603-604(1966).

PHASE QUADRATURE TECHNIQUES IN HOLOGRAPHY

WINSTON E. KOCK

DIRECTOR, THE HERMAN SCHNEIDER LABORATORY
and VISITING PROFESSOR
THE ENGINEERING COLLEGE
THE UNIVERSITY OF CINCINNATI

ABSTRACT

Various extensions of Keating's complex (sine-cosine) hologram procedure are discussed. These include phase-quadrature zone plates and gratings and extensions of the use of quarter-wave plates in applications involving circularly polarized waves.

INTRODUCTION

For simplicity we here give the name <u>phase quadrature</u> to those procedures which involve quarter-wavelength additions. In acoustic holography, the use of two sets of hologram information, a cosine set and a sine set, has been described by Keating and his associates at the Bendix Research Laboratories (1). Referred to as a <u>complex on-axis</u> hologram without conjugate image, this use of a phase-quadrature technique avoids the requirement for an off-axis reference beam in obtaining, on reconstruction, a separated, single, image.

We here discuss, as one extension of phase quadrature, a technique for obtaining circularly polarized waves, where, as has been noted (2), metal phase-quadrature plates were successfully used (Fig. 1) in generating circularly polarized microwaves (3,4). The success of microwave radars using circular polarization in seeing through rain has been discussed (5,6), and, at the first U.S.-U.S.S.R. Symposium (7), in 1975, a similar procedure for light waves was described by the author (8) for seeing through fog and for holography uses. We here describe further experimental results in this program (9) and discuss an extension in which two phase-quadrature structures are used.

We also discuss the application of Keating's phase quadrature technique for improving the efficiency of one of the simplest holograms, the zone plate. A comparable procedure for improving the efficiency of a grating is described.

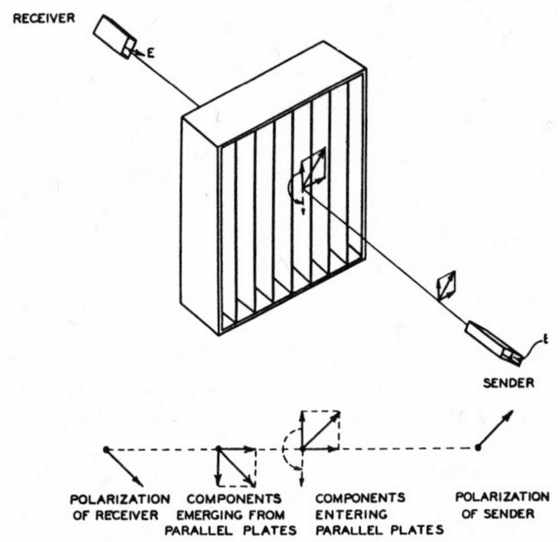

Fig. 1 Rotation of polarization by means of a metal plate waveguide structure. If the structure is made half as thick, the incident linearly polarized waves emerge as circularly polarized waves.

Phase Quadrature Holography

An important extension of the use of holography in detection systems was devised by P. N. Keating and his associates at the Bendix Research Laboratories (1). Through the utilization of a complex (phase quadrature) technique, a hologram sonar detection system was enabled to reconstruct, even with an on-axis reference signal, only the real image, eliminating both the conjugate imagery and the D.C. (undeviated) term.

In hologram sonar, arrays are generally used, and if the information at the array elements is simply mixed with a reference signal (as in an optical hologram), optimum utilization is not realized. This is because (with an off-axis reconstruction) half of the spatial bandwidth is used to accomodate the conjugate image, or, (with an on-axis reference), the two images are inextricably mixed (as in an on-axis optical hologram).

In the earlier Bendix underwater viewing system (Fig. 2), as described (10) in March 1969, each element of an array comprising 400 receivers was sampled, yielding a cosine term for each receiver. An off-axis reference was synthesized to separate the real and conjugate images, following Leith and Upatnieks (11). In the more recent phase-quadrature technique, both sine and cosine terms were utilized (Fig. 3); this permitted the use of an on-axis reference with no conjugate image or D.C. term being generated in the reconstruction. Figure 4 portrays the reconstruction of a small source. The zero order term, had it been present, would have been above and to the right of the white square image. Also, no conjugate image contribution is observed.

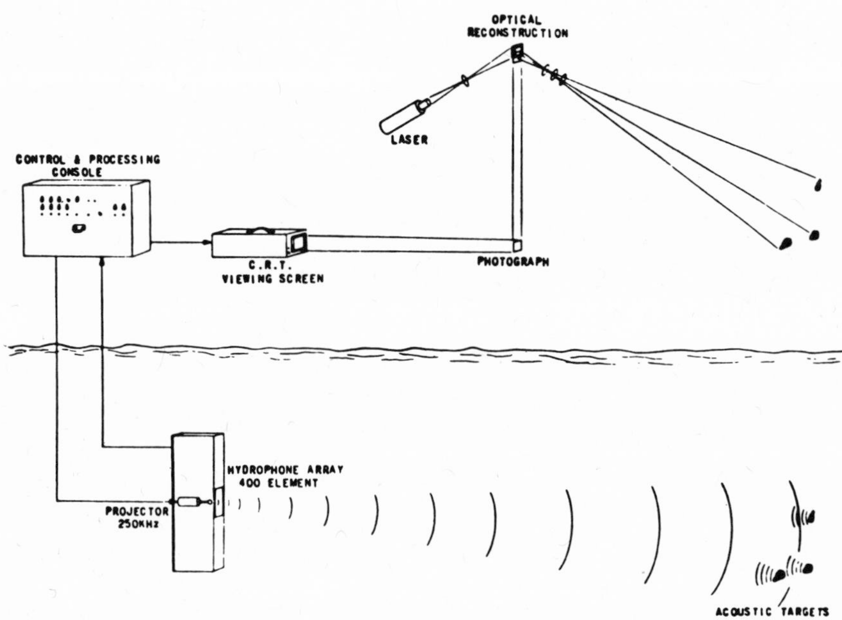

Fig. 2 Elements of the Bendix holographic sonar system.

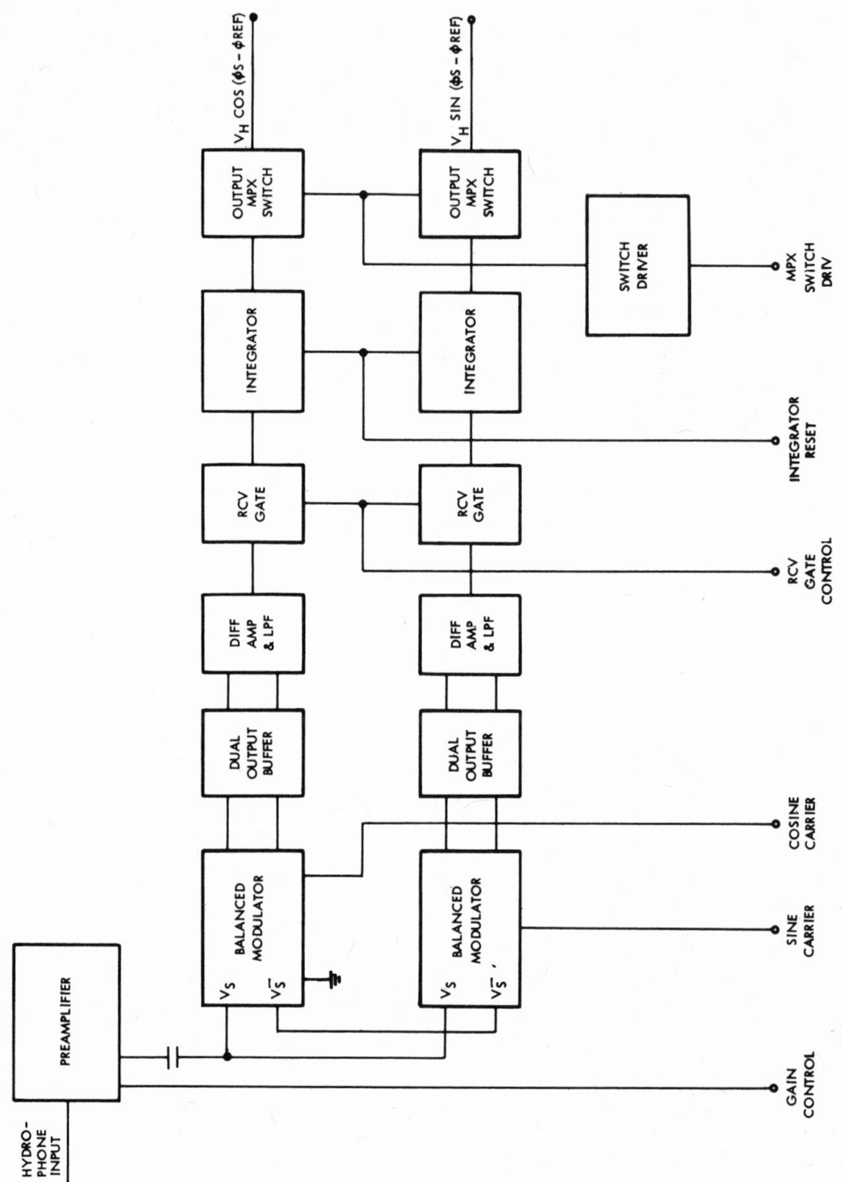

Fig. 3 Alternate (Sine/Cosine) Channel Processor

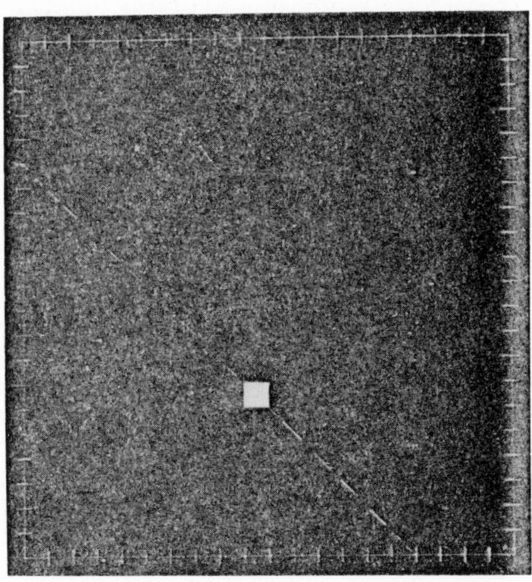

Fig. 4 Reconstructed image of a small source. No zero order contribution of conjugate image contribution is seen, even though an on-axis reference beam was used.

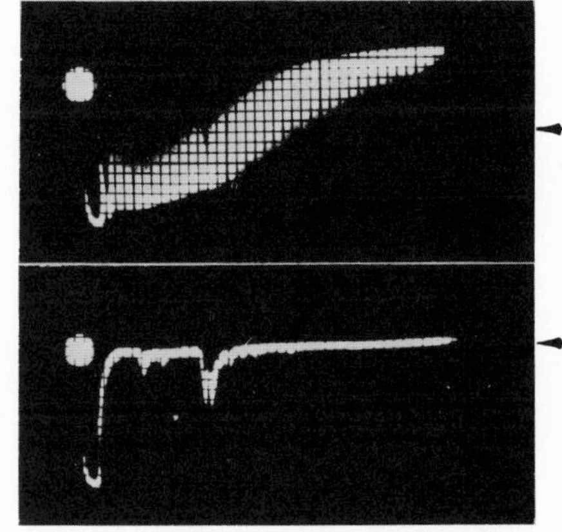

Fig. 5 Suppression of rain echoes is accomplished in a microwave radar by changing the outgoing signal polarization from linear to circular. The plot is amplitude (vertically downward) vs. range (horizontally to the right); top, linear polarization; bottom, circular polarization.

This result is particularly important to the system's detection sensitivity or gain, since all high frequency sonar systems are faced with first-circuit noise as their sensitivity limiting factor (rather than external noise). By avoiding the diversion of received energy into the conjugate image and the D.C. term, the detection gain of a hologram system is made to match closely that of high gain detection systems.

Circular Polarization

It is known that a radar set operating with circular polarization can be used in such a way that the unwanted echoes from symmetrical objects such as raindrops can be suppressed. Presently, circular polarizers are also used in optics where they are valuable in reducing specular backscatter. Linearly polarized waves are converted, by a phase-quadrature (quarter-wave) plate, to circularly polarized waves. Waves reflected from symmetrical objects (such as raindrops) return to the transmitter with the wrong sense of circular polarization and therefore they cannot enter the receiver. Signals reflected from non-symmetrical objects, will, on the other hand experience, on reflection, varying degrees of depolarization; they therefore will possess a component which can enter the receiver. This effect is shown, for a microwave radar, in Figures 5 and 6.

At light wavelengths, this use of circular polarization could prove useful in seeing through fog or mist (12). Thus, if the ratio of the size of fog droplets to rain drops is comparable with the ratio of the size of light wavelengths, the use of circular polarization here should yield a result comparable with that shown in Figures 5 and 6. This would permit objects otherwise obscured by fog or mist to be made more visible.

Depending on the effectiveness of this procedure at light wavelengths, it may be possible for circular polarization to provide improved nighttime visibility for pocket flashlights, automobile headlights, etc., during fog or mist conditions. The availability of several different wavelength narrowband optical filters would permit the transmitted light to match more closely the droplet size that exists in a particular situation, thereby providing the maximum visibility.

In an early experiment using a sheet of Polaroid circular polarizer material HNCP37, placed in the path of

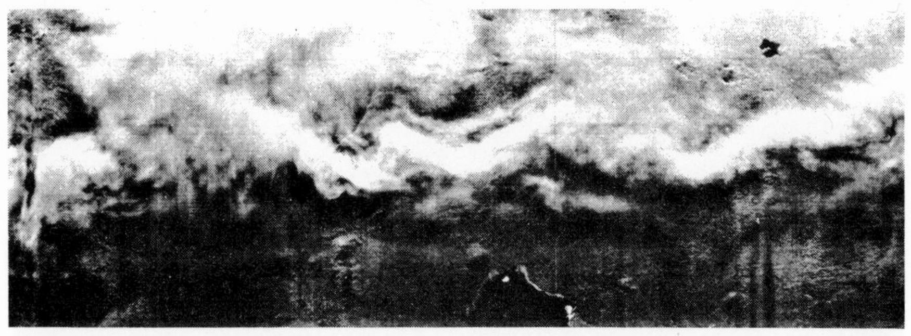

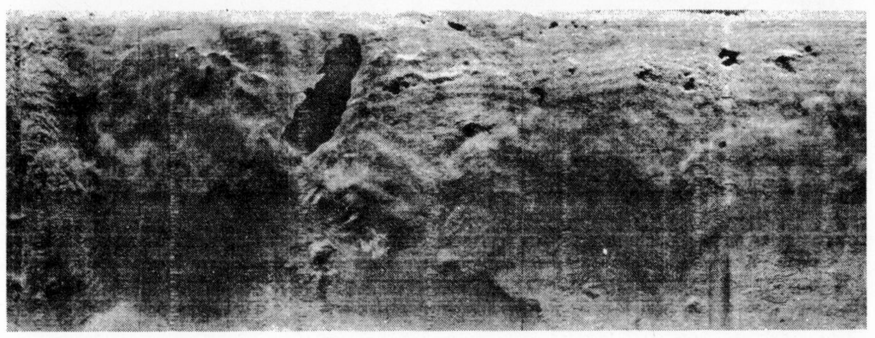

Fig. 6 Precipitation penetration obtained by the use of an opposite polarization sense receiver. Top: same (linear) polarization used on transmit and receive, bottom: opposite sense (Courtesy Center for Research, Inc., University of Kansas).

a narrow band light source illuminating a volume of steam, a significant reduction in the light reflected from the steam relative to that from other (nonspecular) objects was observed. In a more recent experiment, a reflection from mist was observed using ordinary illumination and the reflection was reduced when circular polarization was used. I am indebted to Professor Joseph Boyd at the University of Cincinnati for the report on these experiments. (Fig. 7)

Should the automobile headlight use of circular polarization as described above prove promising, measures would have to be taken to redesign reflectorized road

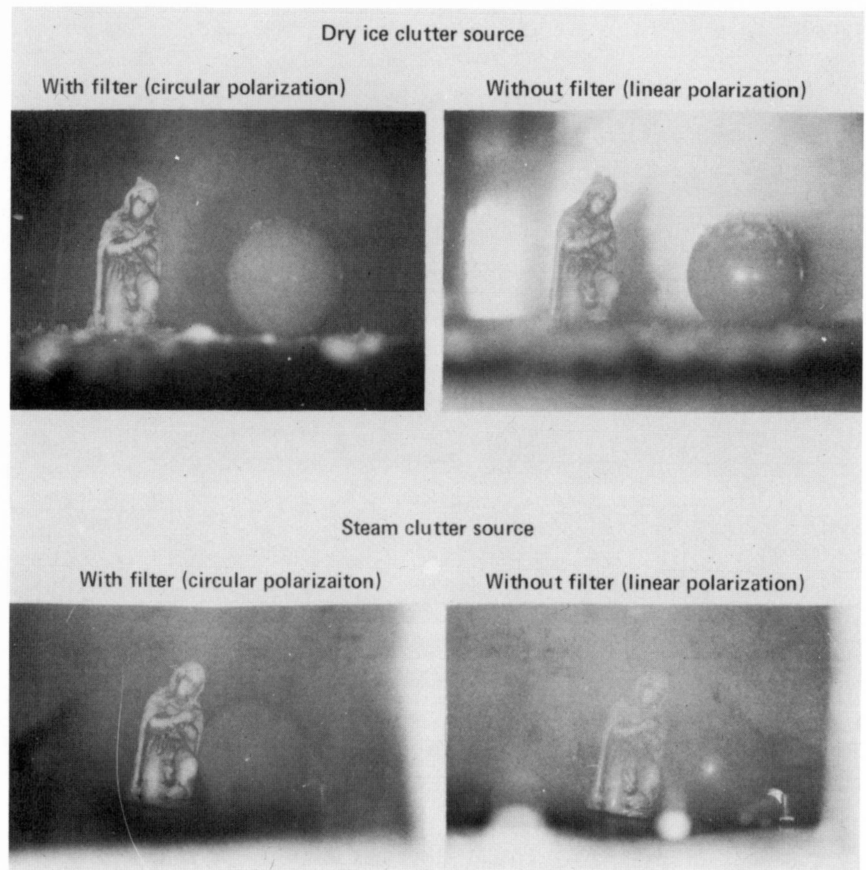

Fig. 7 Two sets of photos showing the effects of circularly polarized waves at optical wavelengths. (Courtesy Professor J. Boyd, University of Cincinnati).

markers for night driving. Most of these are presently equivalent to specular reflectors and their reflections would accordingly also be strongly reduced, along with the reflections from the mist droplets. This situation also exists at radar wavelengths, causing circular polarization radars to receive very weak echoes from standard corner reflector markers. As with raindrops and specularly reflecting surfaces, the echoes from corner reflectors would have the wrong sense of circular polarization. Various reflector designs have been examined at microwaves in an attempt to correct this situation (13).

In the optical case, it was found that the same circular polarization Polaroid material could be used to restore the desired strong reflections from such road markers (or from specularly reflecting material). In this experiment, the polaroid material was affixed to the reflector, with the opposite side of the material facing the illuminating source to obtain the original circularly polarized illuminating waves. The passage of the circularly polarized light through this reversed polaroid twice, once during its passage through to the reflector and once during its passage back following reflection, causes the reflected waves to possess the proper sense of circular polarization. Reflections from these reflectors could thus be highly visible through mist, with the reflections from the mist droplets being suppressed.

The use of circularly polarized light can also be useful in holography. One application (7) discussed at the 1975 U.S.-U.S.S.R. Seminar, would be in the analysis of droplet spray patterns. Photos of the spray pattern of a liquid propellant in a rocket engine are found useful because the size, sphericity, and distribution of propellant droplets are important factors governing the performance of the engine, and they are difficult variables to measure and predict (Fig. 8). Since two sets of circularly

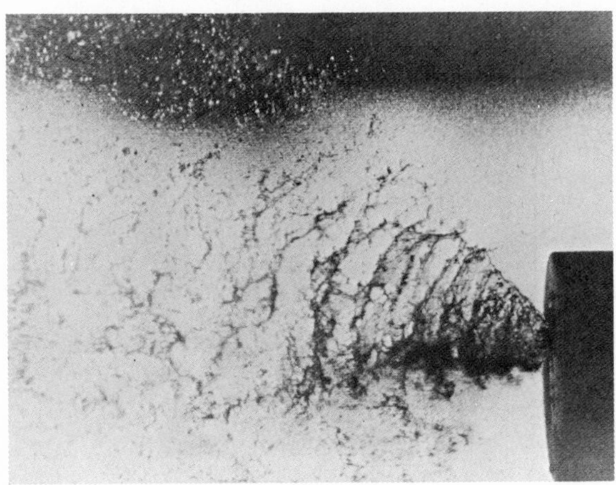

Fig. 8 Reconstructed hologram of a liquid propellant spray in a rocket engine (Courtesy U.S. Air Force Systems Command).

polarized waves having opposite rotation directions cannot generate an interference pattern, the use of circularly polarized waves for the illumination can cause symmetrical objects (such as small droplets) in the recorded scene to be either suppressed or accentuated. In either case, those objects which are displayed will appear, as in a normal hologram, in full three dimensions (7).

In forward-looking, real-time, synthetic aperture radars (14), the reduction in rain reflection made possible by circular polarization would permit shorter microwave wavelengths to be used with their accompanying higher resolution.

Phase-Quadrature Zone Plates

The standard high gain radars or sonars employ, because of the first circuit noise limitations, highly efficient receiving units such as lenses or parabolic reflectors, both of which form only one (focussed) image. Although zone plates have often been considered as substitutes for lenses, their lower efficiency has usually made them less attractive.

Quarter-wave techniques have been used to further improve the efficiency of microwave lenses. Thus, in Fig. 9, the addition of quarter wavelength (phase quadrature) steps cause the reflections from the high points to be 180° out of phase with the reflections from the low points, the result being that reflection losses were largely eliminated. This technique is the equivalent of the use of coatings on optical lenses, a procedure similarly used to increase their light transmissivity, that is, their efficiency or gain.

Because this way of reducing reflection losses in lenses is a quarter-wavelength procedure, the similarlity to Keating's phase-quadrature hologram technique was noted, and it was recognized that an analagous use could significantly improve the efficiency of one of the simplest holograms (15), the zone plate.

If two zone plates are constructed with both designed to focus waves at the same focal point, but with both separated in space by a quarter wavelength of the design frequency, one finds that same zones of the two zone plates overlap (Fig. 10). This overlapping occurs in such a way as to cause one of the zone plate images, (e.g. the real image) to be favored by both. It also results in a

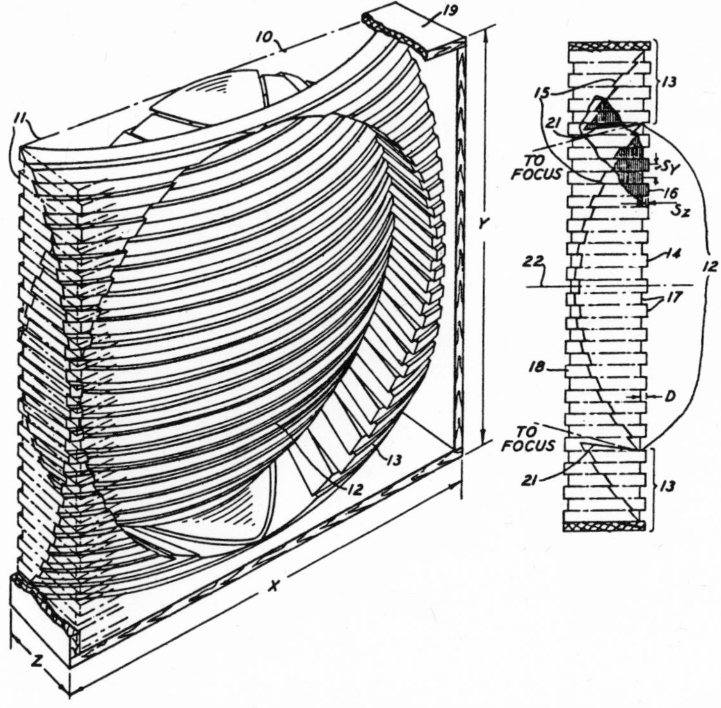

Fig. 9 The use of quarter wave steps on microwave lenses largely eliminates reflection loss.

blocking of the straight through, undeviated, D.C. term. As can be seen in Fig. 10, these effects increase for the outer zones, and for the zones of an offset zone plate (Fig. 11), the effect is very pronounced. In addition, reflected components are also drastically reduced, because the two zone-plate phase-quadrature reflections are 180° out of phase. It is seen that a similar, two-structure,

Fig. 10 Two zone plates, spaced a quarter wavelength apart, enhance the real focal image relative to the conjugate image. The two structures also reduce the zero order component and the reflected components.

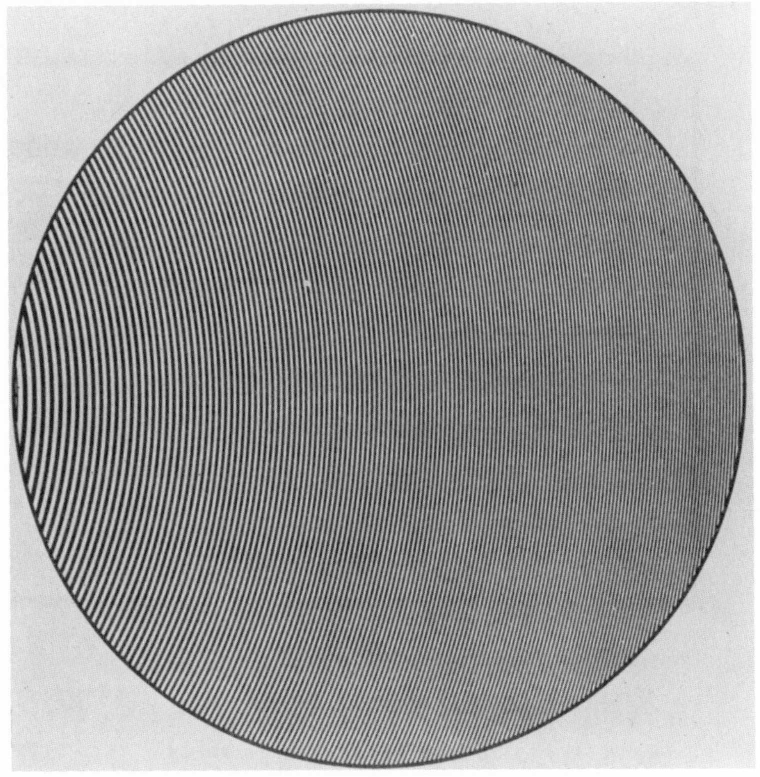

Fig. 11 An off-axis zone plate, showing the close spacing of the zones.

grating, with the structures also separated by a quarter wavelength, will cause one of the two diffracted orders to be favored, the D.C. term reduced, and the reflection loss reduced.

References

1. P. N. Keating, R. E. Koppelmann, R. K. Mueller and R. F. Steinberg, "Complex-On-Axis Holograms and Reconstruction without Conjugate Images", ACOUSTICAL HOLOGRAPHY, Vol. 5 (Plenum Press) 1974.

2. W. E. Kock, "Engineering Applications of Lasers and Holography", (Plenum Press, New York) 1975.

3. W. E. Kock, "Wave Polarization Shifter Systems", U.S. Patent #2,588, 249, filed January 26, 1946.

4. W. E. Kock, "Related Experiments with Sound Waves and Electromagnetic Waves, "Proc. I.R.E. vol. 47", 1959, pp. 1192-1201.

5. J. Brown, "Microwave Lenses", Methuen Publishing (London) 1953.

6. W. E. Kock, "Sound Waves and Light Waves", (Doubleday), 1965, pp. 71-72

7. W. E. Kock, "Extensions of Synthetic Aperture Radar Information Processing" in OPTICAL INFORMATION PROCESSING, Editors, Yu. E. Nesterikhin, George W. Stroke, Winston E. Kock, PLENUM PRESS, 1976.

8. W. E. Kock, "Circular Polarization in Certain Laser and Holography Applications", APPLIED OPTICS, July 1975.

9. This program, under the auspices of the University of Cincinnati Herman Schneider Laboratory, is under the direction of Professor Joseph Boyd.

10. H. R. Farrah, E. Marom, and R. K. Mueller, "An Underwater Viewing System Using Sound Holography", and L. Larmore, PLENUM PRESS, 1970.

11. E. N. Leith and J. Upatnieks, J. Opt. Soc. Am., 53, 1377 (1963).

12. A. M. Nathan, J. Opt. Soc. Am. 48, 285 (1958).

13. See for example, D. R. Wortendyke and R.C. Rudduck, "Metallic-Post Corner Reflectors for Same-Sense Circular Polarization", The Microwave Journal, December, 1970, pp. 61-66.

14. W. E. Kock, "A Real Time Parallel Optical Processing Technique", IEEE Transactions on Computers, vol. C-24, No. 4, April 1975, pp. 407-411.

15. G. L. Rogers, "Gabor Diffraction Microscopy: The Hologram as a Generalized Zone Plate", NATURE (London), 166, 237 (1950).

SYNTHESIS OF KINOFORMS IN REAL TIME

V.P. Koronkevitch, A.E. Meerson, V.G. Remesnik
G.P. Cheido, A.M. Stcherbatchenko

Institute of Automation and Electrometry
and Computing Centre, Novosibirsk, USSR

The multi-step process of kinoform synthesis may be considerably simplified in case of using thin films of chalcogenide vitreous semiconductors as a material for recording. In the films under light influence the refraction index is changed by the value of the order of $\Delta n = 0.1$. Great value of Δn allows us to obtain phase shifts up to 2π at the film thickness $10\,\mu m$. High resolution allows us to reproduce a thin structure of the phase profile. The phase profile can be plotted by displacing the film relative to a laser beam with the help of a computer-controlled precise cross-bar table. All the operations connected to the reduction of the transparency and converting the amplitude profile into a phase one become unnecessary. Chalcogenide materials do not require development. Thus, routine work on the production of kinoforms is completely carried out by the computer and multi-level laser photoplotter.

Our experiments were stimulated by the problems of seismoholography. All mathematical models have approach to the situations taking place in the practice of seismic shooting. The present paper describes the model of the design of a kinoform giving images in the Fourier plane and Fresnel area.

F o u r i e r k i n o f o r m. Consider the plane transparency $p(u, v)$ simulating the initial object, which is given in the following way

$$p(u, v) = t(u, v) \cdot exp[\varphi(u, v)] . \tag{1}$$

Herein $t(u, v)$ is the transparency binary transmission, i.e. 0 corresponds to the opaque element, 1 - to the transparent one; the phase values $\varphi(u, v)$ simulate diffuse scattering which in optics corresponds to illumination through the ground glass.

Distribution of the complex amplitude of the field in the plane (x, y) parallel to the transparency can be represented by the relationship [1]

$$p(\xi, \eta) = \iint_{-\infty}^{\infty} p(u, v) \cdot exp(j2\pi)(\xi u + \eta v) \, du \, dv \tag{2}$$

To the accuracy of the factor Fourier-kinoform is a space Fourier spectrum of the object function $p(u, v)$ calculated from the coordinates ξ, η of the scale $\xi = x/\lambda z$, $\eta = y/\lambda z$ where u, v are the space coordinates, ξ, η are the space frequencies, λ is a wavelength, z is the distance between the transparency plane and the kinoform element.

Having changed the functions $p(\xi, \eta), p(u, v)$ for their disrete analogs and set the condition on the steps of discretization along the coordinates from (2) we shall obtain

$$p(m, n) = \sum_{k=0}^{N-1} \sum_{\ell=0}^{N-1} p(k, \ell) \cdot exp(2\pi j/N)(km + \ell n) \tag{3}$$

where $m, n = 0, 1, \ldots, N-1$.

When realizing the algorithm (3) with the computer M-220 we use the Cooley-Tukey technique [2]. Simulating the initial transparency of the size of 128x128 elements we determine the real and imaginary parts of the complex amplitude

$$\begin{aligned} Re[p(m, n)] &= t(k, \ell) \cdot Cos(RAND \times \pi) \\ Im[p(m, n)] &= t(k, \ell) \cdot Sin(RAND \times \pi) \end{aligned} \tag{4}$$

where RAND is a material procedure in the language utilizing the characteristic features of the computer M-220 generates random numbers uniformly distributed within the interval $[0,1]$.

To provide the control over the multi-level laser photoplotter we take out the phase of the field $p(m,n)$ scattered by the transparency to the punch tape as a teletype code M-2

$$\varphi(m,n) = \arcsin\{Re(p)/[Re^2(p)+Jm^2(p)]^{1/2}\} \quad (5)$$

from which the information multiple 2π is eliminated.

Before computing the inverse Fourier transformation for computer reconstruction of image the field amplitude $p(m,n)$ is cancelled to the unit by division of $Jm(p)$, $Re(p)$ over $mod(p)=[Re^2(p)+Jm^2(p)]^{1/2}$.

F r e s n e l k i n o f o r m. To simulate the Fresnel kinoforms the technique mentioned in [3] was used. All parameters of design have approach to the experiments taking place in the practice of seismic shooting. The diagram of the experiment is represented in Fig. 1.

The diffuse point sources (ξ_i, η_i) radiating monochromatic spherical waves provide in the registration plane (x,y) amplitude complex distribution

$$F(x_j, y_j) = \sum_i u_i(\xi_i, \eta_i) \quad (6)$$

where $u_i(\xi_i,\eta_i) = exp[(2\pi j)(\omega t - K r_{i,j})]$ is the field of the ith oscillator with the coordinates (ξ_i, η_i) ; $r_{i,j} = [(x_j-\xi_i)^2 + (y_j-\eta_i)^2 + \ell_i^2]^{1/2}$ is the distance from the ith oscillator up to the point (x_j, y_j) in the plane of registration; $K = 2\pi f/V$ is the wave number; f = 50 cps is the radiation frequency; V = 2 kmps - the velocity of waves; ℓ_i is a distance between the plane of registration and radiating point. Phase value in the point $m(x_j, y_j)$ from the ith oscillator is computed according to the formular

$$\varphi_j = R_i \pi + K r_{i,j} , \quad (7)$$

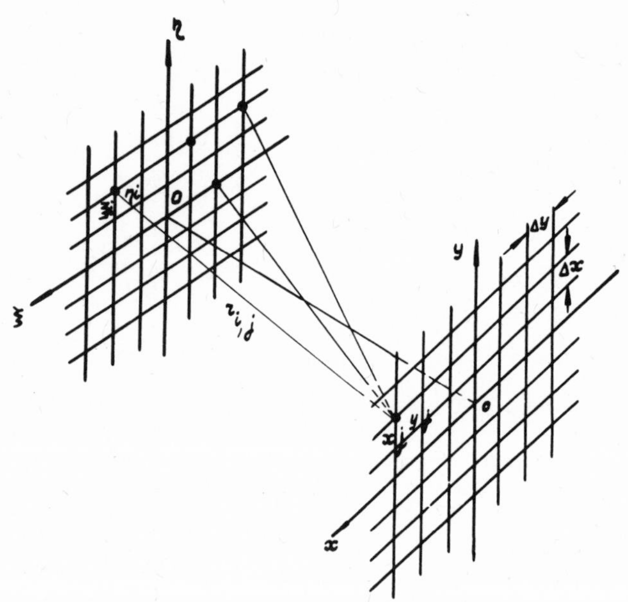

Fig. 1. The scheme of kinoform calculation.

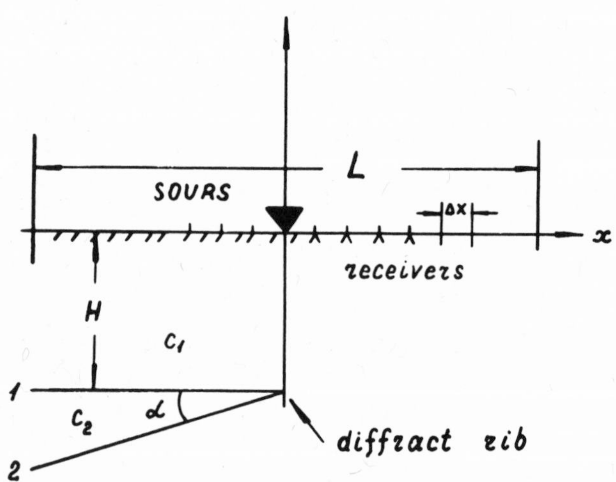

Fig. 2. The model of the medium. Vertical section. The black triangle is the source placed at the origin of coordinates.

where R_i is a random number from the intercept $[0,1]$. The real and imaginary parts of the field are determined from

$$Re = \sum_i \ell_i \cdot Cos(\varphi_i)/z_{ij},\ Jm = \sum_i \ell_i \cdot Sin(\varphi_i)/z_{ij} \quad (8)$$

The aperture consists of 64 x 64 elements which at the step between the counts $\Delta x = \Delta y$ = 15 m corresponds to the size on the ground equal to 945x945 m; the depth of the initial point object is of the order of the aperture sizes ℓ = 750 m.

The third example represents the simulation of the most interesting for the practice case, i.e. seismic wave diffraction on the edge [4].

Seismic shooting of geological structures of the type of wedging out layers and contiguities of different foundaries is of great importance for the search of minerals. The simplest models of geological structures (Fig. 2) are wedge-shaped homogeneous media with different elastic properties having the common top (diffracting edge).

According to the principles developed in [4,5] the wave pattern for the model of Fig. 2 can be considered as a superposition of geometrooptical waves providing the energy transfer in accordance with the classical ray theory, and the diffracted waves resulting from the cross diffusion from the boundaries of shadows of the geometrical field. Each geometrooptical wave (direct, reflected, passing) may be represented by a zero approximation of the ray set [6], and it will differ from the zero only in its "illuminated" area, which is separated from the rest of the space with the shadow sharp boundary.

A diffracted wave corresponding to each boundary of the shadow compensates the break of the respective part of the field on this boundary. The asymptotic method of calculation of the given waves within the limits of the Helmholtz homogeneous equation is described by the formulars

$$f = \sum_i (f_i + f_i^{D}),\ f_i = R_e \int_{\omega_0 \gg 1}^{\infty} x \cdot \Phi_i \cdot e^{-j\omega(t - t_i)} d\omega,\ \Phi_i = \frac{K_i\, S_i(\omega)}{L_i} \quad (9)$$

$$f_i^{\mathcal{D}} = \pm Re \int_{\omega_0 \gg 1}^{\infty} x_i^{\mathcal{D}} \cdot A_{Ti} \cdot W(\omega_i) \cdot e^{-j\omega(t-t_i^{\mathcal{D}})} d\omega, \quad A_{Ti} = [\Phi_i(r,\theta)]_{\theta=\theta_i}$$

$$W(\omega_i) = \frac{1}{2\sqrt{\pi}} \cdot e^{-j\frac{\pi\omega_i^2}{2}} \cdot \Gamma\left(\frac{1}{2}, -j\frac{\pi\omega_i^2}{2}\right), \quad \omega_i = |\theta - \theta| \sqrt{\frac{\omega \cdot r_{oi} \cdot r}{\pi c_i (r_{oi} + r)}},$$

(10)

where $j = \sqrt{-1}$, f_i and $f_i^{\mathcal{D}}$ are, respectively the ith geometrooptical and related to it diffracted waves; t_i, $t_i^{\mathcal{D}}$ and x_i, $x_i^{\mathcal{D}}$ are hodographs and the coefficients of conversion of these waves; r, θ are polar coordinates with the origin at the angle point; K_i, L_i, $S(\omega)$, C_i, θ_i, R_{oi} are, respectively, the direction function, geometrical divergence, complex spectrum of the form, propagation velocity, angular coordinate of the shadow foundary, and radius of the front curvature at the angular point of the ith geometrooptical wave; $\Gamma(\frac{1}{2}, x)$ is incomplete gamma function; the minus sign in the formular for f_i will be in the shadowed area, and plus sign will be in the illuminated area.

When calculating the kinoform we used the program realizing a particular case of the algorithm (9), (10) for determination of the field of diffraction of a spherical wave on the ideally reflecting boundary in a stationary case, when $K_i(\theta)$ = const and reflection coefficients don't depend on the angle. At the points of the space removed from the edge the wave field can be represented as a sum

$$f = f_{inc} + f_{refl} + f_{diff} + f_{backg} . \qquad (11)$$

where f inc. is the field of the wave being radiated by the source; f refl. is the field of the wave reflected into the half-space $\theta > \theta$; f diff. is the part of the diffracted wave field compensating the break of the diffracted wave on the boundary of the geometrical shadow $\theta = \pi - \theta_0$; f backg. is the unbroken part of the diffracted wave field on the shadow boundary.

Kinoforms obtained for the case when all the terms of the formular (11) are summarised, and for the case when two first terms are not taken into account. The aperture size in calculations is $L \times L$ =1260 m x 1260 m;

the frequency is Ω = 62,5 cps; the velocity is V = 2500 mps; the edge depth is H = 500 m; steps along the axes are $\Delta x = \Delta y$ = 20 m; α = 0.

E x p e r i m e n t a l t e c h n i q u e s f o r k i n o f o r m r e c o r d i n g. Kinoform recording was carried out with the help of a multi-level laser photoplotter. The principle of recording is based upon displacing the photoactive material (chalcogenide film) about the writing beam. Displacement of the film and laser beam intensity are controlled by the computer.

The bulk of the multi-level laser photoplotter are the mechanical, electronic, and optical units of the photogrammetric automatic machine developed at the Institute of Automation and Electrometry [7]. The photoplotter (Fig. 3) comprises the carriage with photoactive material 2 mounted on it which moves in horizontal plane along the guides mounted on the massive base 3.

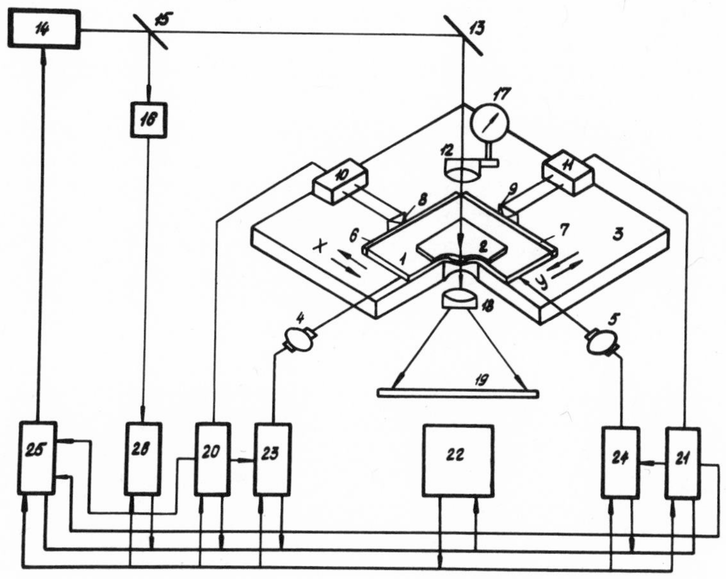

Fig. 3. The structural scheme of the multi-level laser photoplotter.

The carriage is set in motion by two direct-current motors 4,5. Along two sides of the carriage the measuring guides 6,7 with which the angle reflectors 8,9 come into contact of two independent laser interferometers 10,11 are arranged. Interferometer axes along which the corner reflectors move are normal to the measuring guides.

The optical system 12 focuses the argon laser beam 14 upon the chalcogenide film surface. To control the power a part of radiation branches on the photoreceiver 16. The optical system is mounted on air bearings providing its displacement along the axis of sight. A special objective mounting rests upon the plate 2 with photosensitive material via an air pillow. Follow-up system comprising a pneumatic amplifier provides maintenance of the preset distance between the objective and material. Control over displacement of the optical system along the optical axis is performed using the micron indicator 17. Image of the laser luminescent spot is projected on the screen 19 via the optical system 18. The carriage 1 displacement is converted into digital binary code with the help of laser interferometers 10,11 and units 20,21. Each unit of conversion of displacement value-to-digital code comprises three reversible counters. The first counter registers the code of the carriage position relative to the zero reference point, the second - the code of the carriage displacement value relative to the coordinate being registered by the first counter. And finally, the third--quantiges the displacement value with the given quantizing step. The computer Minsk-22 performs recording and reading of the codes from the first and second reversible counters and gives the required factor of division for the third counter. The units of converting the displacement value to digital code 23,24 are connected with the units controlling the electric motors at which signals of the error magnitude and laser interferometer signals proportional to the speed of carriage motion are fed.

The units controlling the electric motors set the carriage at any given position in two modes, i.e. in the mode of movement with constant speed and in the mode of positioning. In the mode of positioning setting the carriage into the given position is carried out with 0.32 m error.

The mode of movement of the carriage with constant speed is realized with a proportional control computer. The speed of the carriage movement corresponds to the numerical code being derived from the computer to the second reversible counter which is a register in this mode as its input is blocked. The carriage stops on the computer command which, when the carriage moves, performs asking the counters registing the code of carriage coordinates and computes the value of the magnitude of displacement relative to the initial point taken as the origin of coordinates.

The unit controlling the laser modulator 25 operates in continuous or pulsed modes. The number of the intensity quantization levels is equal to 1024. The maximum frequency of laser switching is 5 Kcps.

The unit 26 converts the laser signal being received by the photoreceiver 16 to a digital code. The number of quantization levels is equal to 128. The maximum fast response is 20 Kcps. The converting error is 1%.

For phase profile recording the photosensitive material is mounted on the carriage which is set at the given point taken as the origin of coordinates. At this point setting of all the counters of the units 20,21 to the initial state is carried out, then the carriage starts moving with constant speed in the X axis direction according to the computer program. As the carriage moves along the X coordinate in given length increments pulse laser switching is carried out laser radiation intensity being given by the computer. Having exposured a line the carriage moves along the Y axis in the mode of positioning. Then with the laser off the carriage returns along the X axis at the beginning of the next line. The time of the phase profile recording with the element number 64 x 64, the size of a single element being 20 x 20 m^2, is 3 min.

M a t e r i a l f o r k i n o f o r m r e c o r d i n g. Thin films of chalcogenide vitreous semiconductors which previously have been used for recording kinoform optical elements [8] served as photosensitive material for kinoform recording. The films were prepared by electron-beam evaporation of previously synthesized material As_2S_3 in vacuum, on a polished glass substrate. Thickness of evaporated films was equal to 10 μ m. When exposuring the films by argon laser radiation ($\lambda \approx 0.51 \mu$ m) some photostructural changes occur in them. It results in the shift of the basic absorption edge to longer wavelengths and in considerable increase of the refractive index ($\Delta n \sim 0.1$) within the wide range of spectrum. For He-Ne laser radiation ($\lambda \approx 0.63 \mu$ m) the films are transparent and practically pure phase modulation is observed in the recorded areas. The films do not need additional development, i.e. they allows to operate in a real-time operation mode, since their structural changes occur at the moment of exposure.

To determine characteristics of film photoresponse the dependence of the phase shift obtained on the magnitude of the exposure was taken. At the constant duration of a recording pulse the dependence of the phase shift magnitude on the intensity of the recording beam given by control voltage from the computer was set up. The phase shift was determined with the help of a microinterferometer. The relation obtained was well approximated by the function

$$\varphi = A_0 \left[1 - exp\,(\alpha - \beta U) \right] \qquad (12)$$

where φ is a phase shift in the region exposured; A_0, α, β are constants; U is a control code voltage.

E x p e r i m e n t a l r e s u l t s. The multi--level laser photoplotter allows us to fulfil continuous and discrete optical information recording to a high accuracy. As the example of photoplotter possibilities. Fig. 4(a) represents a microphotograph of a grating recorded on the thin aluminium layer, evaporated on a glass substrate. Recording was performed by evaporating the aluminium with the help of a focused argon laser beam. The grating step is 20 μ m. On the photograph it is possible to see the character of film movement: movement along the line, returning, and transition to the next line. Discontinuity of lines depends on the uniformity of table movement and on the stability of recording beam focusing.

Fig. 4(b) represents a microphotograph of a film fragment with a recorded phase grating. Recording was carried out in the mode of uniform film movement. The grating step is 5 μ m. The film thickness is 6 μ m.

The width of recorded phase strokes does not exceed 1.5 μ m.

In the discrete recording mode the photoplotter allows us to record phase distribution according to the preset law on chalcogenide films. In such a way discrete phase holograms, kinoforms can be synthezised and space phase modulation can be recorded in a given manner. A microphotograph of the chalcogenide film with discrete phase recording is shown in Fig. 4(c). For this purpose a mask with a square hole is arranged in the way of the argon laser beam (see Fig. 3). The image of the square is projected with reduction on the chalcogenide film. Control over focusing is performed in accordance with the image on the screen 19. Sizes of the squares recorded made 10, 20 and 40 m. Phase shift in the square is given by the computer and varies from 0 to 2 π .

To check up the correctness of the phase recording when synthesizing kinoforms on the film the test phase triangle was recorded. In case of ideal recording the phase within the triangle must linearly increase and sharply break when it reaches 2 π . Fig. 5(a) represents a microphotograph of one of the experimental test triangles received with a shearing-microinterferometer.

As the first object for synthesis of Fourier-kinoform the image shown in Fig. 5(b) was used. The image was given by an array of 64 x 64 points. Fig. 5(c) represents inverse image reconstruction using only phase carried out by the computer. A microphotograph of the external view of the synthezised Fourier-kinoform consisting of 64 x 64 elements is given in Fig. 5(d). The kinoform sizes are 0.64 x 0.64 mm^2. Optical image reconstruction from such a kinoform gives a multiple image (Fig. 5(e)). Multiplication arises due to diffraction of light on the mosaic structure of a kinoform. A fragment of the image reconstructed is given in Fig. 5(f).

Fig. 6 (a,b,c) shows the image of another object inserted into the computer, the image reconstructed by the computer using only phase, and the general view of optical reconstruction, respectively. The number of elements when calculating and recording Fourier-kinoform was equal to 64x64. Figs. 7(a) and 7(b) represent the result of image reconstruction with the computer

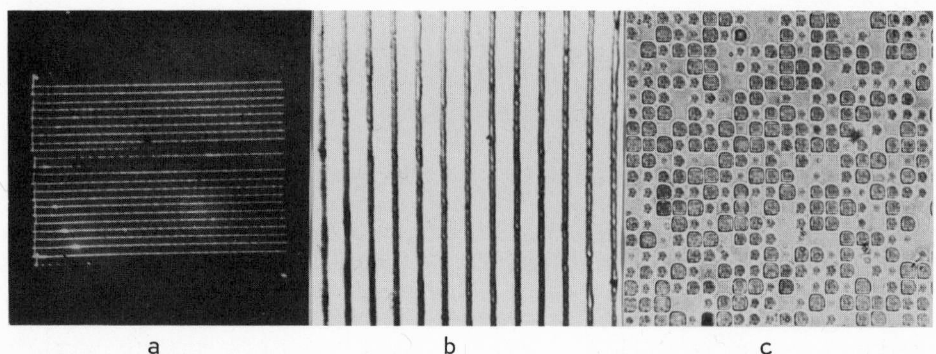

a b c

Fig. 4. Microphotographs of the objects recorded with the photoplotter.

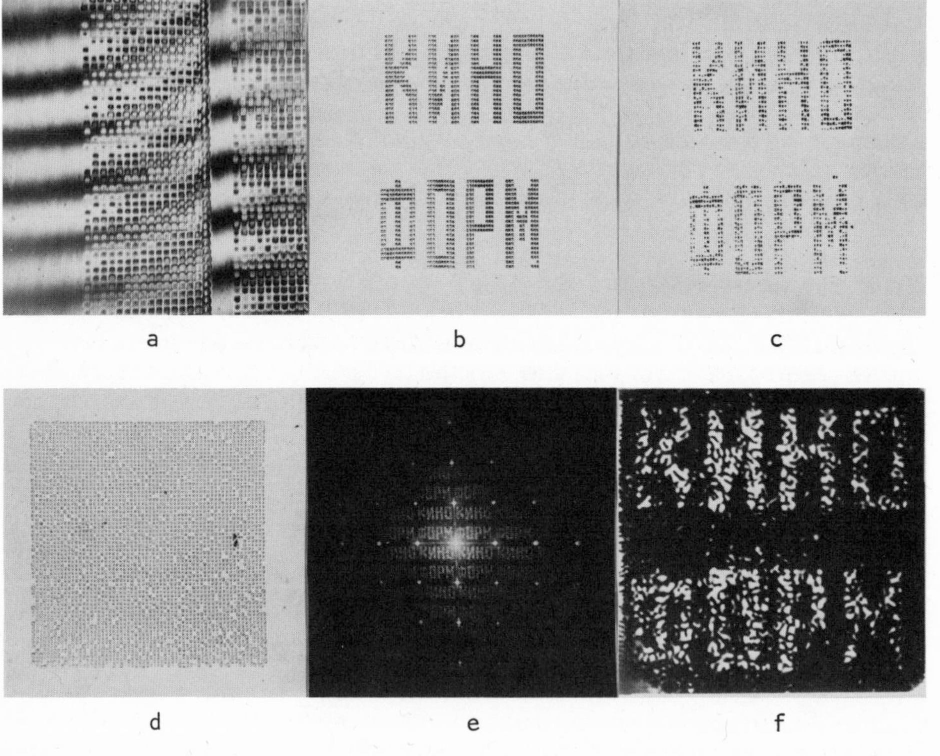

a b c

d e f

Fig. 5. Kinoform synthesis on chalcogenide film.

from the phase and that of optical reconstruction of the image for the Fourier-kinoform consisting of 128 x 128 elements.

Image reconstruction with kinoforms in the Fresnel area is given in Fig. 7(c). The calculation model suited the real conditions of seismic shooting. The object consisted of 32 point oscillators, the number of kinoform elements was equal to 64 x 64, the distance between the aperture and the object was of the order of the aperture sizes.

Fig. 8 represents the results of optical image reconstruction according to the model of the edge diffraction of acoustic waves which is of interest for long wave holography. The aperture of kinoforms was equal to 64 x 64 elements. When reconstructing images lense was used which allowed to obtain on the screen a magnified virtual image being formed by kinoforms. The image in the plane corresponding to the position of the edge is shown in Fig. 8(a). At the double depth the image of a virtual source (Fig. 8(b)) resulting frone mirror reflection from the interface is reconstructed. Both photographs relate to the case when all the terms of the formular (10) are taken into account. Fig. 8(c) represente the case of optical reconstruction when direct and reflected waves are not taken into account. Birfurcated image arising for the case of Sommerfeld edge diffraction is observed. Fig. 8(a) has no such birfurcation, and it is explained by the fact that *frefe* » *faill* . At the double depth in case of elimination of direct and reflected waves caustics of the diffracted wave is observed instead of the virtual source (Fig. 8(d)).

As comparison, Fig. 9 represents the results of numerical image reconstruction with the use of amplitude and phase for the model of edge diffraction which V.G. Khaidukov and G.M. Tsibultchik kindly gave us. Great similarity of numerical and optical image reconstructions is seen to be observed.

D i s c u s s i o n. Turning to the discussion we should like to emphasize that the suggested method of kinoform synthesis is a new one and we embarked on only first test experiments in this field. In the given work the images reconstructed by kinoforms consisting of 64x64 elements and only in one case the aperture was

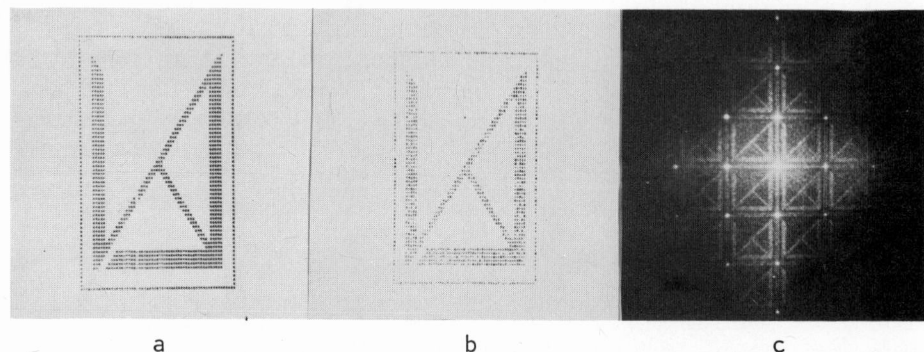

a b c

Fig. 6. A Fourier-kinoform consisting of 64x64 elements. a) The initial image. b) Numerical reconstruction using the phase. c) Optical image reconstruction.

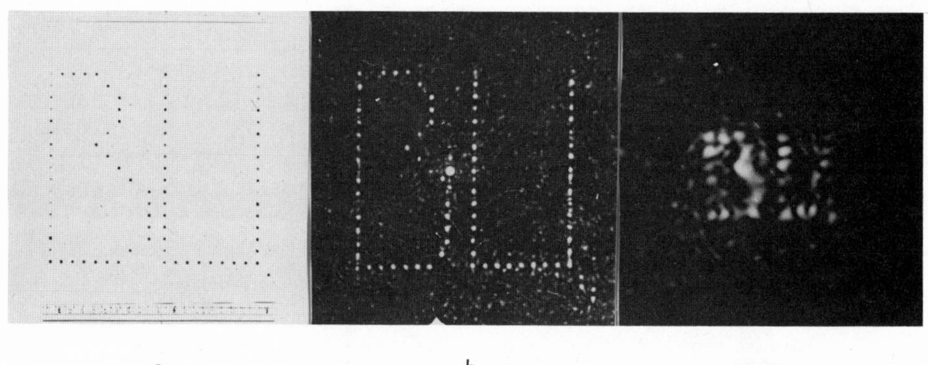

a b c

Fig. 7. a) Numerical image reconstruction using the phase from the Fourier-kinoform consisting of 128x128 elements. b) Optical image reconstruction from the Fourier kinoform consisting of 128x128 elements. c) Optical image reconstruction from the Fresnel-kinoform of 64x64 aperture. The initial object is placed at the depth of the order of aperture sizes.

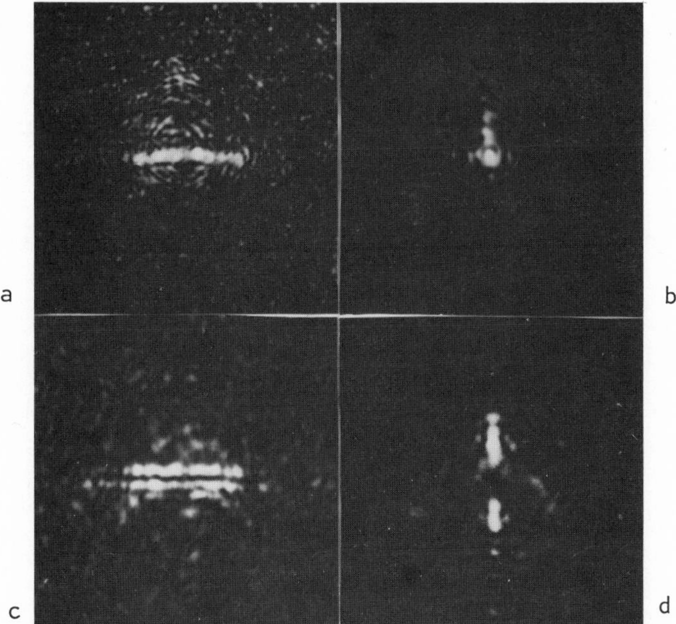

Fig.8. Optical image reconstruction using the model of edge diffraction of acoustic waves.

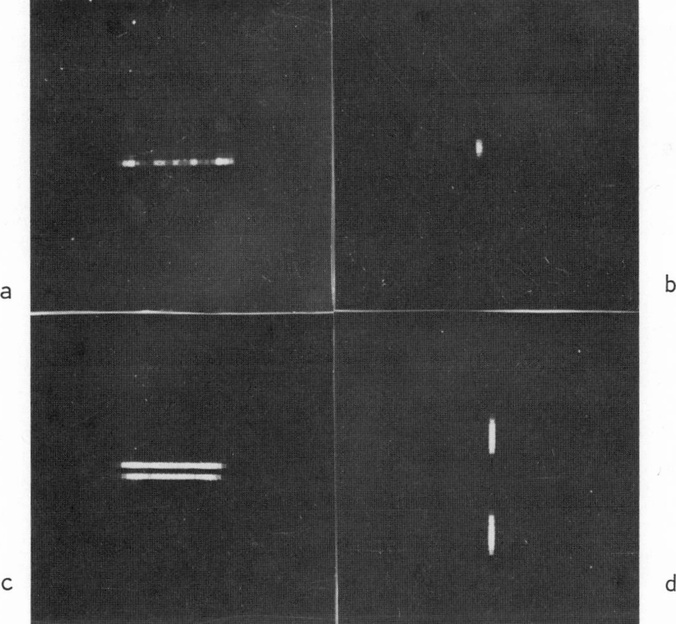

Fig.9. Numerical image reconstruction with the use of the amplitude and phase corresponding to Figure 8.

equal to 128x128 were considered. In spite of the lack of redundancy in our kinoforms the quality of reconstructed images is quite acceptable. For comparison it will be recalled that, for example, in the work [9] when synthesizing a kinoform the redundancy in 2500 times was introduced.

Besides, we shall note the errors arising due to the fact that because of a number of reasons we have not managed "to keep phases" when recording kinoforms (see Fig. 4(a)), though this part is quite controllable.

C o n c l u s i o n. The opinion that optical reconstruction has no prospects compared to the numerical one has recently spread among those investigating the problems of reconstruction non-optical images. The pessimism is thought to be related to technological difficulties of synthesized phase hologram registration on the known media. At the same time the advantages of optical reconstruction over numerical one are well-known and self-evident. The possibility of obtaining three--dimentional images which is necessary for a number of problems, e.g. when indicating three-dimentional images at the computer output, the possibility of processing mass data which is limited by the memory and fast response of modern computers in case of numerical reconstruction, and, finally, the possibility of flexible combining hologram and kinoform synthesis with subsequent optical processing should be related to these advantages. When solving the problems of seismic and acoustic holography optical reconstruction can be indispensable under condition of wave field registration on the great enough aperture and using wavelengths short enough. This work is hoped to provide some progress to investigations in the field of optical processing of non--optical information.

In conclusion the authors wish to express their gratitude to corr. member Yu.E. Nesterikhin and corr. member A.S. Alekseev for the interest shown and stimulation of this work, and besides to V.G. Khaidukov, L.A. Vorontsova, and O.M. Karpova for their assistance in the experiments.

REFERENCES

1. T.S. H u a n g. Digital holography. Proc. of the IEEE v. 59, N 9, p. 1335-1345.

2. J.W. C o o l e y, J.W. T u k e y. - Math. Comp., 1965, v. 19, p. 297.

3. A.E. M e e r s o n, G.M. T s i b u l t c h i k. Nekotorie chislennie experimenti po sintezu seismo-akustitcheskikh gologram. - Sb. "Akustitcheskie metodi i sredstva issledovania okeana", Vladivostok, 1974.

4. K.D. K l e m-M u s a t o v, G.L. K o v a l e v s k i, V.G. T c h e r n j a k o v, K.D. M a k s i m o v. Matematitcheskot modelirovanie difraktsii seismitcheskikh voln v uglovikh oblastjakh. - Geologia i geofizika, 1975, N 11, s. 116-124.

5. K.D. K l e m-M u s a t o v, Printsip Yunga v teorii difraktsii seismitcheskikh voln. - Sb. "Seismitcheskie volni v slozhnopostroennikh sredakh". Novosibirsk, 1974, s. 4-63.

6. A.S. A l e k s e e v, V.M. B a b i t c h. O lutchevom metode vitchislenia intensivnosti volnovikh frontov.- Izv. AN SSSR, seria geophiz., 1958, N 1.

7. L.V. B u r i i, V.P. K o r o n k e v i t c h, Yu.E. N e s t e r i k h i n, A.A. N e s t e r o v, B.M. P u s h n o i, S.E. T k a t c h, A.M. S t ch e r b a t c h e n k o. Pretzizioznii fotogrammetritcheskii avtomat. - Avtometria, 1974, N 4, s. 83-89.

8. V.P. K o r o n k e v i c h, G,A. L e n k o v a, I.A. M i k h a l t s o v a, V.G. R e m e s n i k, V.A. F a t e e v, V.G. T s u k e r m a n. Kinoform optical elements. - Optical information processing. Plenum Press. N.Y. and London. 1976, p. 153-170.

9. L.B. L e s e m, P.M. H i r s c h, J.A.J o r d a n. The kinoform: a new wavefront reconstruction device.- IBM J.Res. Dev., 1969, v.13, N 2, p. 150-155.

SPACE LIGHT MODULATORS
(Effect of Space Light Modulators Characteristics on Information Transmittance of Coherent Optical Processing Systems)

S.B.Gurevich

A.F.Ioffe Physical-Technical Institute
Academy of Sciences of the USSR
Leningrad, USSR

I. Introduction

Many characteristics of systems including information transmittance of coherent optical processing systems depend on space light modulators (SLM) characteristics to a considerable degree. Therefore it is necessary to have a satisfactory match between SLM characteristics and parameters of a system including SLM. Some of these parameters were evaluated in [1-3], the other papers [4-6] give characteristics of SLM. However, it is evident that the problem of matching SLM with other parts of systems of processing, recording and reconstruction of information needs studying.

One of various possible schemes of an optical processing system is shown in Fig.1. At least two parts of the system contain SLM: information input and information optical processing. Holographic memory system is schematically shown in Fig.2. In this figure SLM is a matrix of input data (page composer) and a matrix of holograms. In both systems SLM execute different functions and different parts of the system interact with them.

If SLM is used as an input device, it can be applied to the problem of transformation of two-dimensional fields or series of one-dimensional signal distributions simultaneously processed. In such SLM information is introduced and recorded as a result of various effects: optical incoherent or coherent image of visible light or light of other wavelengths, thermal, acoustic and other images, electric signals. The latter case is very often realised for information processing. In this case one-dimensional signal is transformed into a two-dimensional optical image by means of modulators with electrical commutation or by means of series of one-

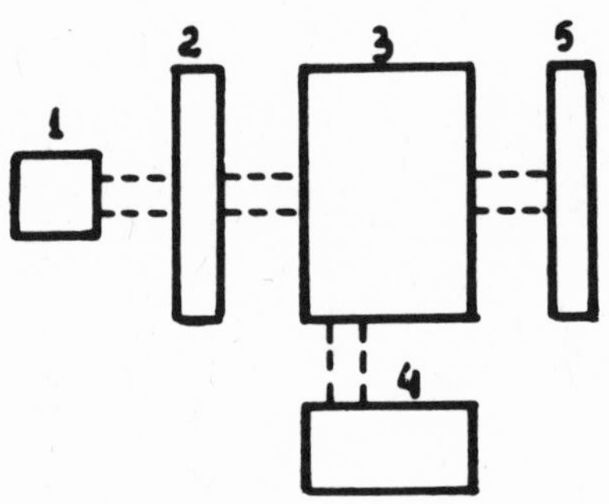

Fig.1. Coherent optical processing.
1 - a source of coherent light responsible for transmission, processing, record or reconstruction of information;
2 - information input;
3 - optical processor (information processing); 4 - operating of memory processing; 5 - information output.

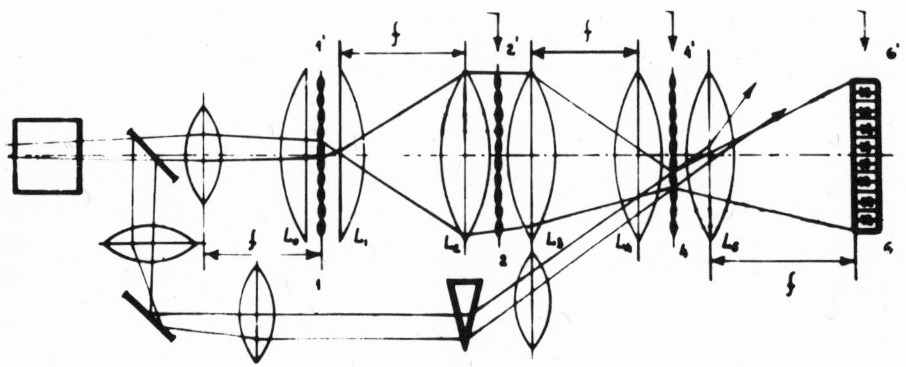

Fig.2. The scheme of holography memory system.
1 - laser beam; 2 - reflector;
3 - splitter; 4 - lens raster;
5 - matrix of data; 6 - matrix of holograms; 7 - matrix of photodiodes.

dimensional acousto-optical transformations. SLM - controlled transparents are used in system of holographic memory, matrix electrode structure being used for an input of digital information.

Other SLM have no differences of principal from the SLM mentioned above, but they meet other requirements. They are elements of processing, storage and reconstruction of information. In a system of the first type generally they are amplitude-phase filters that can be SLM operating in real-time and having different characteristics. There is a matrix of holograms in a system of the second type. In this case processing is choosing of necessary information using appropriate addressing system and reconstruction of this information at the appropriate detectors while passing a coherent beam of the given direction through SLM - hologram matrix.

II. SLM characteristics

The totality of characteristics used for examination of possibilities of SLM applications depends greatly on a type of SLM, effects used and a problem considered. In addition to this the characteristics are related to particularities of SLM operating cycle, in turn consisting of input of information-recording, information storage, read-uot and erasing of information. One may indicate the following principal characteristics of SLM:

1. Output signal-to-noise ratio; variations of signal-to-noise ratio from an input to an output.
2. Transmission of spatial information: resolution per unit area and size of a modulator operating surface or resolution per whole operating area; the corresponding frequency-contrast and frequency-gradational characteristics.
3. Characteristic of a transmission of a space light modulator; deviation from linearity and dynamic range; presence of parasitic (that is carrying no information) background; contrast and contrast sensitivity.
4. Transmission of temporal information (its variations in time): time of operating cycle and its components, delay of recording and erasing (degree of quickness of responce, degree of fullness of erasing and of shifting of recorded information in time. Time of conservation of recorded information without erasing.

5. Sensitivity: value of input energy providing the given output data of spatially modulated light.
6. Transmission of spectral information: spectral characteristics of a transformer, modulator and device as a whole.
7. Spatial non-uniformity and presence of various regular interferences.
8. Clearance, energetic (not connected with information) and economic characteristics.

The contribution of each characteristic can very essentially with variation of a problrm solved by means of SLM. Some of these characteristics are interconnected. The signal-to-noise ratio determines the number of transmitted halftones; it affects the SLM resolution. Therefore it is an important parameter for determination of modulator information capacity. The magnitude of input energy required to record the introduced information is essentially related to a SLM resolution. High resolution of a material is generally associated with the necessity of supply of considerable amount of energy of information carrier when recording. Such factors as degree of non-linearity and dynamic range of a signal transmitted characterize transmission. These characteristics are very important for SLM applications as filters and for holographic recording. Though the dependence of SLM characteristics on phenomena used its operating needs studying, one may note, for instance, that maximum resolution can be achieved at the molecular level of effect of information carrier (for example, photochrome effects under effect of photons). It can be sufficiently high at variation of optical path difference by means of creating of spatial relief (that can be achieved on thin films). When electro-optic effects are used, where sufficient variation of refractive index and polarization plane rotation requires considerable thickness of an active material, resolution appears to be lower. When it is necessary to calculate the characteristics and to choose the SLM suitable to solve the given problem, it is useful to take into account a relation of various characteristics and their dependence on the effect used for producing of spatial distribution of optical parameters.

III. Information loss and matching of SLM characteristics

Since the system considered above have SLM and lenses as main elements, they are the principal cause of information loss in the system. The experience in information loss estimation in systems consisting of successive series of elements is that it is necessary to consider not only the transmittance and information loss of a single element of a system, but those of the system as a whole. Let us limit an effect of lenses by aperture and by main parameters determining the geometry of a system. Thus, information losses due to aberrations will not be taken into accaunt. Let us consider a version of the scheme of optical processing shown in Fig.1. The scheme is represented in Fig.3. The optical

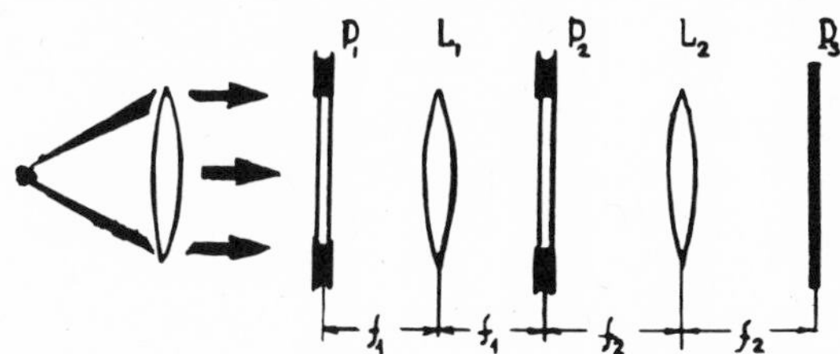

Fig.3. The scheme of optical image filtering.

filtering of images is produced here by means of two SLM (for image input and filtering) and by two lenses producing Fourier-transform. The first lens produces an image spectrum multiplied by the function determined by the second SLM (filter). The second lens produces an image that is reconstructed with appropriate correction. There is a definite relation between the required resolution and field sizes of two SLM and lens parameters.

The maximum volume of information recorded on SLM is determined by the following expression:

$$I_c = N \log_2 (m+1) ,$$

where N is the number of spatial elements and m is the number of distinguishable half-tones. Tn the case of holographic recording (for example, in certain versions of a memory system) $N = N_\ell \cdot N_\alpha$, where N_ℓ is the number of elements at the plane, N_α is the number of distinguishable angle directions in each element. When information is introduced in the first SLM of the system in Fig.3. $N = N_\ell$ and $N_\alpha = 1$ (difference in directions is related to object frequency characteristics only). It was noted that the maximum density of information recorded by light cannot exceed 10^9 bit/cm^2 (usually it appears to be much less) notwithstanding the fact that SLM characterized by the maximum information capacity (chalcogenide vitreous semiconductor films, for example) allow to record densities greater than 10^{10} bit/cm^2 [8].

In order to calculate the relation between information density and the required resolution of the first and the second SLM let us note that two lenses of equal apertures D and focal distances F are used.

Let sizes of an object recorded on SLM be and information capacity of SLM be

$$I = N\log_2(m+1) = 4\nu_{x\,lim}\nu_{y\,lim}\,l_x l_y \log_2(m+1) =$$

$$= 4S\nu_{x\,lim}\nu_{y\,lim}\log_2(m+1).$$

If we neglect the part of information that is transmitted by means of the number of half-tones, then the information density I' of SLM information input is (Fig.3)

$$I' = I_{m=1}/S = 4\nu_x\nu_y .$$

If an objiect recorded on SLM is centered relatively to an exis of the lens producing Fourier transform and if it is known to be situated in the front focal plane of the lens, then the following relation will be fulfilled (we consider one-dimensional case):

$$(D/2 - x)/F = tg\alpha_{lim} = \sin\alpha_{lim} = \lambda\nu_{x\,lim} .$$

It leads to the expression:

$$\nu_{x\,lim} = (D/2 - x)/\lambda F .$$

One can see that frequencies that are transmitted in such a system without losses inside it are limited as shown in Fig.4, with maxumum value $\nu_{x\,lim} = D/2\lambda F$ at $x = 0$ (in the centre of an objiect image) and $\nu = 0$ at $x = D/2$. If it is required to transmit a certain frequency ν_α all over an image field of a length l_a , then an aperture D must be determined from the

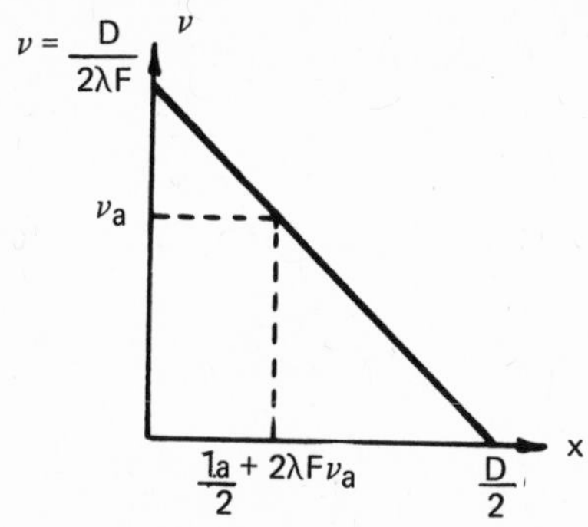

Fig.4. Dependence of frequency on coordinates of object points.

condition:

$$D = l_a + 2\lambda F \nu_a .$$

Now let us consider the ratio of information densities in an input SLM and in a processor SLM.

At the plane P_2 frequencies of a Fourier-image will be given by the values $\nu_{\xi lim} = \frac{D/2 - \xi}{\lambda F}$ and $\nu_{\eta lim} = \frac{D/2 - \eta}{\lambda F}$. Taking into consideration scale coefficient $1/\lambda F$ we have: $\nu_{x lim} = \xi/\lambda F$ and $\nu_{y lim} = \eta/\lambda F$

Then

$$x_l = D/2 - \xi_l \; ; \quad y_l = D/2 - \eta_l .$$

Thus we have the following expression for the ratio of information densities:

$$\varphi = \frac{I'_{in}}{I'_p} = \frac{4\,\nu_{x lim}\,\nu_{y lim}}{4\,\nu_{\xi lim}\,\nu_{\eta lim}} = \frac{(D/2 - x_l)(D/2 - y_l)}{x_l\, y_l}$$

The density of information recorded in both SLM will appear to be the same if field sizes of an object are chosen as follows: $l_x = 2x_l = l_y = 2y_l = D/2$. In this case information field sizes at the planes P_1 and P_2

will apparently be of the same value. If at the given D and the given resolution of the first SLM it is required to increase the volume of introduced information, magnitudes of ℓ_x and ℓ_y must exceed $D/2$. Then the information density recorded on the second SLM must exceed the density of the input information. For instance, if the resolution of the first SLM is 20 lines/mm and if we have to transmit 600 x 600 elements and aperture is 2 cm, we have $\ell_x = \ell_y = 1,5$ cm.

$$\varphi = (D-\ell)^2/\ell^2 = 0.11$$

that is the information density in the second SLM exceeds that in the first by one order while the area occupied by it is by one order less than that in the first SLM. Hence, the second SLM must resolve not less than 60 lines/mm in order to take part in optical processing.

The examples of non-agreement between SLM characteristics are shown in Fig.5. Fig.5a, case 1, shows

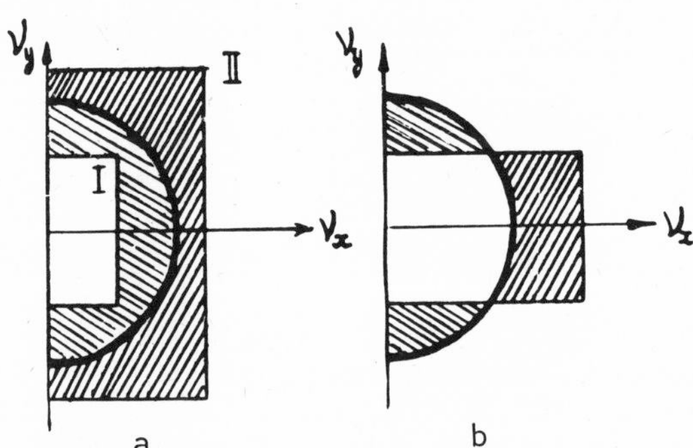

Fig.5. Matching between SLM information input and output characteristics; cases of isotropy and anisotropy of objiect field.

the case when the binary information density of the field formed by an effect of the first SLM along two components (x and y) is less than the binary information density that can be recorded on the second SLM; the latter has certain abundance in this case, its capacity is not fully filled in this system. Case 2 characterizes the situation when the field binary information density exceeds the density that can be recorded on SLM. In this case it transmits only a part of lower-frequency information present in a light field, while another part of it is lost. Fig.5b shows the case when x and y-components of information density at the planes P_1 and P_2 have different densities: x - component of the density in the field is greater than in the filter while y - component is less though the information density as a whole is the same. In this case a part of information (of an x - component) is lost while a part of the filter capacity (y - component) is unfilled. It can occur in the case of different field sizes in the x - and y - directions that produce different information density in the field in x - and y - directions, while the density of recorded information in a material of the filter is anisotropic as a rule.

The scheme of Fig.3 is not always unique and optimum scheme for optical filtering. As was shown by A.Vander Lugt [2], information density of the frequency plane can be considerebly increased if the scheme given in Fig.6 is used. In this case a plane wave modulated by an input information SLM is replaced by a con-

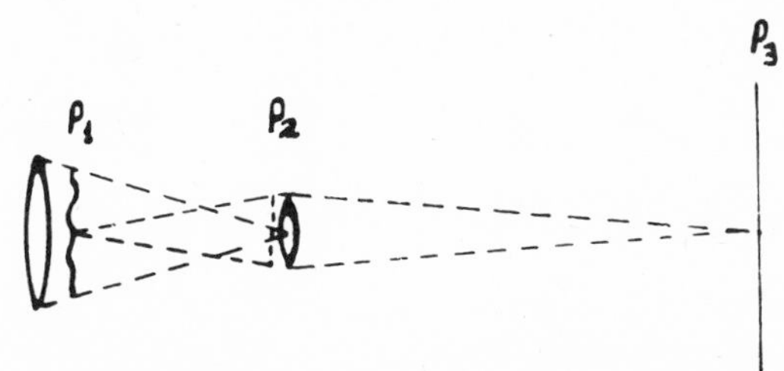

Fig.6. The scheme of optical filtering using convergent beam passing SLM.

verged spherical wave. Nevertheless it should be noted that the above-mentioned considerations are referred to systems with aberrationless objectives. Real objectives produce Fourier transform with less accuracy comparing to objectives without aberration, therefore they provide worse resolution of an image both in the center and at the edges. it will affect the conditions of matching between SLM characteristics and those of lenses in optical processing systems. Other factors that must be considered when matching the characteristics are additional information losses due to limited dynamic range and noises introduced by a SLM-filter. One may read about possible effect of these factors in [9,10,3], though accurate calculations must be related to concrete conditions.

References

1. A. V a n d e r L u g t, Design Relationships for Holographic Memories, Applied Optics, vol.12, p.1675, 1973.

2. A. V a n d e r L u g t, Packing Density in Holographic Systems, Applied Optics, vol.14, p.1081,1975.

3. S.B. G u r e v i c h, Informational Capacity of Coherent Optical Processing Systems, Optical Information Processing, G. S t r o k e et al Eds, Plenum Press, New York, 1976.

4. D. C a s a s e n t, Materials and Devices for Optical Information Processing. G. S t r o k e et al. Eds, Plenum Press, New York, 1976.

5. A.I. K o s a r e v, V.K. S o k o l o v, Space-Temporal Light Modulators, Zarubeznaya Radioelectronica, N°8, p.56, 1974 (in Russian).

6. A. V a n d e r L u g t, Coherent Optical Processing, Proc. of IEEE, v.60, p.1300, 1974.

7. S.B. G u r e v i c h. Effectivity and sensitivity of television systems, "Energy", Moscow-Leningrad, 1964 (in Russian).

8. S.B. G u r e v i c h, V.K. S o k o l o v, On the maximum information capacity of holographic systems, Z.T.Ph., v.43, p.645, 1975 (in Russian).

9. I.W. G o o d m a n, Film-Grain Noise in Wavefront Reconstruction Imaging, JOSA, v.57, p.493, 1967.

10. R.I. C o l l i e r, C.B. B u r c k h a r d t, L.H. L i n, Optical Holography, A.P. New York, 1971.

INFLUENCE OF PROBABLE CHARACTER OF DETECTION PROCEDURE AND LIGHT FIELD FLUCTUATIONS UPON PHOTORECEPTION QUALITY

V.M. Efimov, A.M. Iskoldskii

Institute of Automation and Electrometry
Novosibirsk, USSR

ABSTRACT

The image, being detected, is a rather slowly changing in time and space transparency, modulating the intensity of a stationary light field. Such reproduction error components as an accidental one, resulting from light field fluctuations, "shot" noise of a detector, and dark current additive noise, and a systematic one resulting from smoothing effect of a readout aperture, are taken into account.

Relationships for the mean-square error at high enough reproduction accuracy have been obtained, they allow to optimize a readout aperture size.

Transparencies were classfied according to behavior of the transparency "correlation function" under three types, nondifferentiated (highcontrast), once differentiated (mean-contrast), twice and more differentiated (low-contrast) transparencies.

A "determinated" model is often used when discussing the problems of image recording and processing; in the model some function [1] is correlated with an object and its image. On the other hand, more adequate discription can be obtained within the bounds of classical (or quantum) statistical optics, where a light field is considered to be accidental [2].

Further we shall regard a combined model preserving the main features of determinated and probable descriptions.

High-speed photographic recording deals with time--dependent objects, that is a light field of these objects is not stationary according to the definition; and it makes difficult to get constructive theoretical results. This obstackle can be overcome if consider the time and space stationary accidental light field $v(x,t)$, being modulated by the relatively slowly changing scalar function $\sqrt{f(x,t)}$ - the transparency $(0 \leq f(x,t) \leq 1)$. We are interested in the function $f(x,t)$, (x-are the coordinates of a point on a plane).

Taking modulation into account, the function of a light field mutual coherence will be:

$$\Gamma_f(x_1,x_2;t,t+\tau)=\sqrt{f(x_1,t)f(x_2,t+\tau)}\langle v(x_1,t)v^*(x_2,t+\tau)\rangle,$$

where

$$\langle v(x_1,t)v^*(x_2,t+\tau)\rangle=\Gamma(x_1,x_2,\tau)$$

is the function of the mutual space-time coherence of a stationary light field.

If we can neglect the changes of the function $f(x,t)$ in the intervals of time and space field coherence $v(x,t)$ then the function of this field munual coherence

$$\Gamma_f(x_1,x_2,\tau)=f(x_1,t)\,\Gamma(x_1,x_2,\tau),$$

and its mutual spectrum density

$$\tilde{\Gamma}_f(x_1,x_2,\nu)=f(x_1,t)\,\tilde{\Gamma}(x_1,x_2,\nu),$$

where $\tilde{\Gamma}(x_1, x_2, \nu)$ is the mutual spectral density of the field $v(x,t)$.

The model being under consideration, complies with the cases when macroscopic "movements" of the object don't practically influence upon a spectral formulation of illumination being generated or reflected by it .

When discussing a detecting process, we shall consider only two main detector properties, that is the photocurrent dependence on the incoming illumination frequency $\mathcal{E}(\nu)$ and an accidental nature of photocounts.

The first detector property can be taken into account if we shall introduce the efficient light field $v_э(x,t)$ with the mutual spectral density *

$$\tilde{\Gamma}_э(x_1,x_2,\nu)=\mathcal{E}(\nu)\tilde{\Gamma}_f(x_1,x_2,\nu)=f(x_1,t)\mathcal{E}(\nu)\tilde{\Gamma}(x_1,x_2,\nu) \quad (1)$$

and the mutual coherence function

$$\Gamma_э(x_1,x_2,\tau)=f(x_1,t)\int_0^\infty \mathcal{E}(\nu)\tilde{\Gamma}(x_1,x_2,\nu)\exp(-i2\pi\nu\tau)\,d\nu ,$$

and if we shall correlate this field with the Poisson photocount flow ("electronic" image) of the intensity

$$\lambda(x,t)=v_э(x,t)\,v_э^*(x,t), \quad (2)$$

where $\lambda(x,t)$ is the average specific number of photocounts, being generated at the point x in the time t . Relation (2) represents a non-linear operation being realized at a photocathode, that is square-low detection.

In connection with the fact that the efficient wave field is accidental, $\lambda(x,t)$ is an accidental

---

* Such a description corresponds to the fact that $\mathcal{E}(\nu)$ is a module square of a linear filter amplitude-frequency characteristic, which influences upon the field $v(x,t)$ [2].

process, as well. Its average value is connected with statistical characteristics of the efficient field by the following relationship

$$\langle \lambda (x,t) \rangle = \Gamma_э (x,x,0) = \alpha f(x,t) \quad , \qquad (3)$$

where $\alpha = \int_0^\infty \mathcal{E}(\nu)\, \tilde{\Gamma}(x,x,\nu)\, d\nu$ characterizes the efficient illumination level.

Let us consider an idealized model - a photocounter in the volume $V = ST$, where S is an eye square, T is cumulation time. The integration operation over the volume V can be described when introducing the weighting function of a readout device $w(x-\xi, t-\theta)$, which is equal to one, when variable integrations are within V , and which is, otherwize, equal to zero, that is

$$w(x-\xi, t-\theta) = \begin{cases} 1, \text{if } & x-\xi\, ;\, t-\theta \in V, \\ 0, \text{if } & x-\xi\, ;\, t-\theta \,\bar{\in}\, V. \end{cases}$$

The number of photocounts $n(x,t)$, registered within the volume V (the centre is at the point x, t) according to which one should estimate the value of the required modulating function $f(x,t)$, is the result of measurement. The number of signal photocounts $n(x,t)$ was distributed according to the Poisson law with the following parameter

$$\Lambda = \iint \lambda(\xi,\theta)\, w(x-\xi\, ;\, t-\theta)\, d\xi\, d\theta \; , \qquad (4)$$

that is

$$P(n) = \frac{\Lambda^n}{n!} \exp(-\Lambda) .$$

Then the average value of the number of signal photocounts

$$\langle n(x,t) \rangle = \langle \Lambda \rangle = \iint \langle \lambda(\xi,\theta) \rangle\, w(x-\xi, t-\theta)\, d\xi\, d\theta \, . \qquad (5)$$

From (5) with the account of (3) it follows that

$$\langle n(x,t)\rangle = \alpha \iint f(\xi,\theta)\, w(x-\xi,\, t-\theta)\, d\xi\, d\theta\,. \qquad (6)$$

The formula (6) takes account of the fact that the average signal value is proportional to convolution of modulating and weighting functions of the readout device, i.e. counter.

We should evaluate the required modulating function $f(x,t)$ according to the results of measurement $n(x,t)$. If $f(x,t)$ is practically constant on characteristic eye sizes of the readout device, then

$$\langle n(x,t)\rangle \cong \alpha V f(x,t)$$

and we can assume

$$\hat{f}(x,t) = (1/\alpha V)\cdot n(x,t)\,. \qquad (7)$$

as an estimation of the value $f(x,t)$ at the point (x,t).

The use of estimation (7) leads to arising of systematic and accidental error components.

The mean-square error can be the characteristic of the accuracy of such an estimation

$$\delta^2 = \sigma^2 + \varepsilon^2, \qquad (8)$$

where

$$\varepsilon^2 = \Big[f(x,t) - (1/V)\iint f(\xi,\theta)\, w(x-\xi,\, t-\theta)\, d\xi\, d\theta\Big]^2\,. \qquad (9)$$

The first addend in (8) characterises an accidental error component resulting from two factors, i.e. light field fluctuations and a probable mechanism of photoelectron generation. The second addend describes a systematic error caused by a smoothing effect of a weighting function of the counter.

According to the Poisson character of distribution the dispersion of an accidental error component

$$\sigma^2 = (1/\alpha V)^2 \{\langle\Lambda^2\rangle + \langle\Lambda\rangle - \langle\Lambda\rangle^2\} =$$
$$= (1/\alpha V)^2 \{\langle[\iint \lambda(\xi,\theta)\, w(x-\xi; t-\theta)\, d\xi\, d\theta]^2\rangle +$$
$$+ \iint \langle\lambda(\xi,\theta)\rangle\, w(x-\xi; t-\theta)\, d\xi\, d\theta - \tag{10}$$
$$- [\iint \langle\lambda(\xi,\theta)\rangle\, w(x-\xi; t-\theta)\, d\xi\, d\theta]^2\} .$$

If neglect the changes $f(x,t)$ within the aperture limits V we shall have

$$\sigma^2 \cong (1/\alpha V)^2 \{f^2(x,t) \iint d\xi\, d\theta\, R_\lambda(\xi,\theta)\, w_0(\xi,\theta) + \tag{11}$$
$$+ \alpha V f(x,t)\} ,$$

where

$$w_0(\xi,\theta) = \iint w(\xi+\xi_1; \theta+\theta_1)\, w(\xi_1,\theta_1)\, d\xi_1\, d\theta_1$$

is an autoconvolution of the counter weighting function. $R_\lambda(\xi,\theta)$ is a correlative function of $\lambda(x,t)$ intensity, when $f(x,t)=1$.

One should obviously reduce V , to reduce the systematic error in the estimation of a required function, according to (9); however, in this case the accidental error component characterized by (11) increases indefinitely.

Really, the first component of error dispersion caused by an accidental nature of a light field

$$\lim_{V\to 0} [(1/\alpha V)^2] f^2(x,t) \iint d\xi\, d\theta\, R_\lambda(\xi,\theta)\, w_0(\xi,\theta) =$$
$$= [(1/\alpha V)^2] f^2(x,t) R_\lambda(0,0) \iint w_0(\xi,\theta)\, d\xi\, d\theta =$$
$$= (R_\lambda(0,0)/\alpha^2)\, [f(x,t)]^2$$

increases, but remains limited, while the second one, caused by a discret property of a signal behaves like an inversely proportional one to the volume V. According to an alternative behaviour of error components it follows that there exist optimum parameters of a weighting function for a readout device, and an error of measurement can't be made an infinitesimal quantity. It is this error which characterizes "limited" (in a sense of a mean-square error of restoration) photorecording possibilities.*

From the obtained relationships it follows that error accidental components are multiplicative in reference to the modulating function $f(x,t)$. When the illumination α is small, a "shot" component makes the main contribution to an accidental error. Relative to a systematic error one can note that it depends on the degree of the modulating function $f(x,t)$ "smoothness".

Let us further consider the additive noise influence, supposing it to be initiated by the Poisson flow of noise photocounts (dark current) with the space and time stationary intensity $\mu(x,t)$. In this case the component

$$\delta_{ш}^{2} = (1/\alpha V)^{2}\Big\{\iint d\xi\, d\theta\, R_{\mu}(\xi,\theta)\, w_{0}(\xi,\theta) + \langle\mu\rangle V + \langle\mu\rangle^{2} V^{2}\Big\}. \qquad (12)$$

will be added to a mean-square error δ^{2}. The first and second addends (12) are analogous to the addends (11). The last addend (12) results from an average value of a dark current, producing a systematic shift (increase) of the counter readings. If we correct for this shift then the additional component

$$\sigma_{ш}^{2} = (1/\alpha V)^{2}\Big\{\iint d\xi\, d\theta\, R_{\mu}(\xi,\theta)\, w_{0}(\xi,\theta) + \langle\mu\rangle V\Big\}. \qquad (13)$$

---

* In general, the error magnitude can be smaller if a weighting function shape is optimum, i.e. it allows for a priori light field and modulating function data.

The behaviour $\sigma_{ш}^2$ according to V is analogous to that of σ^2. This error component prevents the decrease in the counter aperture when V is small, as well

$$\sigma_{ш}^2 \cong \frac{1}{\alpha^2} R_\mu (0, 0) + \frac{\langle \mu \rangle}{\alpha^2 V} .$$

Besides, this error component is more sensitive to an illumination level.

Let us consider a situation when reproduction errors are small enough, supposing that coordinates of each photocount within the range of existance V_0 of the function $f(x, y, t)$ are determined; and the reproduction operation is carried out by successive scanning V_0 with the aperture $V = \Delta_x \Delta_y \Delta t$, and calculation of the photocount number within this aperture ranges.

The error is small when Δx, Δy, and Δt substantially exceed the corresponding correlation intervals of the intensities $\lambda(x, y, t)$ and $\mu(x, y, t)$. In this case relationships (11) and (13) will be of the form:

$$\sigma^2 = (V_\lambda / V)(R_\lambda(0,0,0)/\alpha^2) f^2(x,y,t) + (1/\alpha V) f(x,y,t), \quad (14)$$

$$\sigma_{ш}^2 = (V_\mu / V)(R_\mu(0,0,0)/\alpha^2) + (1/\alpha^2 V)\langle \mu \rangle , \quad (15)$$

where

$$V_\lambda = \frac{1}{R_\lambda(0,0,0)} \iiint d\xi\, d\eta\, d\tau\, R_\lambda(\xi, \eta, \tau) ,$$

$$V_\mu = \frac{1}{R_\mu(0,0,0)} \iiint d\xi\, d\eta\, d\tau\, R_\mu(\xi, \eta, \tau) .$$

are correlation "volumes" of fluctuations of the intensities $\lambda(x,y,t)$ and $\mu(x,y,t)$. From (14) and (15) it follows that when a readout aperture "volume" V is great enough, the summary dispersion of an accidental error component of a reproduction decreases inversely proportionally to V :

$$\sigma_\Sigma^2 = \sigma^2 + \sigma_w^2 = \beta / V ,$$

where $\beta = V_\lambda \, (R_\lambda(0,0,0)/\alpha^2) f^2(x,y,t) +$

$$+(1/\alpha) f(x,y,t) + V_\mu (R_\mu(0,0,0)/\alpha^2) + (1/\alpha^2)\langle\mu\rangle .$$

The mean-square error according to the region V_0 of the function existance $f(x,y,t)$*

$$\langle \delta_\Sigma^2 \rangle = \langle \sigma_\Sigma^2 \rangle + \langle \varepsilon^2 \rangle = \langle \beta \rangle / V +$$

$$+[\langle f^2\rangle - \langle f\rangle^2]\{1 - 2(1/V)\iiint d\xi\, d\eta\, d\tau\, \rho_f(\xi,\eta,\tau)\, w(\xi,\eta,\tau) +$$

$$+(1/V^2)\iiint d\xi\, d\eta\, d\tau\, \rho_f(\xi,\eta,\tau)\, w_0(\xi,\eta,\tau)\}, \tag{17}$$

where the normalized "correlation function" of the image

$$\rho_f(\xi,\eta,\tau) = (1/[\langle f^2\rangle - \langle f\rangle^2]) \times (1/V_0) \times$$

$$\times \langle (f(x,y,t) - \langle f\rangle)(f(x+\xi, y+\eta, t+\tau) - \langle f\rangle)\rangle \tag{18}$$

---

* Herein the averaging sign means

$$\langle \psi \rangle = (1/V_0) \iiint \psi(x,y,t)\, dx\, dy\, dt .$$

Let us further consider transparencies with

$$\rho_f(\xi,\eta,\tau) = \rho_f(r) = \rho_f\left(\sqrt{(\beta_x\xi)^2+(\beta_y\eta)^2+(\beta_t\tau)^2}\right) \tag{19}$$

i.e. the correlation function depending on the "distance"

$$r = \sqrt{(\beta_x\xi)^2 + (\beta_y\eta)^2 + (\beta_t\tau)^2}$$

The factors β_x, β_y, β_t characterize the "speed" of the transparency change along corresponding coordinates.

Let the aperture of a readout device be

$$V = \Delta_x \Delta_y \Delta_t$$

In this case it follows from (17) that a minimum of the summary error being realized if

$$\beta_x \Delta_x = \beta_y \Delta_y = \beta_t \Delta_t = \Delta_0 \tag{20}$$

This result corresponds to a simple reason, i.e. typical size of a readout aperture must be larger in the direction, in which $f(x,y,t)$ changes more slowly.

For further analysis it is convenient to classify transparencies into three types:

1. Nondifferentiated (high-contrast) transparencies. For these transparencies the "correlation function" at small argument values

$$\rho_f(r) \cong 1 + \rho'(0+)\, r\,, \tag{21}$$

and a systematic error square is

$$\langle \varepsilon^2 \rangle \sim \Delta_0 \;. \tag{22}$$

2. Once differentiated (mean-contrast) transparencies. For them

$$\rho_f(r) \cong 1 + \frac{\rho^{II}(0)r^2}{2!} + \frac{\rho^{III}(0_+)r^3}{3!}, \tag{23}$$

and a systematic error square is

$$\langle \mathcal{E}^2 \rangle \sim \Delta_0^3 . \tag{24}$$

3. Twice and more differentiated (low-contrast) transparencies, for which

$$\rho_f(r) \cong 1 + \frac{\rho^{II}(0)r^2}{2!} + \frac{\rho^{IV}(0)r^4}{4!}, \tag{25}$$

and

$$\langle \mathcal{E}^2 \rangle \sim \Delta_0^4 . \tag{26}$$

Thus, at high reproduction accuracy

$$\langle \delta_\Sigma^2 \rangle = \langle \sigma_\Sigma^2 \rangle + \langle \mathcal{E}^2 \rangle = \frac{\langle \beta \rangle \beta x \beta y \beta t}{\Delta_0^3} + C_m^{III} \Delta_0^m, \tag{27}$$

where $m = 1, 3, 4$ according to a degree of the transparency contrast; C_m^{III} is a constant.

From (27) it is easy to switch over to the cases of less argument number. E.g., for time stationary transparencies

$$\langle \delta_\Sigma^2 \rangle = \frac{\langle \beta \rangle \beta x \beta y}{T \Delta_0^2} + C_m^{II} \Delta_0^m, \tag{28}$$

where T is the recording time; and for $f(t)$

$$\langle \delta_\Sigma^2 \rangle = \frac{\langle \beta \rangle \beta t}{L_x L_y \Delta_0} + C_m^{I} \Delta_0^m, \tag{29}$$

where L_x, L_y – is a transparency size. When Δ_0 is optimum, a share of a summary error accounting for an accidental component

$$\frac{\langle \sigma_\Sigma^2 \rangle}{\langle \delta_\Sigma^2 \rangle} = \frac{m}{n+m} , \qquad (30)$$

where n = 3, 2, 1, according to a number of arguments.

A share of a summary error, accounting for a systematic component is, respectively,

$$\frac{\langle \varepsilon^2 \rangle}{\langle \delta_\Sigma^2 \rangle} = \frac{n}{n+m} . \qquad (31)$$

To provide a high accuracy of reproduction, a characteristic volume of the readout aperture is required to be much greater, than the correlative volume V_λ ; and a mean number of photocounts within the aperture limits $(\langle N \rangle = \alpha V)$ to be great enough.

If we neglect dark current, then

$$\langle \sigma_\Sigma^2 \rangle = \langle \sigma^2 \rangle = (V_\lambda / V)(R_\lambda(0,0,0)/\alpha^2)\langle f^2 \rangle + (1/\alpha V)\langle f \rangle . \qquad (32)$$

E.g., for the Gaussian light field $(R_\lambda(0,0,0) = \alpha^2)$ the display accuracy is of the order of 10% $(\langle \sigma^2 \rangle \sim 0{,}01)$ can be achieved, when $V_\lambda / V \leq 0.01$ and $\alpha V \geq 100$.

The qualitative conclusions, obtained above, also remain true for the case, when a counter apparatus function differs from the parallelepiped.* In general, in case of a linear filtration

$$\langle \sigma_\Sigma^2 \rangle \geq \int \frac{\sigma_f^2 S_f(\omega)[V_\lambda \langle f^2 \rangle + (1/\alpha)\langle f \rangle]}{\sigma_f^2 S_f(\omega) + V_\lambda \langle f^2 \rangle + (1/\alpha)\langle f \rangle} d\omega \qquad (33)$$

where $\sigma_f^2 = \langle f^2 \rangle - \langle f \rangle^2$, $S_f(\omega)$ is a Fourier analysis from $\rho_f(\xi, \eta, \tau)$, i.e. "spectrum density" of the transparency.

---

* For other counter weighting functions

$$v = \left[\int w(\xi,\eta,\tau)\, d\xi\, d\eta\, d\tau\right]^2 \Big/ \int w^2(\xi,\eta,\tau)\, d\xi\, d\eta\, d\tau .$$

Relationship (33) corresponds to the case of an optimum filtration of an additive mixture of a signal and white noise with the spectrum density $C = V_\lambda \langle f^2 \rangle + \langle f \rangle / \alpha$. It is obvious, that a high-quality photoreception corresponds to small C, i.e. to large α and small V_λ.

REFERENCES

1. A r r i e l, R o z e n f e l d. Picture processing by computer. Transl. from Engl. M. "Mir", 1972.

2. J.R. K l a u d e r and E.C.G. S u d a r s h a n. Fundamentals of quantum optics. Transl. from Engl. M. "Mir", 1970.

OPTICAL DATA TRANSFORMATION AND CODING IN ELECTRO-OPTICAL PROCESSORS

A.A.Vasiliev, I.N.Kompanets, V.N.Morozov

P.N.Lebedev Physical Institute, Ac.Sci.

Lenin prospect 53, Moscow, USSR

1. CONTROLLED PHASE TRANSPARENCIES IN COHERENT OPTICAL SYSTEMS REALIZING WALSH AND HILBERT TRANSFORMATIONS

To create coherent optical processors /1/ with ample functional opportunities of parallel block data processing in the real time, it is necessary to provide fast data input, as well as the control of optical system impulse responses. These problems can be solved by means of electrically and optically addressed space-time optical modulators or controlled transparencies (CT) termed as "tunable spatial filters" /2/ when being used for the impulse response control.

Previously /3,4/ used were electrical and optically controlled transparencies acting as tunable filters in simple electrooptical processors. One of them realized a modified Walsh transformation /3/. The modified Walsh functions differ from the usual ones by a substitution of "0" for "-1", that reduces the efficiency of this transformation owing to its broken orthogonality. Such a substitution is due to the use of a controlled transparency based on the dynamic scattering effect in nematic liquid crystals realizing only positive values of complex transmittance.

One can give positive values of complex transmittance by using controlled phase transparencies, e.g.

based on the field S-effect in nematic liquid crystals with positive dielectric anisotropy /5/. In the present paper controlled transparencies based on S-effect are described as spatial filters in optical processors realizing and utilizing Hilbert and Walsh transformations /6,7/. These transformations are of great interest since they provide effective methods of two-dimensional optical data processing, which is important for the character recognition and pattern coding systems.

Controlled Transparency

Matrix addressing of elements has been used in experimental samples of controlled phase transparencies as the most suitable addressing technique for the realization of two-dimensional functions with separable variables including Walsh functions and "phase knife" (Hilbert mask). The mixture of nematic liquid crystals has been used as the working substance consisting of 23.3% cyanidephenyl ether of butylbenzene acid, 30% cyanidephenyl ether of hexylbenzene acid and 4.67% cyanidephenyl ether of heptilbenzene acid. The mixture is characterized by positive dielectric anisotropy ($\Delta\varepsilon$ = + 23) and temperature interval of nematic phase existence from 10°C to 70°C /8/.

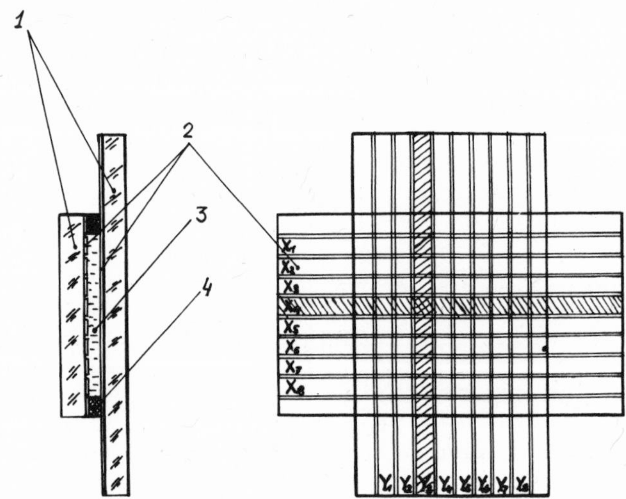

Fig. 1. Experimental model of the controlled phase transparency: 1 - quartz substrates, 2 - transparant electrodes, 3 - liquid crystal, 4 - spacers.

A liquid crystal layer is sandwiched between two quartz substrates 1 (Fig. 1) with transparent strap electrodes 2 mm wide and 50 μm spaced. Thickness of the liquid crystal layer of 12 μm is fixed by mica spacers 4. Initial homogeneous alignment of liquid crystal molecules is accomplished by rubbing the substrates along their vertical strap electrodes.

The addressing scheme of matrix CT is shown in Fig. 2. Initial alignment of liquid crystal molecules is carried out with the help of ac voltage of 20-50 kHz applied via capacitors C_1-C_8 and C_9-C_{16} across all the vertical and horizontal electrodes of transparency. Its magnitude is controlled by potentiometer R_1. Each of X and Y electrodes can be connected with positive and negative poles of dc power supply via $\Pi_1-\Pi_8$ and $\Pi_9-\Pi_{16}$ switches. Additional realignment of molecules occurs only in the areas of a liquid crystal layer located between the electrodes connected with different poles of dc power supply. The potentiometer R_{18} allows us to control the magnitude of dc voltage so that an additional change of the phase delay of monochromatic light passing through the transparency electrodes switched on should be equal to π . If we assume that an initial complex transmittance of the transparency is equal to "+1" then for the light with the electrical vector placed in the same plane with the direction

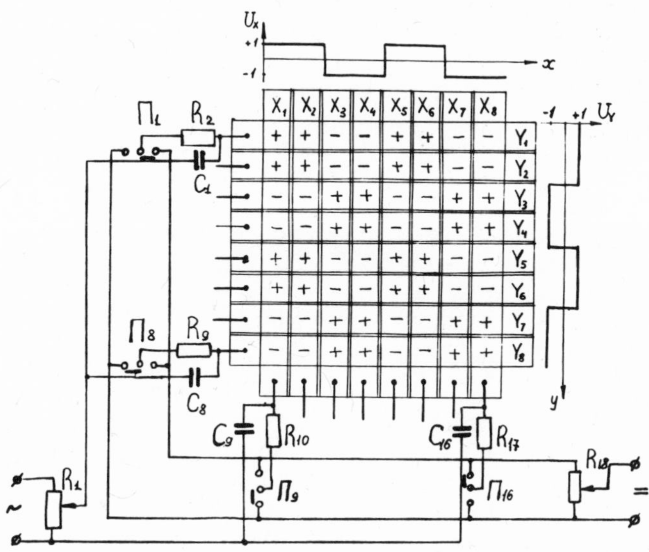

Fig. 2. Matrix transparency addressing scheme.

of propagation and direction of alignment of the liquid crystal molecules, phase delay changing of π corresponds to the complex transmittance equaling "-1".

As is known, two-dimensional Walsh function of (m,n) order is a product of the corresponding one-dimensional functions:

$$\mathrm{Wal}\,(m,n,x,y) = \mathrm{Wal}\,(m,x)\cdot \mathrm{Wal}\,(n,y) \tag{1}$$

where x and y are the spatial variables; m and n are the Walsh function numbers that are equal to the quantity of sign changes of the function per interval. Thus, positive voltage applied to the transparency electrode corresponds to "+1" and the negative one, to "-1". Then to obtain any of 64 two-dimensional Walsh functions it is sufficient to distribute the voltage between transparency electrodes so that is corresponds to one-dimensional Walsh functions. Under these conditions a complex transmittance of each transparency element is the product of signals supplied to the corresponding X and Y electrodes, since the change of liquid crystal alignment forced by dc control voltage, does not depend on the polarity of this voltage (a summary effect of dc and ac voltage is proportional to the rms value of voltage).

For example, Fig. 2 shows schematically the voltage distribution at the transparency electrodes generating Wal (3,3,x,y) Walsh function. Figure 3 illustrates images of the transparency working area generating various Walsh functions in crossed polarizers.

Walsh Transformation. Optical Scheme and Computation of Transformation Coefficients.

To obtain the coefficients of expansion in Walsh functions of two-dimensional function f(x,y) defined over the unit square, we need the dot product of f(x,y) and Wal (m,n,x,y), namely,

$$F(m,n) = \int_0^1\int_0^1 f(x,y)\cdot \mathrm{Wal}(m,n,x,y)\,dx\,dy \tag{2}$$

The optical scheme for computing the dot product of two functions presented in the form of complex transmittances, is shown in Fig. 4. The beam of He-Ne laser 1 with wavelength $\lambda = 0.63\,\mu m$, collimated by a telescope 2, is directed to Mach-Zender interferometer, formed by the mirrors 5,6,3,9. Transparency 7 with the amplitude transmittance $t_1(x,y) = f(x,y)$ (3)

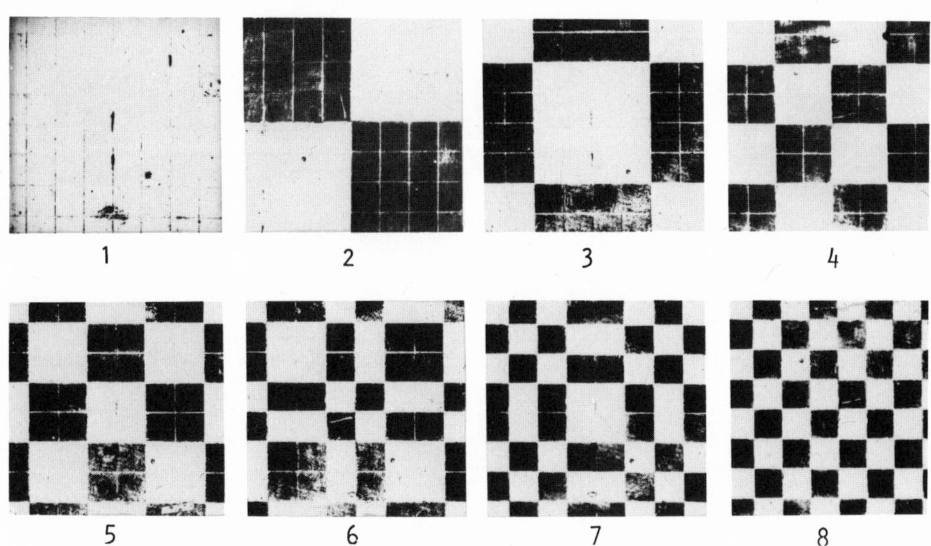

Fig. 3. Photographs (obtained in crossed polarizers) of phase controlled transparency generating Walsh functions of various orders:
1) Wal (0,0,x,y); 2) Wal (1,1,x,y);
3) Wal (2,2,x,y); 4) Wal (3,3,x,y);
5) Wal (4,4,x,y); 6) Wal (5,5,x,y);
7) Wal (6,6,x,y); 8) Wal (7,7,x,y).

and controlled transparency 8, generating 64 Walsh functions:

$$t_2(x,y) = \mathrm{Wal}(m,n,x,y), \quad 0 \leqslant m,n \leqslant 7 \qquad (4)$$

are arranged close to each other in the upper arm of the interferometer.

The light passing through the transparency is focused by a lens 10 on the input photomultiplier pupil 11. If the diameter h of the photomultiplier input pupil is small enough, then the intensity distribution in the pupil is supposed to be uniform. In this case the photomultiplier input V is proportional to the light intensity in the lens focus:

$$V \sim I_0 \sim |E(\xi, \eta)|^2 \qquad (5)$$

when $\xi = \eta = 0$, where $E(\xi, \eta)$ is the light field amplitude, ξ and η are the coordinates of the focal plane.

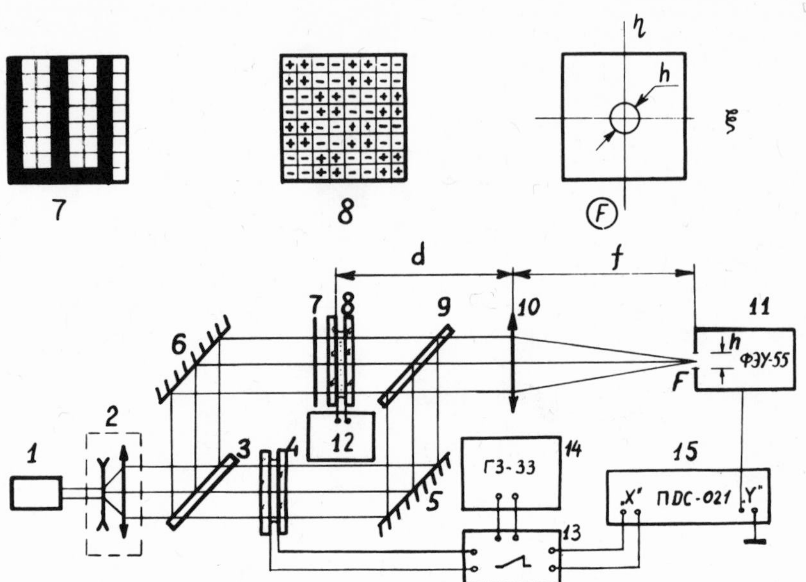

Fig. 4. Schematic diagram of measuring Walsh transform coefficients: 1 - laser; 2 - telescope; 3,9 - semi-transparent mirrors; 4 - liquid crystal cell (phase shifter); 5,6 - mirrors; 7 - transparency with an input signal f(x,y); 8 - controlled transparency; 10 - lens; 11 - photomultiplier; 12 - addressing scheme of the transparency; 13 - time scanning block; 14 - ac generator; 15 - automatic recording potentiometer.

If the lower arm of the interferometer (see Fig. 4) is completely closed then the light amplitude distribution in the focal plane is proportional to Fourier-transform $\mathcal{F}$ of the light amplitude distribution in the plane of transparency 8 /9/:

$$E(\xi,\eta) = \mathcal{F}(\xi/\lambda f, \eta/\lambda f)\exp[j\pi/\lambda f \times (1-d/f) \times (\xi^2+\eta^2)] \qquad (6)$$

where f is the focal length of the lens 10, and d is the distance between transparency and the lens.

In the lens focus ($\xi = \eta = 0$) the phase factor

$$\exp\left[j\pi/\lambda f \cdot (1-d/f)\cdot(\xi^2+\eta^2)\right] \qquad (7)$$

is equal to 1 and

$$\mathcal{F}(\xi/\lambda f, \eta/\lambda f) = A_0 \iint_S t_1 \cdot t_2 \cdot \exp(-j2\pi(\xi x/\lambda f + \eta y/\lambda f))\,dxdy \tag{8}$$

where A_0 is the amplitude of the wave, which illuminates transparency 7; S is equal to 64 a^2, that is the CT working area (a is the size of a controlled transparency element).

At the point with the coordinates $\xi = 0, \eta = 0$ the complex amplitude is

$$\mathcal{F}(0,0) = A_0 \iint_S t_1 t_2 dxdy = SA_0 F(m,n)/jf\lambda \tag{9}$$

Thus, to measure modulus and argument of F(m,n) coefficients, it is sufficient to find the amplitude A_1 and phase φ_1 of the light coming in the photomultiplier input pupil from the upper arm (Fig. 4) of the interferometer (factor $SA_0/j\lambda f$ can be eliminated by calibration). For this purpose the liquid crystal cell with the size of working area equal to 16 x 20 mm was placed in the lower arm of the interferometer and served as a tunable phase shifter.

Initial alignment of the liquid crystal molecules in the phase-shifter cell is accomplished by ac voltage applied to the cell electrodes from the generator 14. Under the action of the sowtooth pulse voltage from the time scanning block 13, there occurs realignment of molecules of the liquid crystal layer in phase shifter cells. As a result, the phase φ_2 of the light coming in the photomultiplier input pupil from the lower arm of the interferometer, is changed. In this case the intensity registered by the photomultiplier depends on the amplitudes and phases of light beams produced in the interferometer arms. When changing voltage at phase shifter electrodes the modulation of light intensity is determined by

$$K = (I_{max} - I_{min})/(I_{max} + I_{min}) = 2A_1A_2/(A_1^2 + A_2^2) \tag{10}$$

where A_1 and A_2 are the amplitudes of light incident on the photomultiplier input pupil, from the upper and lower arms of the interferometer, respectively. The modulation index is maximal if A_1 is equal to A_2. This condition is realized by means of the iris diaphragm as an attenuator placed in the lower arm of the interferometer.

If the input signal is identically equal to one over the total working area of the transparency (7), then

$$f(x,y) = \text{rect}\ (x/8a) \cdot \text{rect}\ (y/8a) \quad \text{and } F = 1 \quad (11)$$

When the lower arm of the interferometer is open and the CT generates the Wal (0,0,x,y) function, the light amplitude A_1 registered by the photomultiplier is assumed to be equal to one. Thus, when A_1 equals A_2, and the CT generates Wal (m,n,x,y) transparency with the expanded signal f(x,y) being available in the input plane, the modulation index is

$$K = 2 \cdot |F(m,n)| \ /(1 + |F(m,n)|^2) \quad (12)$$

wherefrom

$$|F(m,n)| = 1 - (1-k^2)^{1/2}/K \quad (13)$$

Experimental Results

In the experiment the modulus and the argument of the coefficients F(m,n) were derived from the multiplier output dependence on the voltage across the phase shifter cell electrodes. For this the voltage V from the output of time scanning block 13 arrived in the "X" input of the two-coordinate automatic recording potentiometer ПДС-021, and input Y received a signal from the photomultiplier. From the start, a calibration curve was plotted while the controlled transparency was generating Wal (0,0,x,y) Walsh function, and the input signal was identically equal to 1. AC and DC voltages were of such magnitude that the calibration curve represented one intensity minimum and two maxima.

The dependence of registered intensity upon the phase φ_2 of the light emerged from the lower arm of interferometer is as follows:

$$I(\varphi_2) = I_{max} \sin^2(\varphi_2/2) \quad (14)$$

(with the voltage corresponding to the intensity minimum, we assume $\varphi_2 = 0$).

Similar plots were registered at the presence of a signal, while the CT was generating various Walsh functions. With the help of these plots the modulation index K was determined, and the coefficient modulus F(m,n) was calculated by using Eq. (13). The argument of the transformation coefficient was determined from the value of voltage V_o across the electrodes of the phase shifter cell, corresponding to the intensity minimum by the following formula:

$$\varphi_2 = \pm 2 \arcsin^{1/2} (I_o(\varphi_2) / I_{max}) \qquad (15)$$

where $I_o(\varphi_2)$ is the light intensity corresponding to voltage V_o. The sign before the radical depends on the direction of the intensity minimum shift.

A minimal width of intensity distribution in lens 10 focal plane depends on the CT size equal to 8a. Thus, the maximum possible size h_{max} of the input pupil in the photomultiplier is determined from the width of the main spectral maximum of the function:

$$t(x,y) = \mathrm{rect}\ (x/8a,\ y/8a) \qquad (16)$$

which is equal to

$$\mathcal{F}_o(\xi,n)=64a^2 \sin(8\ a\pi\xi/\lambda f)\cdot\sin(8\pi a\eta/\lambda f)/(8\pi a/\lambda f)^2\xi\cdot\eta \quad (17)$$

If we assume a possible intensity irregularity in the photomultiplier input pupil to be equal to 5%, then h_{max} can be determined from the condition:

$$\sin(4\pi a h_{max}/\lambda f)/(4\pi a h_{max}/\lambda f))^2 = 0.95 \qquad (18)$$

then

$$h_{max} = 0.031\ \lambda f/a \qquad (19)$$

In the case under consideration $\lambda = 0.63 \times 10^{-3}$ mm, f = 600 mm, a = 2 and then h_{max} = 0.006 mm. In the experiment, however, we used a diaphragm with the pinhole diameter of 0.01 mm, which increased the measurement error up to 14% due to the intensity irregularity in the input pupil of the photomultiplier.

Table I indicates theoretically calculated (numerator) and experimentally obtained (denumerator) values of several Walsh transform coefficients for the simplest amplitude signals, shown in Fig. 5. It is obvious that the measurement error of coefficients F(0,0) does not exceed 18%. It is caused by large dimensions of the input pupil and by inaccuracy of reproducing and positioning of the input transparency. An additional error results from inaccuracies of reproducing various Walsh functions by the CT, and from interferometric instability.

Table 1

m,n	$f_1(x,y)$	$f_2(x,y)$	m,n	$f_1(x,y)$	$f_2(x,y)$
0,0	$\frac{0.4370}{0.3600}$	$\frac{0.2030}{0.1730}$	4,4	$\frac{-0.0313}{-0.0936}$	$\frac{0.0156}{0.0167}$
1,1	$\frac{0}{0.0200}$	$\frac{-0.0156}{-0.0133}$	5,5	$\frac{-0.0313}{-0.0485}$	$\frac{-0.0156}{-0.0156}$
2,2	$\frac{-0.0313}{-0.2100}$	$\frac{-0.0156}{-0.0133}$	6,6	$\frac{0}{0.0180}$	$\frac{0.0156}{0.0156}$
3,3	$\frac{0.0313}{0.1300}$	$\frac{0.0470}{0.0390}$	7,7	$\frac{0}{0.0280}$	$\frac{-0.0156}{-0.0195}$

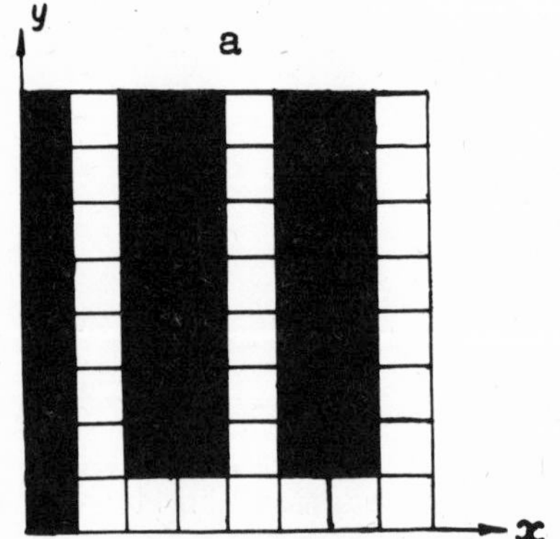

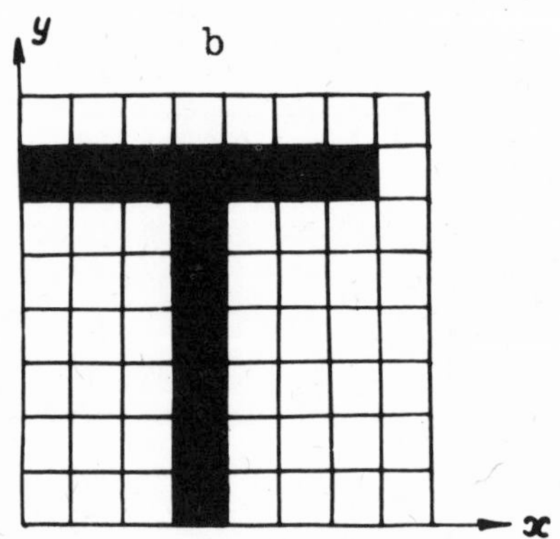

Fig. 5. Photographs of the input signals
a) $f_1(x,y)$; b) $f_2(x,y)$ (dark areas correspond to zero transmittance, the bright ones - to unit transmittance).

The signal transparency $f_2(x,y)$ was made and positioned more carefully in comparison with $f_1(x,y)$ one. While working with $f_2(x,y)$ transparency we made additional arrangements to stabilize the interferometer. This led to decreasing of measurement error, as shown in Table I. Note that improvement of the homogeneity of optical properties and switch responses of the CT materials over the total working area is important for diminishing the measurement error. This can be obtained by improving the initial alignment of the liquid crystal molecules due to the obliquely evaporated thin films /10/.

Hilbert Transformation

It is well known that the Hilbert transformation represents a convolution of the initial function f(x) and the kernel function:

$$g(x) = 1/\pi x \tag{20}$$

Since the Fourier-transform of g(x) equals

$$G(\omega_x) = j \cdot \operatorname{sgn} \omega_x \tag{21}$$

then the Hilbert-transform is usually accomplished with the help of a spatial frequency filter having the following transmittance:

$$G(\omega_x, \omega_y) = -\operatorname{sgn} \omega_x \cdot \operatorname{sgn} \omega_y \tag{22}$$

This expression, with accuracy to a sign, coincides with the complex transmittance of the transparency generating Wal (I,I,x,y) and placed in the space frequency plane of the spatial filtering optical scheme.

The experimental arrangement for the realization of Hilbert transformation is shown in Fig. 6. The laser beam is expanded by a lens 2, then it is focused by lens 3 at the plane 5 of the controlled transparency. The latter is analogous to the transparency considered above, and was used as a spatial filter. The distance d between the input and frequency planes is chosen so that the effective size of the transparency spectrum defined by

$$\Delta = 2\lambda d/a \tag{23}$$

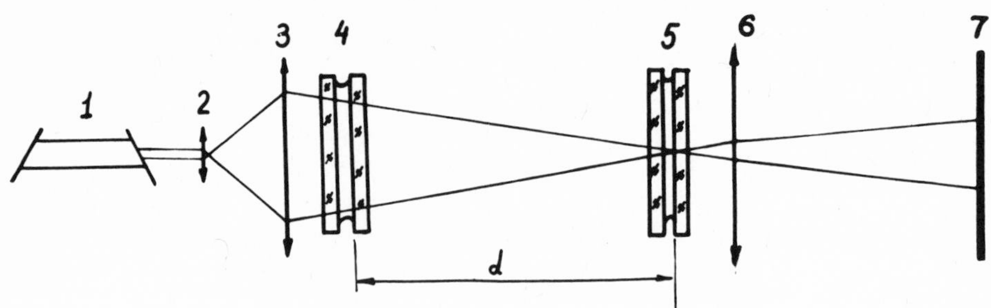

Fig. 6. Schematic diagram of generating Hilbert transforms: 1 - laser; 2,3,6 - lenses; 4,5 -controlled transparencies; 7 - screen (output plane).

(where a is the size of transparency element) is much more than the width of the transient field between the transparency electrodes (~ 20 μm). In the experiment the distance d is equal to 700 mm, which corresponds to the spectrum size of 0.42 mm. Distance d is a factor of twenty larger than the transient field width.

Controlled transparency 4 is used as an input signal. When the filter is switched off, the input signal image is formed at the screen 7 by lens 6. If the filter is switched on, then the Hilbert transform of the input function is formed in the output plane.

Figure 7 shows the image of the output plane of the optical scheme employing various input signals, the spatial filter being in "on" and "off" states. Hilbert transform images of defined functions contain the inherent boundary outlines of the "on" and "off" elements.

2. RECORDING OF HOLOGRAMS WITH REFERENCE BEAM CODING

The problem of information coding is one of the most complex and important problems arising with the creation of coherent electrooptical processors. The choice of optimum coding methods enables us to solve a lot of problems associated with optical data proces-

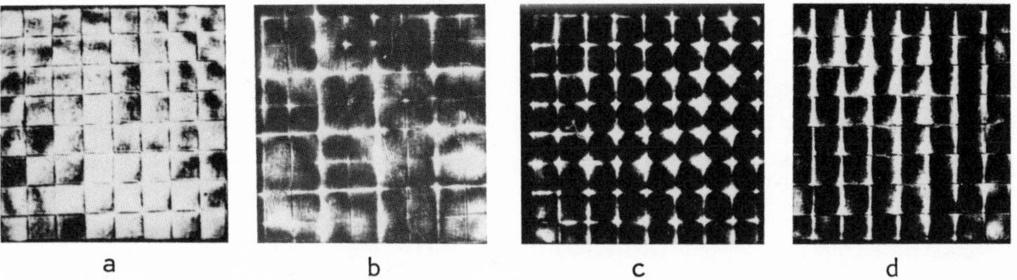

Fig. 7. Output plane patterns of the optical scheme (shown in Fig. 6): a) controlled transparency-filter in the "off" state; b) Hilbert transform of Wal (3,3,x,y) function; c) Hilbert transform of Wal (7,7,x,y) function; d) one-dimensional Hilbert transform of Wal (7,7,x,y) function.

sing. For example, a separate retrieval of superimposed holograms can be accomplished by means of coding of a reference beam /11/.

At present the capacity of holographic storage systems with irreversible storage medium is known to be about 10^8 bits at reasonable parameters of the optical schemes. This corresponds to the information packing density of 10^4 bits/mm^2 /12/. This value is 2-3 orders of magnitude lower than the limit defined by the resolution and proper noises of the storage medium. Under these conditions one can increase the packing density by an order of magnitude by means of the storage technique using superimposed holographic page in a thick photosensitive medium (see, for example, /13/).

For separate retrieval of superimposed holograms we must encode them by changing the angle of incidence of the reference beam /13/, or by using extensive reference sources (reference objects) /14/. A significant complication of the optical scheme is a shortcoming of the first method. In the second method various areas of the frosted glass (diffusor) having almost ideal encoding properties are used. Unfortunately, a replacement of reference objects was realized, up to now, by mechanical movement of the diffusor.

One can take advantage of the second method by making use of the controlled phase transparencies as reference objects. In /15/ the controlled transparencies, which realize binary pseudo-random signals, were proposed as reference objects. The binary pseudo-random signals are widely used in communication and radar techniques /16/. This section deals with the possibility of using various pseudo-random signals for the multiple-exposure hologram coding.

Reference Beam Encoding with the Help of Binary Pseudo-Random Signals

The analysis of Fourier hologram recording and retrieval using intensive reference objects shows that the amplitude distribution in the image reconstructed from the hologram is proportional to the convolution of the input signal and autocorrelation function of field distribution in the reference beam /17/. It means that undistorted separate retrieval of images from superimposed Fourier holograms is only possible if the autocorrelation functions of various reference sources are close to δ -functions, and the cross-correlation functions are close to zero.

The complex transmittance function of the frosted glass appearing as optical analog of white noise, is most suitable for this purpose. But nonlinearity of the modulation response of the majority of phase CT impedes an exact reproduction of a large number of phase quantization levels. This also results in a significant complication of addressing of multiple-element CT. Thus the binary pseudo-random signals are the most suitable means for phase controlled transparencies. Among them the Huffman sequences possess the best correlation properties /18/. For these signals the ratio of the autocorrelation function sidelobe magnitude to the magnitude of autocorrelation main maxima does not exceed the value:

$$(0.7 \div 1,25)/\sqrt{N} \qquad (24)$$

where N is the number of sequence elements. A relative magnitude of sidelobes of the autocorrelation functions is not greater than

$$(1.4 \div 5.1)/\sqrt{N} \qquad (25)$$

Referring to Eqs. (24,25) the controlled transparencies with quite a great number of elements are necessary to obtain binary pseudo-random signals possessing good correlation characteristics. Therefore, in our experiment bleached photographic plates were used as phase masks for the reference beam encoding.

To make a phase mask, the LOI-2 plates possessing high-spatial resolution and low noise, are used. The plates are exposed by means of contact photoprinting.

The exposure was so that the optical thickness of exposed areas exceeded that of the unexposed ones by $\lambda/2$. The plates were processed by D19 developer and a $CuCl_2$ bleacher, and stabilized by inorganic dyes /19/.

To obtain two-dimensional pseudo-random signals six Huffman sequences having 31 elements are employed. A two-dimensional function is obtained by direct multiplication of two one-dimensional functions. The pictures of unbleached masks are shown in Fig. 8.

The uniformity and phase contrast ratio of bleached masks were measured by Mach-Zender interferometer. But a principle quality of such a mask is its correlation function. Auto-correlation and cross-correlation functions of complex transmittances of the mask are measured in the arrangement of Vander Lugt's holographic matched filtering.

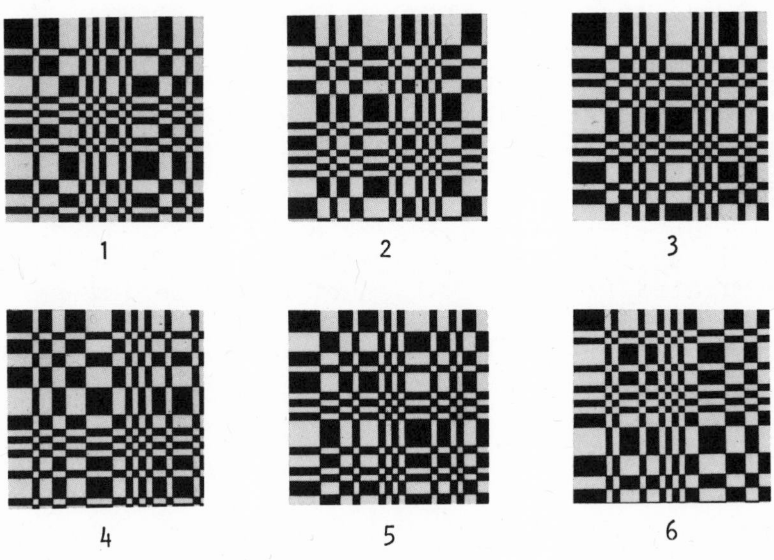

Fig. 8. Photographs of amplitude masks producing pseudo-random codes.

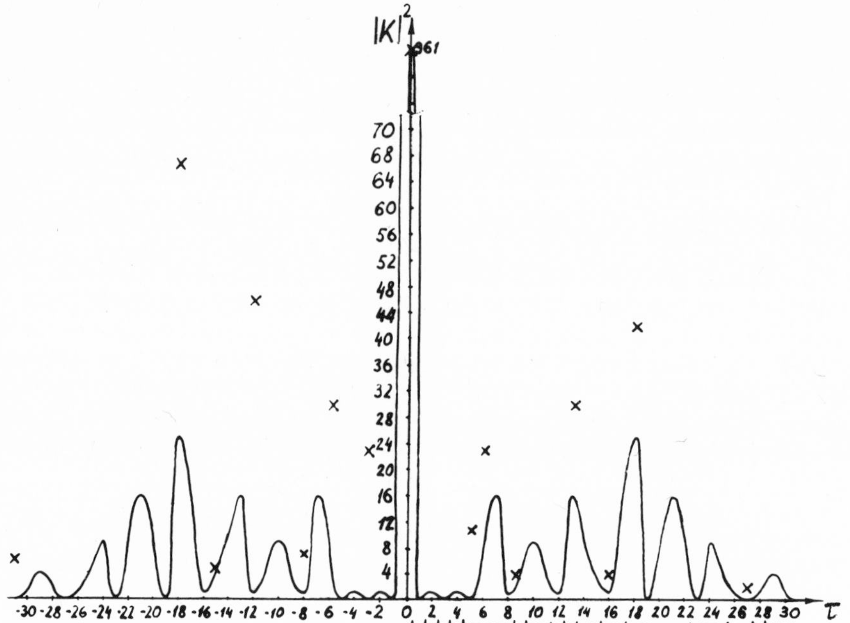

Fig. 9. Experimental results and theoretical calculations of the modulus squared of the autocorrelation function of a pseudo-random code:
—— theoretically calculated curve,
× experimentally measured results.

Figure 9 shows an auto-correlation function plot of one of the pseudo-random sequences having the basis N=31 in comparison with the experimental values of the auto-correlation function sidelobes of the pseudo-random masks, which realize this sequence. Referring to Fig. 9, the sidelobe intensity of the mask is 3-6% of the main maximum intensity. Although the theoretically calculated relative values of sidelobe intensities do not exceed 2.6%, this disagreement, nevertheless, has not practically worsend the mask correlation characteristics, since the sidelobes of auto-correlation function of the masks are less than the theoretically calculated ones. The measurement results show the possibility of employing pseudo-random masks produced by using the photographic film for the reference beam encoding.

Diffuse illumination was used when the superimposed holograms encoded by means of pseudo-random phase masks were recorded. The holograms obtained by this method possess all the properties of Fourier holograms. However, they are characterized by the uniform distribution of intensity for any form of the input signal spectrum /20/. This allows us to avoid saturation of the sensitive medium in the low spatial frequency field.

The recording and retrieval of the superimposed holograms were performed in a scheme shown in Fig. 10. The beam of He-Ne laser 1 passed through the lens 2 and illuminated the frosted glass area sized 5 mm. Lens 4 formed the image of the frosted glass in the hologram plane 7. Transparencies with two windows were inserted to the input plane close to lens 4. A pseudo-random phase mask 5 sized 20x20 mm was put in one window, while the other sized 5 x 5 mm served as an input signal. The transparencies having different masks differed also in a signal window position with respect to the position of the reference signal. Lens 8 generated the image of transparencies in the output plane 9.

Each of the superimposed holograms is exposed until their diffraction efficiencies are equalized, whereas the total optical density of the film is equal to the optimum one (corresponding to the maximum intensity of a single-exposed hologram). To record the holograms, LOI-2 plates are used. The plates are processed by D-19 developed. We did not use bleaching of the holograms. Under these conditions the diffraction efficiency of holograms did not exceed 1%.

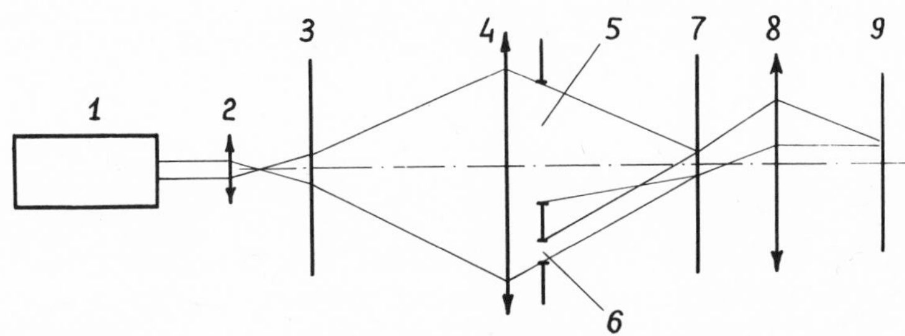

Fig. 10. Schematic diagram of hologram recording using reference beam coding by means of phase masks producing pseudo-random function: 1 - laser; 2 - lens; 3 - diffuse plate; 4,8 - lenses; 5 - the mask producing pseudo-random code; 6 - input signal transparant; 7 - hologram; 8 - restored image plate.

When reconstructing the holograms the light passed through the transparency with a corresponding phase mask, while the signal window was closed.
Figure 11 illustrates the images reconstructed from three superimposed holograms. One can see that the light intensity of images interferring with the useful ones does not exceed the intensity of diffuse noises, while the SNR of reconstructed images was about 5 (the calculated SNR values were equal to 3-5 with regard for the sidelobe intensities of the auto-correlation function). In the case of 5 superimposed holograms the SNR of restored images was about 1.5 (theoretical values were equal to 1.2-1.6).

Application of Walsh Functions for Coding of Superimposed Holograms

The Walsh functions represent a complete orthogonal system of functions, which can be easily realized by using controlled transparencies. However, their correlation functions have high intensities of sidelobes (up to 40%).

To improve the correlation properties of Walsh functions we can multiply each of them by just the same modifying sequence/21/. Similarly, the reference source having the required correlation properties may be produced by multiplying the CT complex transmittance by a random or pseudo-random function possessing good correlation properties, e.g. by the transmittance function of a diffuse plate (frosted glass).

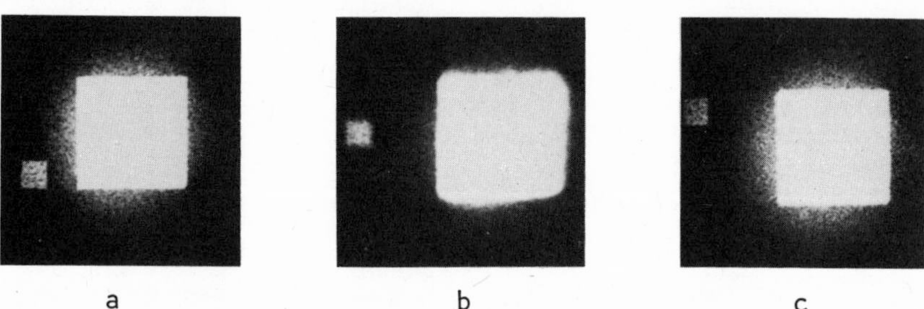

a b c

Fig. 11. Images restored from 3 holograms recorded in the same area of the plate with the use of reference beam coding by various pseudo-random signals.

If a diffuse plate with complex transmittance $t_d(x)$ (we consider one-dimensional case for the sake of simplicity) is placed near CT having somplex transmittance $t_w^i(x)$ and generating the i-th Walsh function the resulting complex transmittance is

$$t^i_{(x)} = t_d(x) \cdot t_w^i(x) \tag{26}$$

To calculate the correlation functions $C^{ik}(x)$ of such reference signals having i and k numbers we used a famous theorem dealing with Fourier transformation of convolution and correlation /9/:

$$C^{ik}(x) = t^{i*}(x) \ast t^k(x) = \mathcal{F}^{-1}\{T^{i*}(\eta) \cdot T^k(\eta)\} =$$
$$= \mathcal{F}^{-1}\{[T_d(\eta) \circledast T_w^i(\eta)]^* \cdot [T_d(\eta) \circledast T_w^k(\eta)]\}, \tag{27}$$

where $\ast$ and $\circledast$ denote, respectively, correlation and convolution operations; $T_d(\eta)$, $T_w^i(\eta)$ and $T_i(\eta)$ are Fourier transforms of $t_d(x), t_w^i(x), t^i(x)$, and $\mathcal{F}^{-1}$ denotes inverse Fourier transformation.

A complex transmittance of a diffusor can be conveniently represented as an ensemble of point sources emitting randomly phased secondary waves having unit amplitudes:

$$t_d(x) = \sum_{n=1}^{N} \delta(x - x_n) \exp(j\varphi_n) \tag{28}$$

where φ_n is the random phase, and N is the number of point sources. Then the Fourier transform of diffuse transmittance is

$$T_d(\eta) = \sum_{n=1}^{N} \exp(2\pi j x_n \eta + j\varphi_n) \tag{29}$$

The halfwidth δ of the transmittance spectrum of a real diffusor is determined by the effective dimension of a diffuse center:

$$\delta \simeq f\lambda/a \tag{30}$$

where f is the focal length of the optical system.

Walsh function spectrum represents an ensemble of sufficiently sharp maxima, having the width Δ ,which depends on the internal length where they are defined, i.e. on the transparency size L, whereas $L \gg a$, and consequently

$$\Delta \ll \delta \tag{31}$$

In this case the expression for Walsh function spectra takes the form:

$$T_w^i(\eta) = \sum_{l=1}^{M} \delta(\eta - \alpha_l^i) \tag{32}$$

where M is the number of a maximum in the spectrum. Then

$$T_d \circledast T_w^i = \sum_l^M T_d(\eta - \alpha_l^i) = \sum_n^N \sum_l^M \exp\left[2\pi j x_n(\eta - \alpha_l^i) + j\varphi_n\right] \tag{33}$$

and consequently

$$\begin{aligned} T^{i*} \cdot T^k = &\sum_{n=n'}^{N} \sum_{l,l'}^{M} \exp\left[2\pi j(x_{n'}\alpha_{l'}^i - x_n\alpha_l^k)\right. + \\ &+ \sum_{n \neq n'}^{N} \sum_{ll'}^{M} \delta(x - x_n + x_{n'}) \exp 2\pi j\left[(x_{n'}\alpha_{l'}^i - x\alpha_l^k) + \varphi_n - \varphi_{n'}\right] \end{aligned} \tag{34}$$

Then the autocorrelation function (when i=k) equals

$$\begin{aligned} C^{kk}(x) = &MN\,\delta(x) + \delta(x)\sum_{n \neq n'}^{N}\sum_{l \neq l'}^{M} \exp 2\pi j x_n(\alpha_{l'}^k - \alpha_l^k) + \\ &+ \sum_{n \neq n'}^{N}\sum_{l,l'}^{M} \delta(x - x_n + x_{n'}) \exp 2\pi j\left[(x_{n'}\alpha_{l'}^k - x_n\alpha_l^k) + \varphi_n - \varphi_{n'}\right], \end{aligned} \tag{35}$$

which practically coincides with the autocorrelation function of the diffusor /17/.

Cross-correlation function (when $i \neq k$) is

$$\begin{aligned} C^{ik}(x) = &\delta(x)\sum_{n \neq n'}^{N}\sum_{l,l'}^{M} \exp 2\pi j x_n(\alpha_l^i - \alpha_l^k) + \\ &+ \sum_{n \neq n'}^{N}\sum_{l,l'}^{M} \delta(x - x_n + x_{n'}) \exp\left[(x_{n'}\alpha_{l'}^i - x\,\alpha_l^k) + \varphi_n - \varphi_{n'}\right] \end{aligned} \tag{36}$$

From Eqs.(35) and (36) the autocorrelation function of all reference signals approaches δ-function, when $i = k$, meanwhile the cross-correlation function represents white noise, whose average intensity is approximately N times less than the autocorrelation function maximum intensity. However, real autocorrelation function always has a finite width limited in this case by effective aperture of the spectral plane in the optical system using these reference sources. It leads to the limitation of spatial resolution and SNR.

The result obtained can be explained by a following qualitative consideration. The convolution of the diffuse spectrum and Walsh function spectrum means the diffusor spectrum shift determined by the maximum position in the Walsh function spectrum.

In our case this value is equal to:

$$\alpha^i = m\lambda f/L \tag{37}$$

where m is the integer (1,2...). When a diffusor is placed in the reference beam, the displacement of the hologram defined by

$$\alpha = \lambda\sqrt{f^2 + L^2/4}\ / 2L \tag{38}$$

leads to disappearance of the reconstructed image /17/. Thus, as a first approximation, a separate reconstruction of the superimposed holograms recorded with the coding of the reference beam by means of diffusor and CT generating the Walsh function, is possible under the condition of

$$m\lambda f/L > \lambda\sqrt{f^2 + L^2/4}\ /2L$$

or

$$m > \sqrt{1 + L^2/4f^2}\ /2 \tag{39}$$

If $L/2 < f$, the condition (39) is fulfilled for any m. Such cases are realized in practical schemes.

The recording of holograms with reference beam coding by means of CT is accomplished in the scheme shown in Fig. 12. The beam of He-Ne laser 1 is split up by a semi-transparent mirror 2 into the signal and reference beams. The reference beam is expanded by lens 4 and focused in the hologram plane 17 by means of the optical scheme consisting of lenses 6 and 7 while the diffuse plate is removed.

CT 11 generating Walsh functions is arranged close to lens 6. The transmittance spectrum of transparency 11 is observed with the use of photodetector 13, measuring the intensity minimum in the region of low spatial frequencies. Diffuse plate 15 is situated close to lens 7 in the plane where the transparency image is formed.

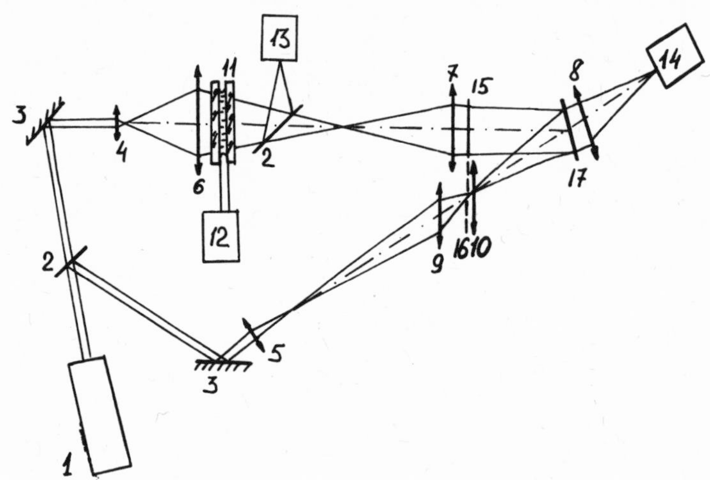

Fig. 12. Schematic diagram of hologram recording by using reference beam coding by means of the controlled transparency, which generates Walsh functions: 1 - laser; 2 -semi-transparent mirrors; 3 - mirrors; 4,5,6,7,8,9- lenses; 10-collective lens; 11 - controlled transparency; 12 - block of CT addressing, 13,14 - photodetectors; 15 - diffuse plate; 16 - diaphragm with a set of pinholes; 17 - hologram.

The signal beam is focused by lens 9 in one of 64 pinholes (0.1 mm in diameter) of the diaphragm 16, which is located in the plane of diffusor 15. Passing through lens 10, the beam illuminates the region (2 cm in diameter) of the hologram.

A pinhole serving as a signal point source with the corresponding code generated by CT, is chosen by displacing lens 9. The displacement of this lens did not practically affect the position of the signal beam in the hologram plane. Lens 8 formed the restored images in the input pinhole plane of the photodetector 14.

The holograms were recorded by using LOI-2 plates and treated with the HP-2 developer. The diffraction efficiency in the case of single-exposed holograms reached 20%.

Using this technique the recording and separate restoration of 8 superimposed holograms have been made. Picture of the point source images reconstructed from

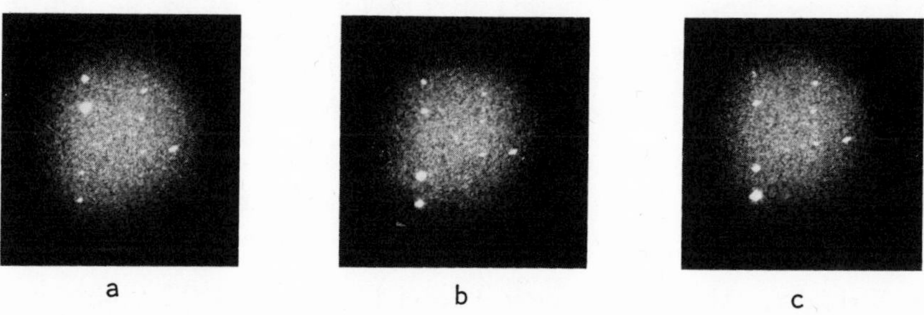

Fig. 13. Photographs of point sources restored from 3 of 8 superimposed holograms: a) CT generates Wal(0,0,x,y) function; b) CT generates Wal (1,1,x,y) function; c) CT generates Wal(2,2,x,y) function.

the holograms are shown in Fig. 13. The ratio of the signal intensity to the interference image intensity exceeds 20, while the diffuse noise level makes up 1-3% of the maximum intensity of the retrieval image.

3. CONCLUSION

The presented theoretical calculations and experimental results have indicated the possibility of using controlled phase matrix transparencies in electro-optical systems, which generate two-dimensional Walsh and Hilbert transformations. The possibility of recording and separate reconstruction of the superimposed holograms using reference beam coding pseudo-random signals based on M-sequences and by controlled transparencies generating Walsh-functions in conjunction with a random phase mask (diffusor) has been shown.

The proof of the second method for the formation of reference objects is the possibility to achieve any of the given correlation properties (by means of selection or synthesis of a single phase mask), while the number of CT elements is limited. The addressing scheme of the matrix transparency is extremely simple (it consists of several flip-flop and "exclusive or" schemes /3/, meanwhile the controlling voltage applied to the electrodes of the transparency, generating Walsh functions, may be formed in a cycle.

When coding the reference beam by means of CT, which generates two-dimensional signals based on M-sequences, one can obtain good correlation properties of the reference objects when a sufficient number of transparency elements is used, but it complicates the addressing design of it. At the same time, the coding of superimposed holograms by M- sequences is more flexible, e.g. it allows us to perform a shift and multiplication of images /13/.

Notice, that besides liquid crystal CT, the controlled phase matrix transparencies based on electrooptical ferroelectric ceramics /22/ having a high speed of operation, may be used in the considered scheme of transforming and coding of optical signals.

The most flexible optical processors may be created with the use of optically controlled spatial light modulators based, for example, on the photosemiconductor + liquid crystal /2/ or PROM /23/ devices.

References

1. V.P.Koronkevich, Yu.E.Nesterikhin, P.E.Tverdokhleb. Coherent optical processors. Autometry, 1972, No. 6, p. 3 (In Russian).
2. A.A.Vasiliev, I.N.Kompanets, A.G.Sobolev. Controlled Transparencies for Optical Information Processing. USA-USSR Seminar on Optical Information Processing (1975, Washington,D.C.), Plenum Press, N.Y. 1976.
3. S.Inokuchi, Y.Morita, Y.Sakurai. Optical pattern processing utilizing nematic liquid crystals. Applied Optics, 11,N10, 2223 (1972).
4. H.F.Harmuth. Transmission of Information by Orthogonal functions. Berlin, Heidelberg, N.Y. 1970.
5. L.M.Blinov. Electrooptical effects in liquid crystals. Usp.Fiz.Nauk, 144,N1, 67 (1974) (In Russian)
6. L.M.Soroko. In "All-Union School on Holography", Novosibirsk, 40 (1973) (In Russian).
7. J.Poncin. Utilization de la transformation de Hadamard pour le codage et la compression des signaux d'images. CNET, Ann.des Tel. 26,N7-8, 235 (1971).
8. M.F.Grebenkin, V.A.Seliverstov, L.M.Blinov, V.G. Chigrinov. Crystallography, 20,N5, 984 (1975)(In Russ.)
9. J.Goodman. Introduction to Fourier Optics. McGraw-Hill Co., 1968.
10. E.Guyon, P.Peiranski, M.Boix. On different boundary conditions of nematic films deposited on obliquely evaporated plates. Lett. on Appl.and Engin.Sciences, 1,N1, 19 (1973).

11. J.T.LaMacchia, D.L.White. Coded Multiple Exposure holograms. Appl.Optics, 7,N1,91 (1968).
12. A.Akaev, C.A.Maiorov, N.A.Smirnov. Zarubezhnaya Radioelektronika,N5, 57;N6, 80 (1975) (In Russian)
13. L.D'Auria, J.P.Huignard, C.Slezak, E.Spitz. Experimental holographic read-write memory using 3-D storage. Appl.Optics 13,N4, 808 (1974).
14. G.W.Stroke. A reformulated general theory of holography. Symp. of Modern Optics, New York, 1967.
15. V.N.Morozov. Kvantovaya Elektronika (to be published) (In Russian).
16. Noise-like signals in the systems of information transmittance (ed.Pestryakov), Sov.Radio, 1973 .
17. R.J.Collier, C.B.Burckhardt, L.H.Lin. Optical holography. Acad.Press. N.Y.,London, 1971.
18. D.A.Huffman. The synthesis of linear sequential coding networks. In: Information theory. Acad.Press, N.Y., 1956, p. 77-95.
19. S.L.Norman. Dye-induced stabilization of bleached holograms. Appl.Optics, 11,N5, 1234 (1972).
20. F.Membry, J.Duvernoy. Reconaissance d'une forme et détermination de ces dimensions au moyen d'un filtre-hologramme à adaption multiples. Nouv.Rev. Optique, 4,N2, 83 (1973).
21. D.D.Stiffler. Synchronization of Telemetry Codes. IRE Trans. Set-8 (1962),No.2,p.112.
22. M.D.Drake. PLZT matrix-type block data composers.. Appl.Opticss, 13, No.2, 347 (1974).
23. P.Nisenson, S.Iwasa. Real time optical processing with $Bi_{12}SiO_{20}$ PROM. Appl.Optics, 11,N12,2760 (1972).

OPTICAL METHODS OF INFORMATION PROCESSING FOR SOME RADIO PHYSICAL PROBLEMS

V.A.Zverev

Radiophysical Research Institute

Gorky,USSR

The solution of a number of radio physical problems requires that multichannel integral transformations should be fulfilled.Such transformations are comparatively simply fulfilled by the optical methods of information processing.

For this purpose different devices were developed which were reviewed in [I].In this monography examples of application of developed optical information processing devices are given. Among them are the problems of radio physics such as measurements of autogenerator linewidth, antenna radiation patterns, the analysis and synthesis of speech signals, etc.

The methods based on the incoherent light application were used for such problems connected with the processing of electric signals. Advantages of these systems over systems using coherent light were analysed in [2] .

It is shown there that the main obstacles of the incoherent light application for the optical information processing is the low accuracy of processing due to the presence of the constant light component in a mask-filter. To improve the accuracy of calculations a new principle is proposed to construct an optical computer using the incoherent light.The main idea of the method proposed is the application of the spatial

spatial and time consistent modulation of the light beam by the mask. To realize the idea, a light modulator consisting of two crossed gratings was used [1,2] . In such a modulator the total intensity of the light beam remains practically constant during time and spatial modulation of the beam. The mask structure provides involving the spatial x and time t coordinates in the form $x \cdot t$. This permits to use the modulation method for the separation of a useful signal from the variation of the light beam intensity. Such a modulator permits to perform two different regimes of information processing - spatial and time integration [2] . In the regime with the spatial integration the processing is performed in space coordinates and the separation of a useful signal - in time coordinates by frequency filters. In the regime with time integration the input signal is presented in the form of the brightness variation of the light source with time. The signal processing in this case is performed in the real time without the preliminary memory of the information processed. The separation of a useful signal is performed in space when reading the processing results [2].

The optical method of information processing is of significance in radiation pattern measurements in Fresnel zone. This possibility was earlier investigated by the method of optical modeling [3]. At present the optical method of processing has been investigated for the real antenna [4] . In this paper to exclude errors connected with the photo-processing regime a dashed record rather than brightness one was used. Inaccuracies related with scanning errors and other factors are investigated. Figure I shows a comparison of one of the sections of the two-dimensional radiation pattern obtained by optical processing in Fresnel zone and by a direct method using the radiator re-focusing. Here the section denoted by crosses is obtained by optical processing. As was noted by the authors, the advantage of the method consists in that it permits the use of transparencies of comapatively larger dimensions that reduces the requirements on the apparatus producing a hologram and on the objectives decreasing the transparencies.

Processing a great of information using simple devices opens the possibilities to apply the developed methods and apparatus to medical research. Some results of optical method of processing in medical studies were given in [I] .The apparatus is comparatively

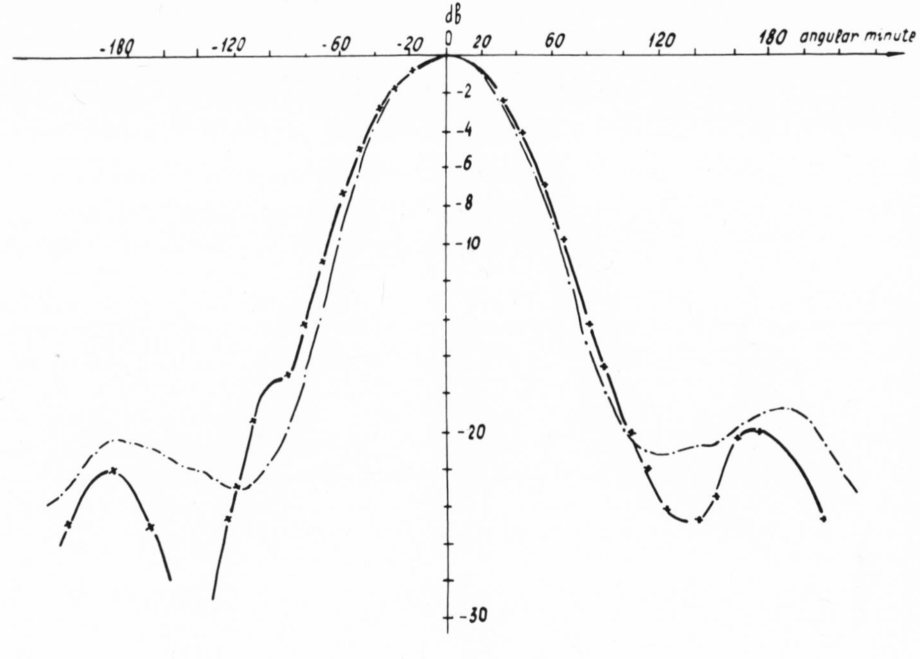

Figure I

simple, compact and employed by a regular medical staff. The most interesting results were obtained using this apparatus in the spectral analysis of ballistocardiograms [5] . A ballistocardiogram represents itself almost a periodic signal the spectrum of which is described by a set of harmonics. If a known periodic force which frequency differs from that of a pulse is applied to a patient,while ballistocardiogramming, it will be seen in the ballistocardiogram spectrum apart from the signal as is shown in Figure 2. Figure 2 a illustrates the spectrum of a normal ballistocardiogram and Figure 2 b shows the same spectrum the applied known external force the harmonics of which are denoted by arrows. From the comparison of the amplitudes of spectral components one may judge about the forces operating inside cardiovascular system and make some conclusions of interest for physiology of blood circulation [6] .

Recently the optical processing methods have been useful in ionospheric investigations [7] . The optical processing is used for observation of frequency varia-

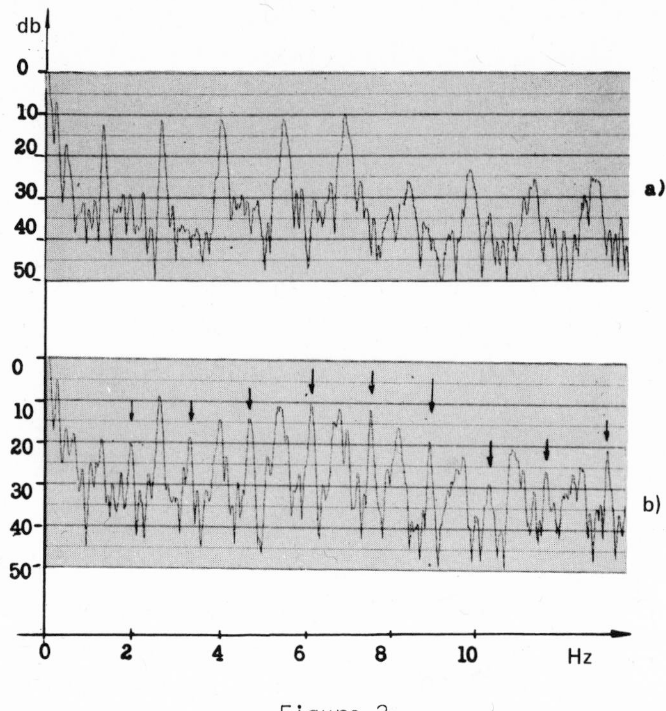

Figure 2

tions of radiowaves reflected from the time-varying ionosphere.This new method of the ionospheric investigations is called in the literature the high-frequency Doppler method. In paper [7] the optical processing was applied in two variants depending on the record regime. In the first variant the change in the Doppler broadening at some fixed ranges was investigated. In the first variant a dashed line which example is given in Figure 3 a was used to record the input signals.In the second variant the frequency variations of signals simultaneously reflected from many ranges were investigated. In this regime a brightness record of information was made by an electron ray tube shown in Figure 3 b.

Figure 4 gives some spectra of reflected signals in the time-frequency coordinates (dynamic spectra). A typical spectrum of signals scattered from the heights of the ionospheric D-region is shown in Figure 4 a. The spectral width is defined by the motion of inhomogeneities filling the scattering volume. The velocity for the spectrum in Figure 4 a amounts to 10 m/sec if the motion is assumed to be chaotic.

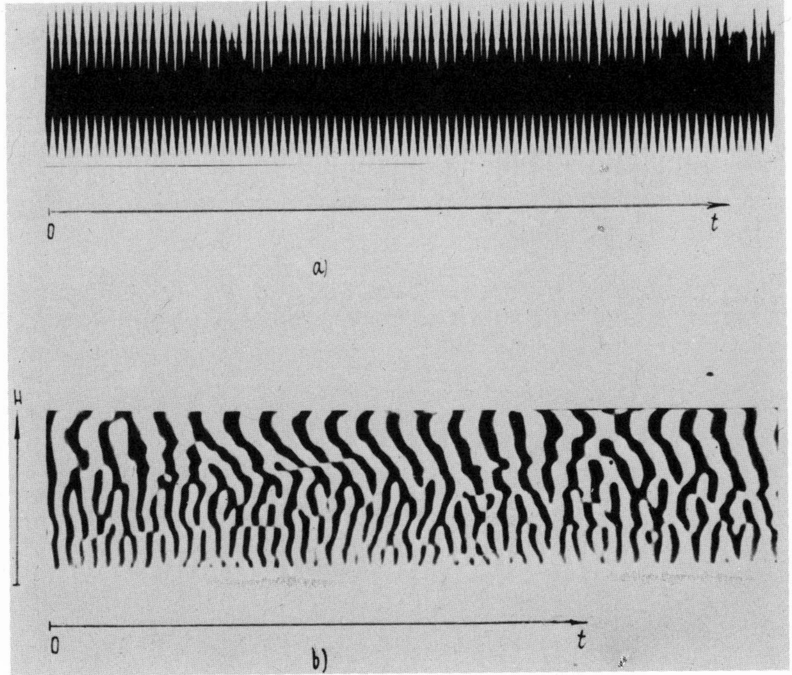

Figure 3

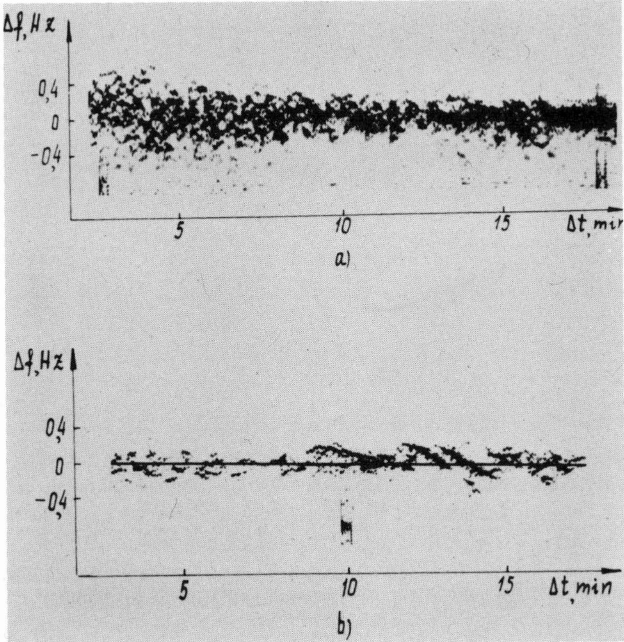

Figure 4

If it is assumed to occur regularly and horizontally, such a spectrum will correspond to the velocity of the order of 80-100 m/sec. A more particular form of the spectrum is represented in Figure 4 b. Typical discrete spectral lines testify to a nonuniform filling of the scattering volume by inhomogeneities. The slope of lines defines unambiguously the velocity of horizontal motion which in this case is equal approximately to 65 m/sec.

The distance between two "clouds" of inhomogeneities moving with the equal velocity is about 4 km.

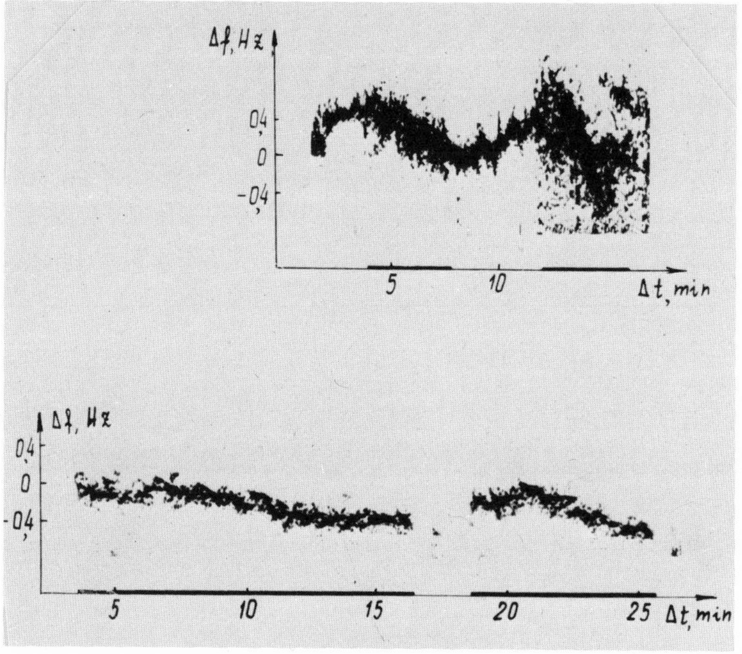

Figure 5

The method permits to determine a lot of important parameters of the ionosphere [7] .Figure 5 shows a dynamic spectrum variation of a reflected signal, caused by power transmitter radiation. The moments of transmitter switching on are marked by thick lines on the time axis. In other cases, such variations of spectrum are not discovered, though the amplitude of the scattered signal is increased many times.The results obtained are in a good agreement with the theoretical considerations concerning the nature of artificial ionosphere inhomogeneities.

Multichannel processing of signals from all ranges permits to obtain the most complete information about the motions in the ionosphere. In particular, this procedure enables one to obtain the visual representation on the dynamics of artificial inhomogeneity origination [7] .

At present, the systematic investigations of processes of cloud-burst sediments and hail-stone formations are carried out in order to increase the efficiency of anti-shower measures. The most complete information on flows and sediments microstructure can be derived by Doppler radar methods with an optical spectrum analysis 8 . In this paper the spectral analysis is performed by the multichannel system with time integration. The system is switched in after a phase detector of pulse coherent decimeter range radar. The result of spectral analysis is presented in a display in frequency-range coordinates. Thus, a spectrum profile of hydrometeor velocities may be observed at a video control system. An envelope of velocity spectrum for any separate range may be observed at a display. An analyzer operates in a real time and provides instantaneous analysis of Doppler signal with a number of filters up to IOO simultaneously in IOO range channels with a dynamic range of 26db. In those cases, when the dynamic range of 26db is insufficient, the analysis system with spatial integration that requires a preliminary record of the signal on the film I is used.Using this system, the signals are analysed successively from all ranges and not simultaneously. In this case, the number of frequency channels may be sufficiently increased. This procedure increases an accuracy of the velocity spectrum measurement. Within the range of 40m/sec the accuracy of the velocity measurements may be up to 0,I5m/sec.

Some typical spectra are given in Figure 6. Radiation is directed vertically. A zero velocity of meteoformation is marked by a vertical line at the frequency of 247cps. Frequencies above 247cps correspond to the meteoformation moving upwards and the frequencies below 247cps cprrespond to the meteoformation moving downwards.

Figures 6a,b show a heavy rain, and Figures 6c,d give a picture of hail process. Spectra of rain clouds and layer-rain clouds differ from those spectra of hail clouds by less velocity dispersion and by some other features [8] .

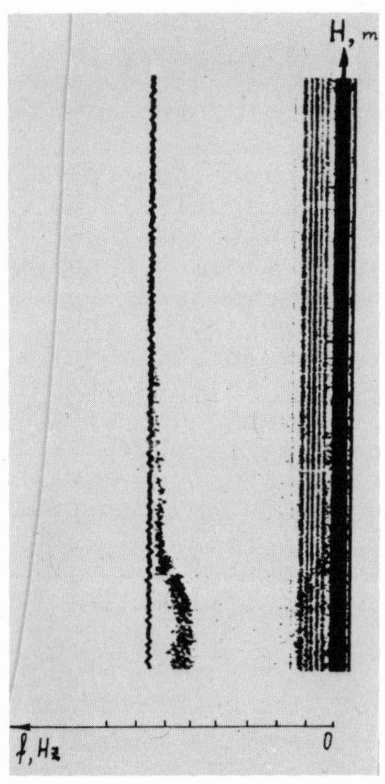

Figure 6a

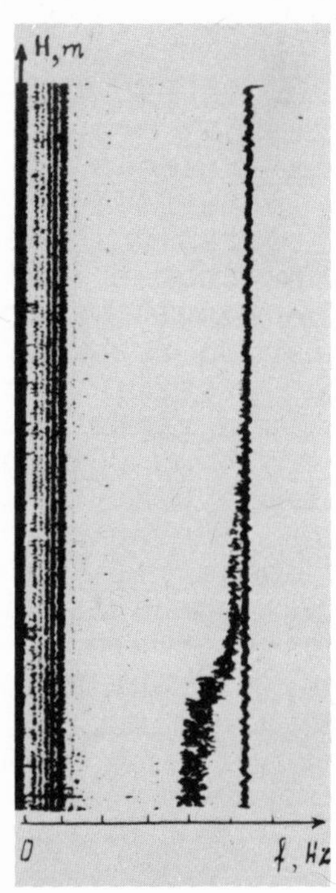

Figure 6b

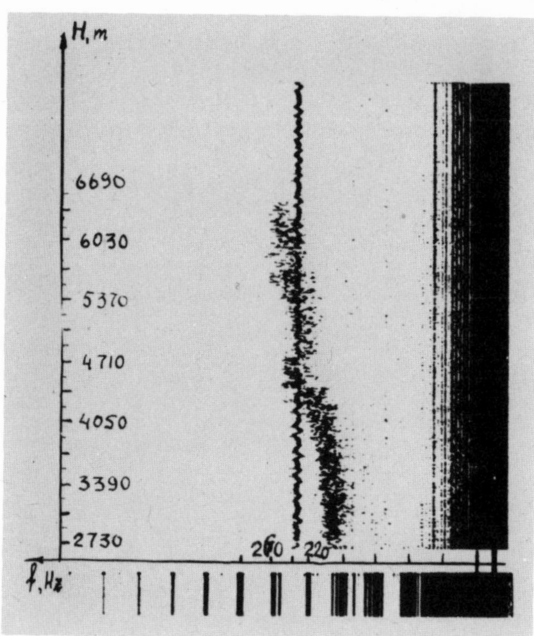

Figure 6c

Upwards flows can be fixed and localized with special accuracy that is difficult to achieve by other methods [8] . The analysis of height-time distribution of spectrum envelopes permits us to determine the microstructure of sediments and other important parameters of hydrometeors [8] . A similar method is used to investigate the reflected acoustic signals from fish shoal [9] . Spectra of reflected hydroacoustical signals provide information on ray velocities of scatterers. These spectra in contrast to radiolocation are measured for each individual pulse, restricting oneself to resolution being equal to an inverse value of radiating pulse range. For radiated frequencies of 20kps at which the most fishing-search stations operate, it turns to be difficult to find a reasonable compromise between the range and velocity resolutions. In [9] to

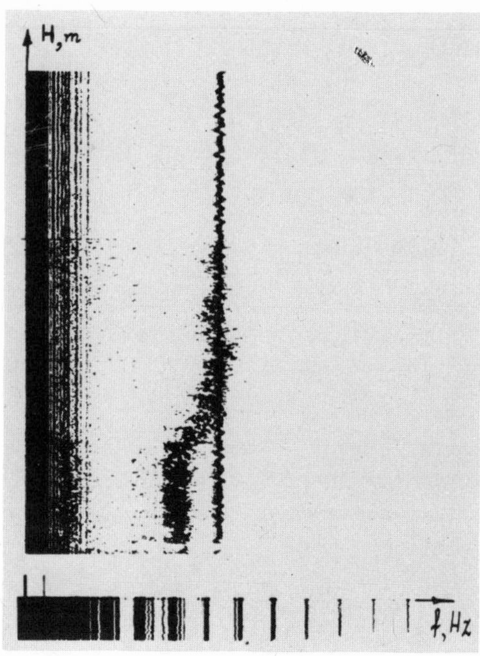

Figure 6d

obtain the scatter ray velocity distribution it is suggested to pass from an ordinary short pulse operation to single necessary-duration pulse radiation and to make a sliding spectrometering of a reverberation response. A spectral analyzer with the time integration analogous to that used in other investigations to obtain a sliding spectrum of the reverberation signal was employed. The spectrum is observed on a display in the frequency-range coordinates. Investigations were made by this apparatus in the Pacific Ocean on research ships TINRO in 1972-1974 [9] .

The possibility was shown to identify echolocated fishes by means of the ray velocity spectra ,to mark out fast and big biological objects,to determine the velocities and direction of migration of fish runs,to reveal their reaction to approaching ships and fishing-tackle. Based upon these investigations the methods of calculation of the quantity of biological sea organisms [9] was essentially refined. Figure 7 shows one of the

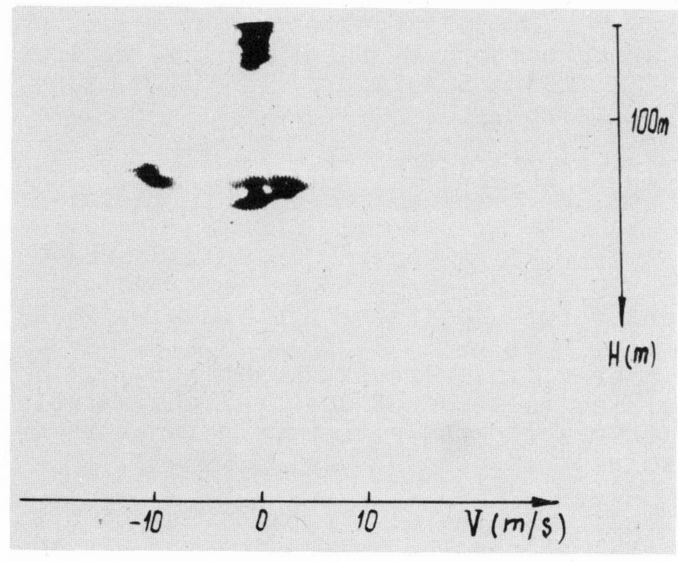

Figure 7

spectrograms. Echolocating was made vertically downwards. The figure also shows the reflection from the bottom and scomber shoal. The latter had a vertical directed-downwards velocity of about IO m/sec.Such a behaviour may explain the reaction to an approaching ship.

The given examples testify to the fact that the optical method of processing the information find ever increasing use in research works.

REFERENCES

I. V.A.Zverev,E.F.Orlov, Optical analizers,M,Sov. Radio,I97I

2. V.A.Zverev,E.M.Zuikova,et.al.Optical methods of information processing using incoherent light. in sb."Optical methods of information processing" Edited by S.B.Gurevich,Leningrad,"Nauka",I972, p.25-32

3. V.A.Zverev,E.F.Orlov,et.al.Measurements of the radiation pattern by the method of optical modelling in incoherent light. Izv.VUZ.Radiophysica, 1969,12,p.1829

4. Yu.I.Belov,E.M.Zuikova,V.I.Turpin, "On the one method of SVU hologram record" (in press)

5. V.A.Zverev,V.A.Antonetz,I.K.Spiridonova,"Spectral approach to the quantitative estimation of ballistocardiogram",Meditzinskaya Tekhnika,N6,1971,p.28

6. V.A.Zverev,K.V.Zvereva,V.A.Antonetz,I.K.Spiridonova,"The relation of spectral characteristics of ballistocardiogram with some parameters of hemodynamics",Vrachebnoe Delo,N3,1974

7. V.V.Belikovich,et.al. "Application of optical spectral processing in the ionospheric investigations. (in press)

8. A.A.Grachev,V.E.Dudin,E.M.Zuikova,et.al."Optical processing of meteo radio echo".(in press)

9. A.V.Zuikov,A.I.Pavlenko,D.A.Selivanovsky,"Spectrometry of reverberational hydroacoustic signals from fish shoals".(in press)

OPTICO-ELECTRONIC METHODS OF INFORMATION PROCESSING IN LASER INTERFEROMETRY

V.P. Koronkevitch, Yu.N. Dubnistchev,
G.P. Arnautov, V.S. Sobolev, V.P. Kiryanov,
and V.A. Khanov

Institute of Automation and Electrometry
Novosibirsk, USSR

Double-beam interferometry experiences a new birth due to the progress in the field of laser and computer art. The interference pattern allows us to get information on the total number of physical values. Fringe width and direction characterize the inclination of wave fronts and serve for determination of angles and directions. The shape of fringes provides information on the object shape. The change in the path difference and shift of fringes are used for absolute measurements of length, refrection index and other physical values which can be expressed by the length (velocity, acceleration, deformation, force, and etc.). By the contrast degree and visibility of fringes one can judge radiation source spectrum and use interferometer as a spectral device.

A new type of interferometers has appeared recently. In these interferometers light, diffracted on the particles crossing the interference pattern, carries information on parameters of movement of flued, gas, and solid. These devices are called laser Doppler velocimeters.

The signal Φ at the output of any double-beam interferometer is the sum of two components

$$\Phi = \Phi_0 + \Phi(\delta), \tag{1}$$

where Φ_0 is a constant proportional to the sum of interfering beam intensities; $\Phi(\delta)$ is a correlation term depending on the phase difference δ and providing information about the parameter being measured. In laser shift sensors the phase of correlation term is defined by the expression

$$\delta = \vec{K}(\vec{x}_1 - \vec{x}_2), \tag{2}$$

where $\vec{K}$ is a wave vector; $\vec{x}_1 - \vec{x}_2$ is a displacement vector being measured. The phase of the correlation term in the output signal of the laser Doppler velocimeter (LDV) is

$$\delta = (\vec{K}_1 - \vec{K}_2)\vec{V}t = \omega_d t \tag{3}$$

where $\vec{K}_1 - \vec{K}_2$ is a differential wave vector which direction is preset by the optical scheme; ω_d is Doppler frequency; $\vec{V}$ is the velocity of the object movement; and t is the time. The problem of the displacement measuring amounts to determination of the absolute phase value change. Doppler frequency equal in time to the phase derivative is determined in the LDV.

In the greater part of interference devices the constant component Φ_0 does not carry any useful information and makes subsequent signal processing more difficult. In modern systems precautions are taken to suppress and discriminate it.

Interference pattern signal is registered photoelectrically. In the first group devices diaphragm is fixed before the photoreceiver. Interference fringes are displaced from the diaphragm. The electric signal frequency is equal to the Doppler frequency and proportional to the interferometer mirror velocity of movement. When the mirror is unmovable the interferogram is stationary. If can be recorded when moving the slit with the photoreceiver along one of the interference field directions. The electrical signal frequency being proportional to the velocity of movement of the slit.

All the variety of the second group systems can be

reduced to two types: 1) differential scheme, 2) scheme with reference beam.

The differencial scheme of the laser Doppler velocimeter can be correlated with an interferometer where a moving slit ahead of the photoreceiver is replaced by a scatterer which velocity should be determined. The stable photoreceiver registers a luminous flux diffracted on the particle. The electrical signal frequency is proportional to the velocity of movement of the scatterer. It is equal to the difference of the Doppler frequency shifts in light scattered by the scatterer from the beams forming the interference field. In schemes with a reference beam selection of the Doppler frequency is carried out by optical heterodyning methods. The analog of this scheme is the double-beam interferometer one mirror of it is changed for a moving scattering objective.

Practical ways of utilization of laser interferometers developed at the Institute of Automation and Electrometry USSR Academy of Sciences, Siberian Branch for physical experements on the definition of global changes of gravitation acceleration, study of mechanism of interaction of light with substance, investigation of turbulence effects will be considered in the given work. Concerning the industrial application the problem of control over accurite characteristics of heavy large-size tools in a machine-building shop is discussed.

L a s e r   G r a v i m e t e r. Operation of the device is based upon the ballistic method of gravitational acceleration [1]. The problem reduces to mesurement of the distance and time of movement of the test body at two intervals at the minimum.

To measure pathway intervals double-beam interferometer with a servo-locked to Lamb dip laser was used. Laser wave length reproduction was not lower than $2 \cdot 10^{-8}$. A crystal oscillator which frequency drift did not exceed $5 \cdot 10^{-9}$ per day, was used as the time standard.

The gravimeter optical scheme (Fig. 1) consists of some basic elements, they are a single-frequency helium--neon laser 1, electron system of stabilization radiation frequency 2, interferometer 3-7, and output photo-

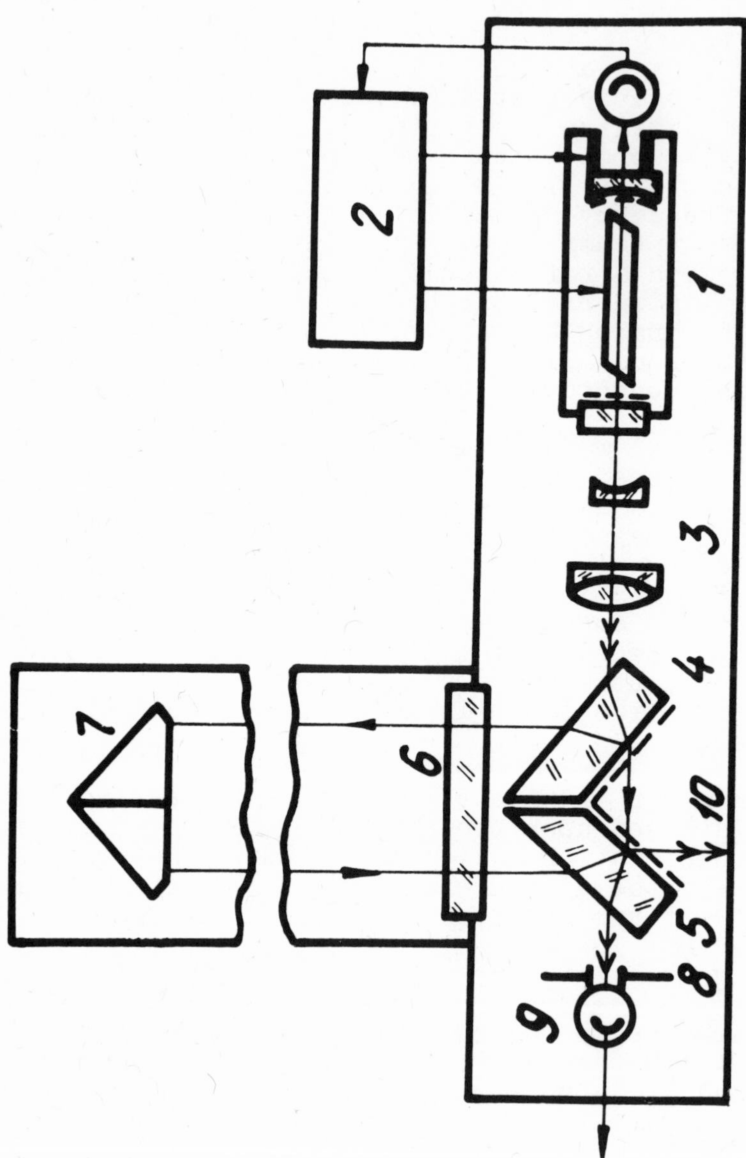

Fig. 1. The optical circuit of laser gravimeter.

electric device 8-9.

The light beam from the laser 1 is transformed by a telescope 3 and strikes a spacer plate 4 of the interferometer. The interferometer consists of two semitransparent plates 4 and 5 of the corner reflector 7 (test body). The photoreceiver 9 is mounted at the interferometer output. Having passed the plates 4 and 5 one of the beams (reference) through the diaphragm 8 is directed to the photoreceiver 9. The measuring beam reflecting from the plate 4 passes through the corner reflector 7 and recombinating with the reference beam on the plate 5 forms an interference pattern between 5 and 9. The field within the area 5 and 10 is used for the adjustment of reference arm of the interferometer in the vertical direction. The geometrical circuit of the interferometer is made with regard to laser polarizing properties. A polarizing vector being preset by the position of the Brewster windows is brought in line with the plane where the edge of the corner reflector and the edge of the plate assembly lie. Such arrangement of the elements allows to increase the contrast of the fringes. The corner reflector 7 moves along the vacuum tube with the protective glass 6.

The value of gravitation acceleration g was determined from the formula

$$g = \frac{\lambda}{2} \left[\frac{N_2 - N_1}{t_1 (t_2 + \tau)} \right] \qquad (4)$$

Herein λ is a laser wave length for the centre of the Lamb dip $N_1 \frac{\lambda}{2}$ and $N_2 \frac{\lambda}{2}$ distances being travelled by the corner reflectors for the time intervals t_1 and t_2, τ is the pause between the first and second intervals with $t_1 = t_2$.

The total time of the gravity drop of the corner reflector was 0.5 s. It corresponds to the distance 1.25 m or $4 \cdot 10^6$ interference fringes. At the beginning of the reflector drop measurements are not made. The delay time is 0.1 s. The interference signal frequency from which the measurements should be started ranges from 1 to 3 MHz. Signal frequency at the finish of the measurement is 16 MHz. Thus, the experiment was carried out in such a way, that the direct component of the in-

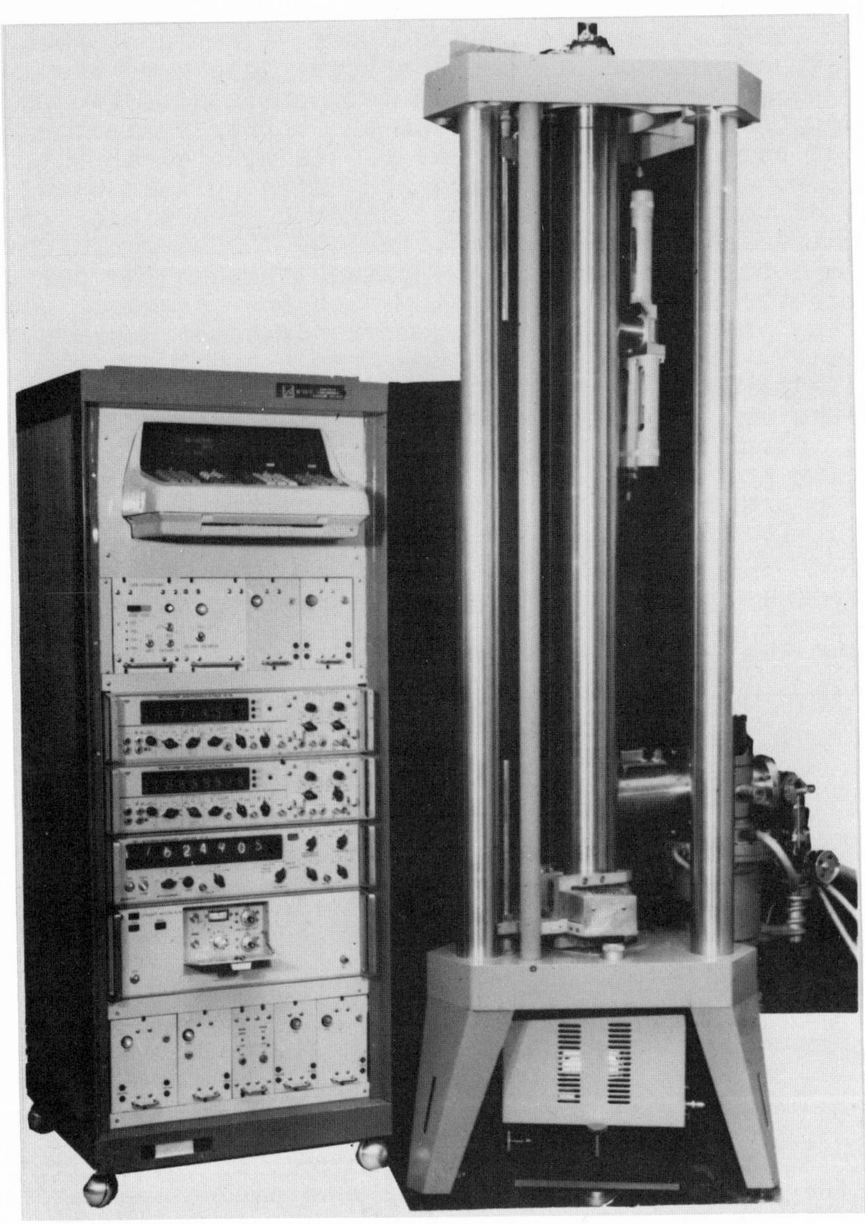

Fig. 2. Laser gravimeter.

terference signal Φ_0 was supressed by the filter, mounted at the photoreceiver output. The transmission band of the filter ranges from 1 to 16 MHz.

The potentialities of the laser gravimeter (Fig. 2) can be illustrated by measurements, obtained during the period from May 1972 to October 1973. To compare the results obtained to the Potsdam system of gravity the station of measurement of our Institute (Novosibirsk) was correlated with the Potsdam system by the aerogravimetric expedition of the Institute of Terrestrial Physics USSR Ac.Scien. with an error of ± 0.051 mGal.

The measurements (see Table) show that the values of gravity obtained differ from one another not more than 0.046 mGal, it is whithin the error limits of our device.

Table

Date	Potsdam system correction, mGal
13 May 1972	-13.891
14 May 1972	-13.875
18 October 1972	-13.912
16 November 1972	-13.921
6 October 1973	-13.909
20 October 1973	-13.902
Average	-13.902 ± 0.006

Such coincidence allows us to conclude that century variations for the given period did not exceed this value at the Novosibirsk station.

Laser gravimeter measurements were run in reference gravimetric stations of Moscow, Tallin, and Tbilisi during one month in 1975. The results of measurements in Moscow at the beginning of the trip and over a month differ by $4 \cdot 10^{-9}$ G. Three series of measurements were carried out in Tallin. The difference of the results did not exceed $4 \cdot 10^{-9}$ G. In Tbilisi measurements were carried out in the high humidity conditions, the difference of the results of the measurements accounted for nearly $2.5 \cdot 10^{-8}$ G.

Interferometer for measuring changes in the refraction index in photomaterials. When studying registering media the necessity of defining local phase changes in the medium affected by radiation arises. This problem is important for synthesis of optical elements of the kinoform type. Heterodyne-type interferometer with a two-frequency laser can be used for this purpose. Application of the two-frequency laser allows to carry out the external (with respect to the interferometer) modulation of signal and eliminate noise resulting from the direct component Φ_0 .

The optical scheme of the device is explained by Fig. 3. The light beam from the two frequency laser 1, having passed through a quarter wave plate 2 and objective 3, is splitted into two beams: reference and measuring ones by a natural quartz plate 4. A phase object 5 being under investigation is mounted in the paths of these beams in a focal plane of the objective 3. Further the beams recombinate with the help of the second quartz plate 4 and pass through a polarizer 6 behind which a photoreceiver 7 is located. Radiation from another end of the laser 1 is splitted with a light splitter into two beams, one of them is used to stabilize the laser difference frequency and the socond one is used to form a reference signal. This beam passes through a polarizer 8 and enters the photoreceiver 9. Electrical signals from the photoreceiver outputs 7 and 9 are supplied to a phasemeter 10 which indications are registered by a recorder 11. Recording phase profile in the medium was carried out with a laser beam 14.

Radiation of the helium-neon laser placed in the magnetic field ($\lambda = 0.63 \mu$ m) consisted of two components with orthogonal circular polarizations and frequencies ω_1 , and ω_2 . The magnitude of the difference of these frequencies $\Delta\omega = \omega_1 - \omega_2$ was defined by laser parameters (the active medium gain, resonator quality factor, generation frequency deflection from the centre of the atomic line) and by the magnetic field

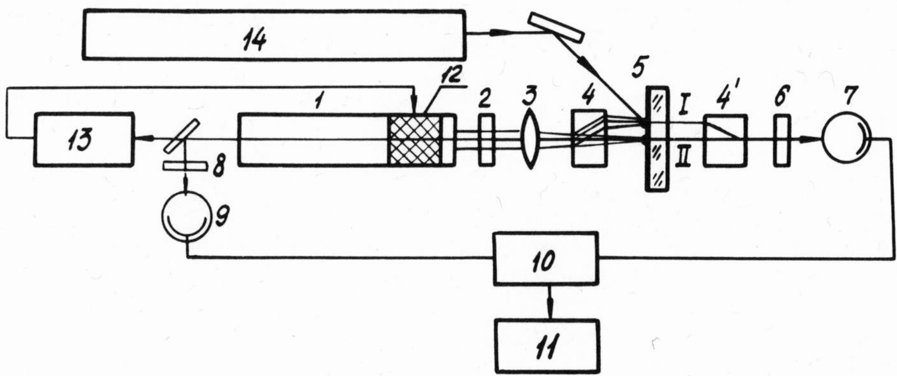

Fig. 3. Optical circuit of the interferometer for determination of the refraction index.

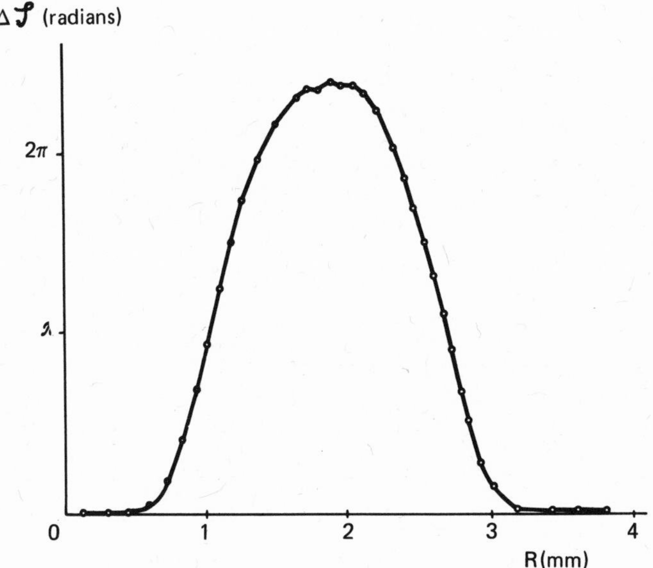

Fig. 4. Phase profile of the point.

magnitude. For our laser construction [2] $\Delta\omega$ was chosen within the range 0.03 + 2 MHz.

The expression of the light intensity being registered by the photoreceiver 7 (Fig. 3) is of the form

$$I_1 = A^2\{1 + \cos[\Delta\omega t + 2\pi(n_1L_1 - n_2L_2)/\lambda]\}, \quad (5)$$

where A is an amplitude, n_1L_1 and n_2L_2 are optical objective depths for the beams I and II (Fig. 3).

In a like manner for the photoreceiver 9

$$I_2 = A^2[1 + \cos(\Delta\omega t - 2\alpha)] \quad (6)$$

where α is the angle of rotation of the polarizer 8. When comparing the signals of the photoreceivers 8 and 9 one can notice that their phases differ by the magnitude

$$\Delta\varphi = 2\pi(L_1n_1 - L_2n_2)/\lambda + 2\alpha \quad (7)$$

When $L_1 = L_2$ the value $\Delta\varphi$ is proportional to the difference of refraction indices $\Delta n = (n_1 - n_2)$ $\Delta\varphi$ is measured by a phasemeter providing an accuracy of phase difference counting equal to 0.05° at the averaging time 10 s. When watching the kinetics of phase changes in photomedia during the process of exposing the averaging time can be reduced to $1\cdot10^{-4}$ s that leads to the reduction in accuracy up to 10°.

Experimental estimation of the interferometer was made with the use of halchogenide glassy semiconductor films (HGS). Fig. 4 represents the results of measurement of the phase profile of the point obtained by exposing of the film As_2S_3 with the help of the focused beam of the argon laser (λ = 0.51 μm) with the Gaussian intensity distribution. The interferometer space resolution was defined by the diameter of the measuring beams I and II in the plane of the sample and accounted for 70 μm in our case.

Investigation of the kinetics of phase changes in the films of the composition As_2S_3 and $As_2S_3 \times As_2Se_3$ was carried out integrally with the measurement of the absorption coefficient (K) for $\lambda = 0.63\,\mu$ m. The choice of the composition was defined according to the absorption condition close to uniform in the depth of the layer.

When the power of the exposing beam was equal to $P = 10^{-1}$ W·cm^{-2} for fresh evaporated films it was discovered that K increases and then reduces while the absorption coefficient is increasing monotonously (Fig. 5,a). The reduction of the film geometrical depth in the result of photostructural transforms did not exceed 1%. The refraction index and the absorption coefficient for the films calcinated at the temperature close to that of softening appeared to be correlated (Fig. 5,b). The experimental results obtained are testimony to the difference of the processes of photostructural transforms in the fresh evaporated and calcinated films.

Interferometer for determination of the displacement. When designing large-size heavy tools and constructing modern unique structures such as nuclear particle accelerators it is necessary to measure length up to 60 m with a relative error of $5 \cdot 10^{-7}$. This problem is solved with the help of homodyne or heterodyne interferometers [3,4]. The device is connected with a counting--computing unit based on a reversible fringe counter.

Interference signal investigation in the paths from 0 to 60 m shows that the air turbulence maks the main contribution to broadening of the signal spectrum. In the conditions of a standard machine-building enterprise the direct signal component Φ_0 ranges within the band 20 Hz with its maximum at 10 Hz. The value of the correlative component $\Phi(\delta)$ when path difference is 120 m (the length of 60 m being measured) is 2-10 times less in the amplitude according to the interferometer geometrical factor. The maximum of noises for radiotechnical circuits is within the 2-5 MHz band, that is why the preferable frequency band for interferometer operation will be 1 KHz - 2 MHz. For the most part, designers produce heterodyne-type interferometers shifting signal spectra to the range 1-3 MHz with subsequent low frequency component filtration [5]. Unfortunately, heterodyne interferometers occupy for information transmitting the band twice as broad as homodyne ones. Thechnological difficulties arising with their

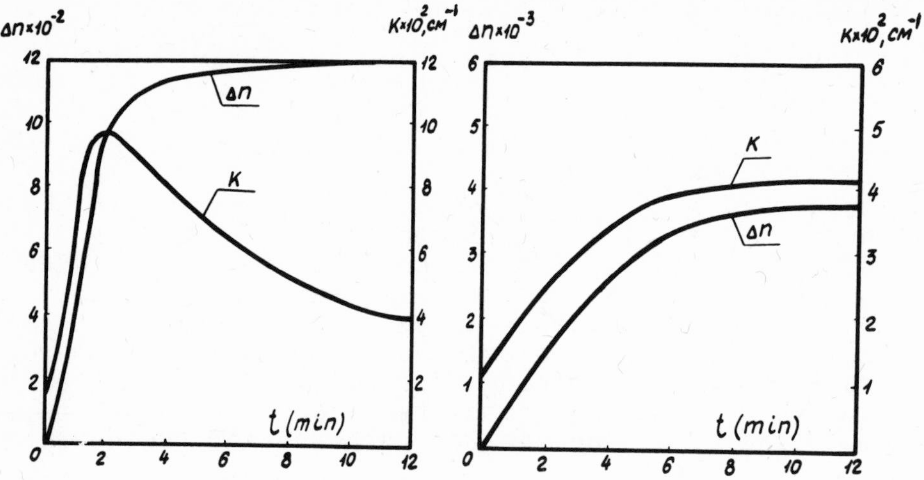

Fig. 5. Dependence of the refraction index and relative transmission on the exposing time.

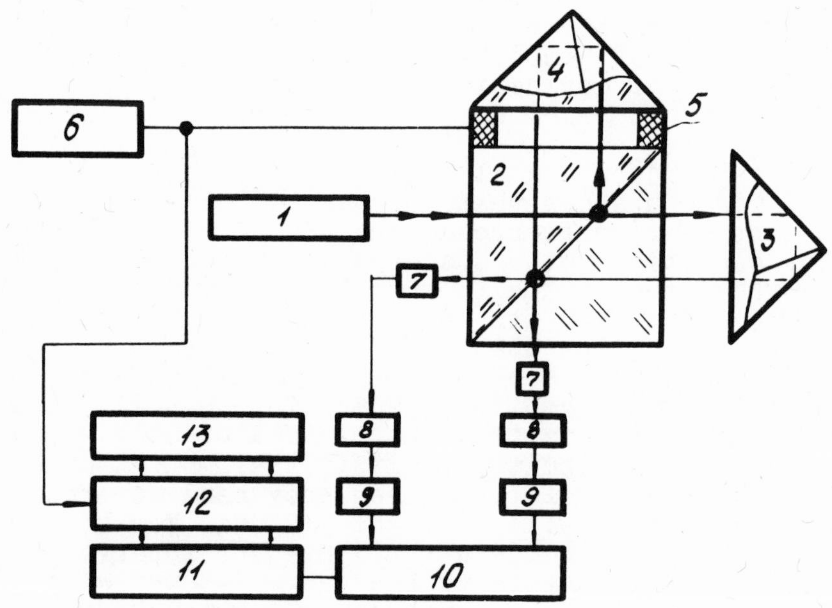

Fig. 6. The block-diagram of the displacement interferometer.

construction relate to design of two-frequency lasers, acoustic and electrooptical cells or polarization modulators. Because of this when developing the interferometer for measuring large lengths we used the internal phase modulation with subsequent signal processing similarly to that of the herodyne-type interferometers. The combination of the principles of processing according to the scheme of the homodyne interferometer with the method of internal phase modulation allows to use the favourable band of medium frequencies, cut off a low-frequency interference background, increase quick operation (the speed of the reflector displacement), having combined this with the ease of the construction.

The device comprises (Fig. 6) the frequency stabilized gas laser 1, interferometer 2, and counting-computing unit. The moving reflector 3 in the measuring arm is connected with the object of displacement, and the reflector 4 in the reference arm is rigidly bound with the 2 via the piezoelement 5 being controlled by the modulation generator 6. The photoreceivers 7 are mounted at the interferometer output. The interference patterns being projected on the 7 are phase shifted by 90°. The photoreceivers 7, wide-band amplifiers 8, and level reconstructors 9 form amplification channels. The read-out pulse shaper 10 produces signals for the operation of the reversible counter 11.

The reversible counter memorizes current information on the changes of the path difference relative to some zero position. This information exactly corresponds to the displacement being measured only at the moments when the modulated signal passes the position at which the chosen zero is adjusted. That is why the connection between the buffer of the indication unit 13 and the reversible counter 11 is accomplished via the gates 12 at which control input the gate from the generator 6 of modulation signal is supplied. The moment of the gate appearance corresponds to the moment of achievement of the zero level.

The mode of operation of the interferometer is explained by Fig. 7. The Figure represents path difference changes according to the character of movement of the reference and measuring reflectors. The sum of these changes is the current value of the path difference. The time signal and the shape of the read-out pulses are given under the Figure. As information reading occurs at the moments when the modulating signal passes its medium position, the state of the reversible counter is always equal to the current value of the path difference, i.e. the displacement being measured.

Low-frequency phase modulation shifts the signal spectrum to the high-frequency band thus allowing to eliminate turbulent pulses of the interference background. Depending on the relationship of the measuring and reference reflector speeds low-frequency components out of the pass band of the input filter can appear. The resulting destortions will produce spontaneous signal drift because of the constant component loss.

To eliminate them it is enough to fix the medium line of the interference signal relative to the threshold levels of the shapers. To do this it is necessary to reconstruct the constant component level. The latter can be carried out if the phase modulation index will exceed π . Then fixing maximal and minimal signal levels the average value can be determined. A distinctive feature of the new interferometer is low-frequency phase modulation with the index exceeding π , the possibility of discrete yielding of results only at the moments when the reference reflector passes the conventional zero, and reconstruction of the constant component level by electronic way.

Utilization of cheap one-frequency lasers, standard piezoelectric modulators, and a minimum of optical elements is characteristic of the technical realization of the device. Taking into account the fact that in case of internal phase modulation the light energy losses do not occur the advantages of the given interferometer for measuring large lengths will be quite evident.

The interferometer was used for the alignment of frame of the heavy large-size machine designed to produce the details weighting up to 250 t. The machine comprised two guides 15 m long. The gantry with tools moved along the guides. Displacement of the gantry was indicated by two scales connected to each quide. The scales were composed of metal sections 300 mm long. The problem of alignment consisted of the joint of scales and their certification.

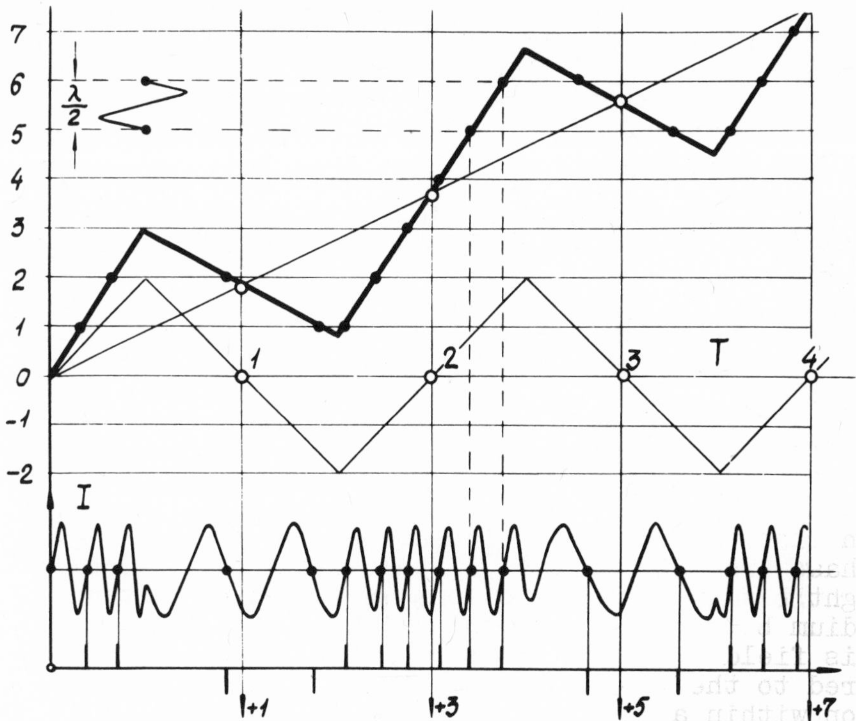

Fig. 7, The interferometer signal.

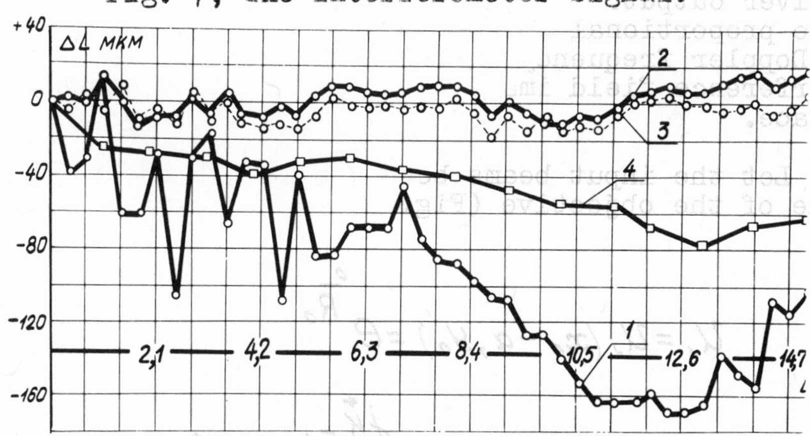

Fig. 8. The frame error of a heavy large-size tool-machine.

1 - the error of adjustment with step by step method with the stroke standard; 2 - the error of adjustment with a laser interferometer; 3 - the error obtained over a month after scale fixing; 4 - the error of the firm tool.

The results are given in Fig. 8. It shows that graduation of the frame by a usual step by step method of the stroke standard did not give a good result. The interferometer allowed to adjust the frame of the tool to an accuracy of 0.02 mm. When the scale was adjusted and finally fixed a reference detail 5 m long and its mirror reflection were produced. Then both halves of the detail were joined. The spread of the centres of four holes spaced at the 5 m distance ranged within 0.02 mm.

For comparison Fig. 8 represents similar results for the gantry tool-machine produced by "Waldrich Corburg". The alignment of the frame for this machine was carried out with the help of a heterodyne laser interferometer produced by "Hewlett-Packard".

L a s e r D o p p l e r V e l o c i m e t e r s. Consider the LDV operation using Fig. 9. The amplitud--phase (real or virtual) interference field from two light beams is known to form in the area of the moving medium being under investigation [5-9]. The image of this field in light scattered by scatterers is transfered to the photoreceiver. The spacial light distribution within a scattering volume is transformed to the time distribution of the electric signal at the photoreceiver output. The electric signal frequency appears to be proportional to scatters velocity and equal to the Doppler frequency difference in light creating the interference field image on the receiver photosensitive surface.

Let the input beams be given in the front basic plane of the objective (Fig. 9) by the expressions

$$\mathcal{U}_1 = \mathcal{U}_1(x_0 - a, y_0) = e^{\frac{j\vec{K}}{R_0}\left[(x_0-a)^2 + y_0^2\right]}$$

$$\mathcal{U}_2 = \mathcal{U}_1(x_0 + a), y_0) = e^{\frac{j\vec{K}}{R_0}\left[(x_0+a)^2 + y_0^2\right]} \qquad (8)$$

where R_0 is the radius of the curvature of wavefront; $\vec{K}$ is the wave vector; a is shift of the beam relative to the optical axis.

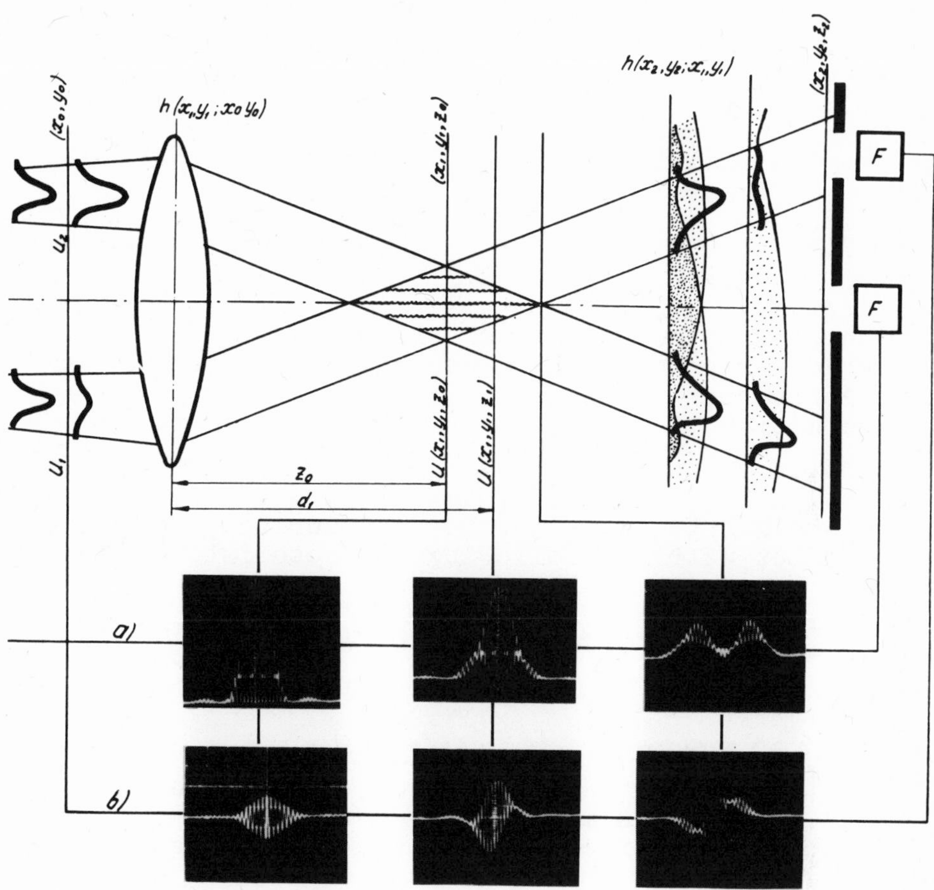

Fig. 9. Generalized optical circuit of the LDV.

a) the differencial circuit output signal with the scatterer moving through various sections of the interference field;
b) output signal of circuit with a reference beam in the mode of autocompensation of the direct constant.

The expression for the field in the plane (x_1, y_1) spaced at d_1 distance from the preset basic plane of the objective is of the form [9]

$$\mathcal{U}(x_1, y_1) = \iint_{D_1} [\mathcal{U}_1(x_0 - a, y_0) + \mathcal{U}_2(x_0 + a, y_0)] h_1(x_1, y_1; x_0, y_0)\, dx_0 dy_0 =$$
$$= G_1(x_1, y_1)\, e^{-j\frac{a\vec{K} x_1}{d_1}(1 - z_1\varphi)} + G_2(x_1, y_1)\, e^{j\frac{\vec{K} x_1}{d_1}(1 - z_1\varphi)}, \qquad (9)$$

where $h(x_1, y_1; x_0, y_0)$ is a pulse response of the shaping objective; $z_1 \varphi$ is an additional phase shift depending on beam parameters; D_1 is the objective aperture, $d_1 = z_0 + z_1$. In the formula (9) the condition $\frac{1}{z_0} + \frac{1}{R_0} - \frac{1}{F} = 0$ has been fulfilled. The latter means that in the plane x_1, y_1, z_0 the images of the laser "neck" are crossed. The scattere moves in the plane x_1, y_1, z_1 it can be described by the spatial filter with the following characteristics

$$P(x_1, y_1, z_1) = circ\left\{\frac{2}{b}\left[(x_1 - x_1^o)^2 + (y_1 - y_1^o)^2\right]^{1/2}\right\} \qquad (10)$$

where $x_1^o = V_x(t - t_0)$; $y_1^o = V_y(t - t_0)$; b is the scatterer diameter; V_x and V_y is the velocity projection on the axis x_1 and y_1; t_0 is the moment of the scatterer entering into the scattering volume. The expression for the field in the plane x_2, y_2, z_2 on the photosensitive surface of the photoreceiver is of the form

$$\mathcal{U}(x_2, y_2) = \iint_{D_2} \mathcal{U}(x_1, y_1)[1 - P(x_1, y_1, z_1)]\, h_2(x_2, y_2; x_1, y_1)\, dx_1 dy_1 \qquad (11)$$

where $h_2(x_2, y_2; x_1, y_1)$ is the pulse response of the receiving optical system, and D_2 is the diameter of its aperture diaphragm. Superposition of the field diffracted on the scatterer and the field of incident beams occurs in the photoreceiver plane. The resulting field contains information on the scatterer velocity. The electric signal at the photoreceiver output is proportional to the intensity

$$I = \eta \iint_{D_3} |u(x_2, y_2)|^2 dx_2\, dy_2 \,, \tag{12}$$

where η is the coefficient taking into account sensitivity and gain of the photoreceiver; D_3 is the size of the aperture diaphragm of the photoreceiver.

In the differential scheme [10] only light diffracted on the scatterer strikes the photoreceiver. Thus, the output electric signal can be represented as following

$$I = \eta \iint_{D_3} \left| \iint_{D_2} u(x_1, y_1) P(x_1, y_1, z_1) h(x_2, y_2; x_1 y_1) dx_1 dy_1 \right|^2 dx_2 dy_2 = I_n(t) + I_д \cdot \cos \omega_д t \tag{13}$$

Herein $I_n(t)$ is the low-frequency component of the output signal, corresponding to the constant component of the interference field in the scattering volume; $I_д$ is the amplitude of the correlation (Doppler) component; $\omega_д$ is the Doppler frequency. Figure 9 represents the photographs of real signals obtained with the single scatterer transit within various cross-sections of the scattering volume.

The similar analysis can be carried out in the LDV scheme with a reference beam [4]. A characteristic feature of the signal in the scheme with a reference beam is the possibility of obtaining of the so-called autocompensation mode of the constant component. It can be acheived with the definite ratio of the reference and signal beam intensities depending on the scatterer size [6,9]. Figure 9 (along the line $б$) represents a real output signal operating in the mode of the constant component autocompensation.

When the concentration of scatterers is great the output signal is considerably complicated. For example, the differential scheme besides the constant and correlation components contains cross or, the so-called, coherent components which frequency spectrum is defined by the whole set of twin differential velocities of scatterers being in the scattering volume at the same time, i.e. by the velocity gradient [10, 11].

To discriminate the direct and cross-components of a signal optical compensation circuits operating with a double-ended photoreceiver are applied [11 - 15]. Two-frequency Doppler devices [16-18] are the most promising ones to determine the magnitude and direction of velocity. The electronic unit of measuring the Doppler frequency of these devices provides suppression of constant and cross-components of the signal through their filtration. The circuits comprising a frequency modulator with a rotating electric field (MREF) can serve as an example. The MREF is equivalent to the phase plate rotating at the angular speed equal to the half-frequency of an exciting electric field.

Consider the inverse differential scheme 16 represented in Figure 10 as an example. The device comprises the laser 1, quarter-wave phase plate 2, frequency MREF 3, objective 4, receiving objective 5, diaphragms 6, quarter-wave phase plates 7 and 8, Wallaston prism 9, and photoreceivers 10 and 11 arranged in series. The photoreceivers are connected in parallel to adding and subtracting circuits, the outputs of which are connected with the frequency meter 16, via the filters and pulse sequence shapers 12 and 13. In the scattering volume of this scheme two space superimposed orthoganally polarized virtual interference fields are shaped. Interference bands of these fields run in opposite directions at the speed proportional to the modulation frequency in the MREF. Interference field images in scattered light are selected according to orthogonal polarizations and projected onto the photoreceiver with the receiving optical system.

After filtration of the Doppler component and transforming analog signal into the pulse sequence of the equal amplitude the pulses of $\Omega + \omega_d$ and $\Omega - \omega_d$ frequences enter the inputs of the adding 14 and subtracting 15 circuits. We have pulse sequence of 2Ω frequency at the output 14. The pulse frequency at the subtracting circuit output is equal to $2\omega_d$. The pulses of $2\omega_d$ frequency enter the frequency meter determining the Doppler frequency values.

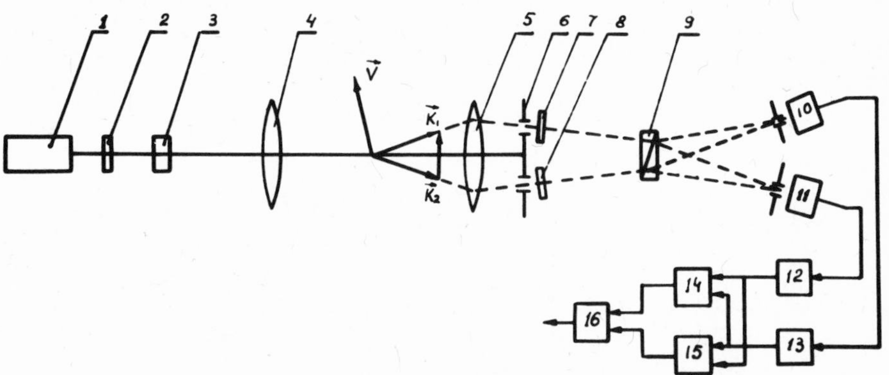

Fig. 10. The LDV optical circuit with a frequency modulator with a rotating electric field (MREF).

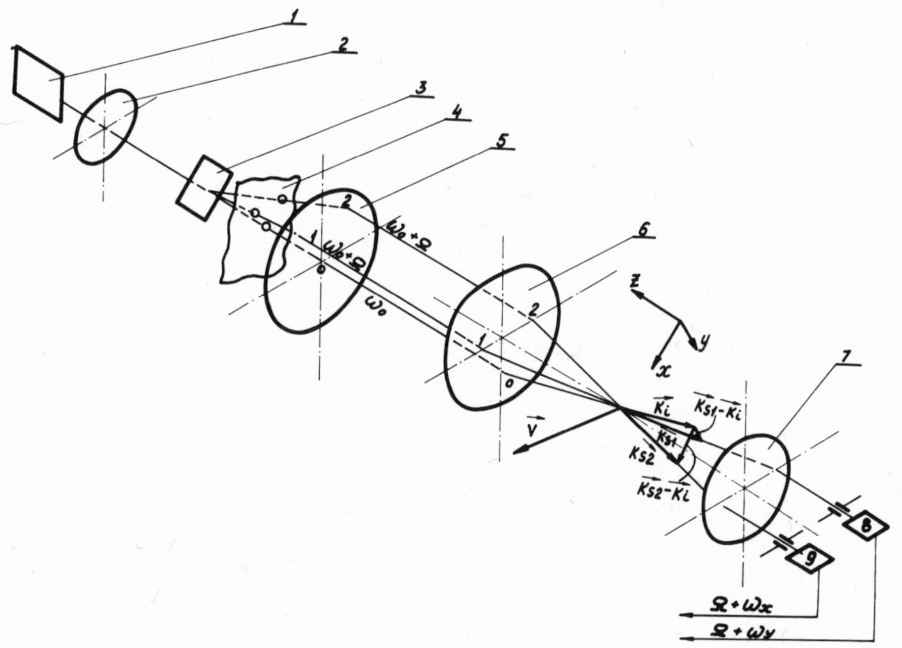

Fig. 11. Optical circuit for simultaneous determination the magnitude and direction of two orthogonal components of the velocity vector.

1 - laser; 2 - objective; 3 - two-coordinate acoustooptical modulator; 4 - screen with diaphragms; 5,6, and 7 - objectives; 8 and 9 - photoreceivers.

The high-frequency 2Ω pulses are fed to the same frequency meter and used as a filling frequency to impove the accuracy of measurements. The direction of the velocity projection being measured is determined according to the phase sign at the output of the circuit 16. Application of the frequency MREF is useful for devices intended for measuring the velocity vector. The examples of such systems are given in the work [21].

The scheme represented in Fig. 11 is intended for simultaneous determination the value and direction of two orthogonal components of the velocity vector. The device comprises the laser 1, objective 2, two-coordinate acousto-optical modulator 3, screen 4 with diaphragms, objectives 5,6, and 7, and photoreceivers 8 and 9 are arranged in series. Two ultrasonic waves running in orthogonal directions are excited in the acousto-optical modulator 3. The laser beam neck is projected with the objective 2 into the acousto-optical modulator into the crossing area of running orthogonally directed ultrasonic waves. As a consequence of the Raman-Nath diffraction at the modulator output 4 light beams diffracted into the first order passing in pairs in orthogonal planes and spreading at the same angle to the beam of the zero order are formed. Frequences of the beams diffracted into the first and zero order differ by modulator frequency (or, which is the same, ultrasonic wave frequency).

$$\omega = \omega_0 \pm \Omega \tag{14}$$

The major optical axis of the objective 5 coinsides with the direction of the light beam of the zero order and the point of splitting the diffracted beams is in its front focus. That is why the diffracted beams are parallel at the objective 5 output. The objective 6 is mounted so that its major axis is at equal distance from the beams of 0, 1, and 2 orders. Having passed the 6 the beams are directed into the moving medium where they cross in the area being under investigation. The diffracted beams 1 and 2 act as reference ones. Scattered ligth beams with the wave vectors $\vec{K}_{s_1}$ and $\vec{K}_{s_2}$ space superimposed with corresponding reference ones enter the photoreceivers 8 and 9. The light beam with a wave vector scattered by a moving particle has the Doppler frequency shift

$$\omega d_1 = \vec{V}(\vec{K}_{s_1} - \vec{K}_i) \tag{15}$$

Conformably, the Doppler shift in a light beam with the wave vector $\overline{K_{S_2}}$ is equal

$$\omega_{d_2} = \overline{V}(\overline{K_{S_2}} - \overline{K_i}) \tag{16}$$

From the geometry of Fig. 11 it is readily seen that the difference wave vectors $(\overline{K_{S_1}} - \overline{K_i})$ and $(\overline{K_{S_2}} - \overline{K_i})$ are orthogonal to each other. Thus they can be corresponded with the orthogonal coordinate axes Ox and Oy, and the Doppler shifts in scattered light beams are proportional to the following velocity projections:

$$\begin{aligned} \omega_{d_1} &= \omega_x = V_x |\overline{K_{S_1}} - \overline{K_i}| \\ \omega_{d_2} &= \omega_y = V_y |\overline{K_{S_2}} - \overline{K_i}| \end{aligned} \tag{17}$$

From this it follows that frequences of the signal Doppler components from the photoreceiver outputs 8 and 9 are in a linear connection with the corresponding orthogonal velocity projections V_x and V_y .

The signals from the photoreceiver output of each measuring channel are fed into electronic units of measuring the Doppler frequency. Such kind of electronic devices are usually the systems with phase or frequency tracking for the input signal [22,23] and they comprise a frequency detector. The output signal of electronic devices is given in the analog form and digital indication of mean velocity is fulfilled during the necessary averaging time. Fig. 12 shows the appearance of the two-frequency laser Doppler velocimeter comprising a unit of electronic signal pocessing, and Fig. 13 represents, as an example, profiles of the velocity longitudinal component in the channel with a rectangular trench.

In the LDV when the concentration of scatterers is high the added measurement error in the form of the analog signal noise at the output of the frequency discriminator appears. This is phase noise and it appears due to superposition of signals from different particles being withing the scattering volume at a time [8]. The measurement error arising due to the phase noise can be eliminated in the correlation LDV. The optical scheme

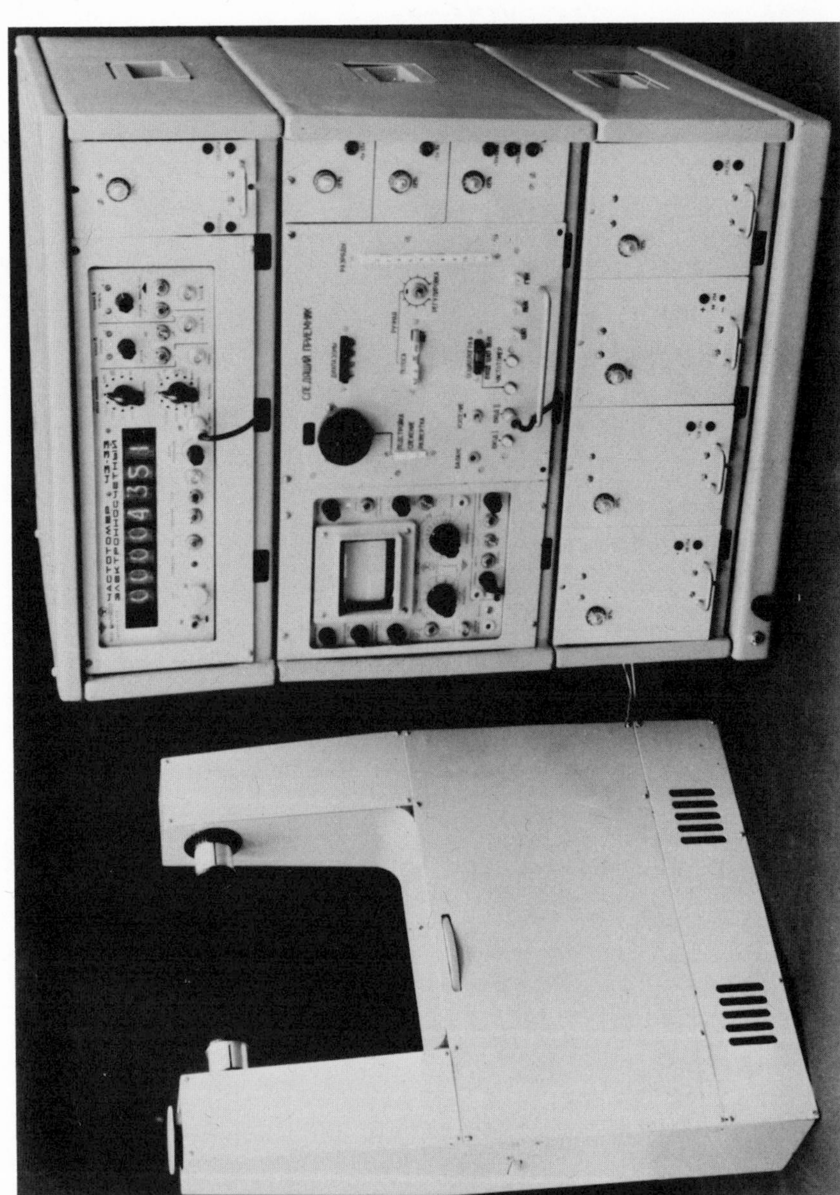

Fig. 12. The appearance of the two-frequency LDV.

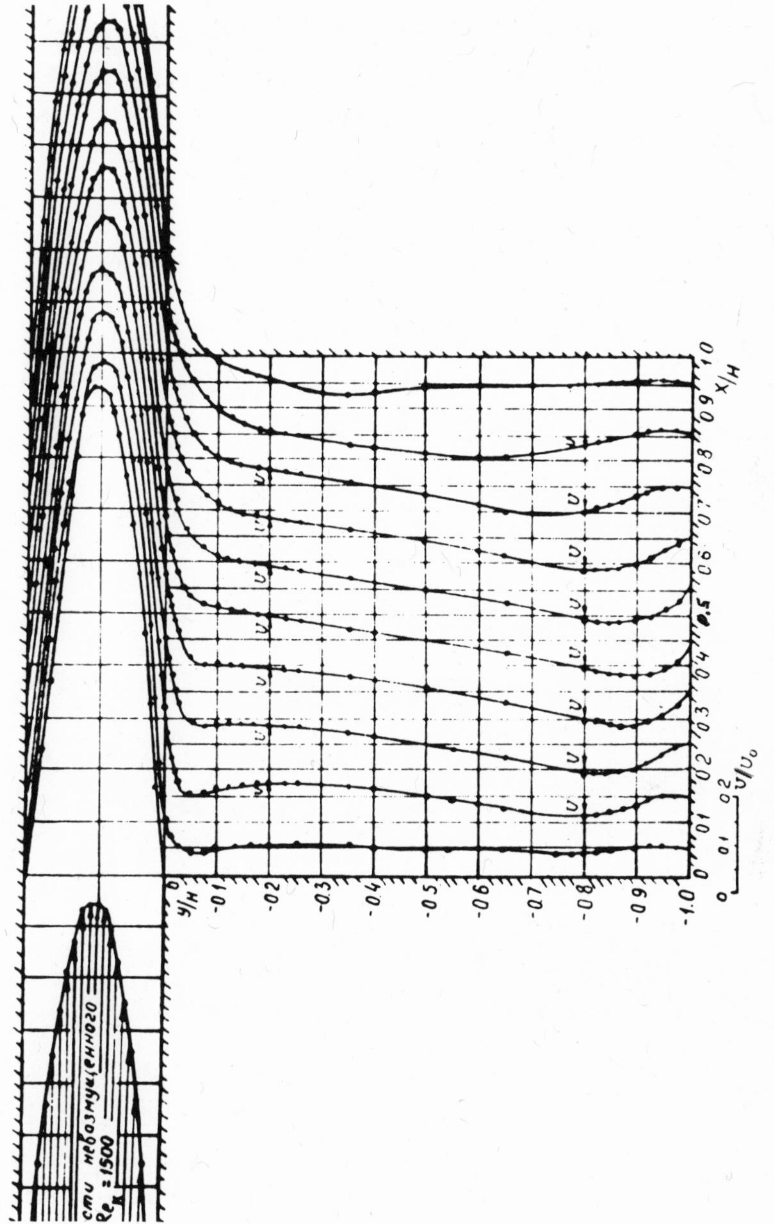

Fig. 13. Velocity profiles in the channel with a rectangular trench.

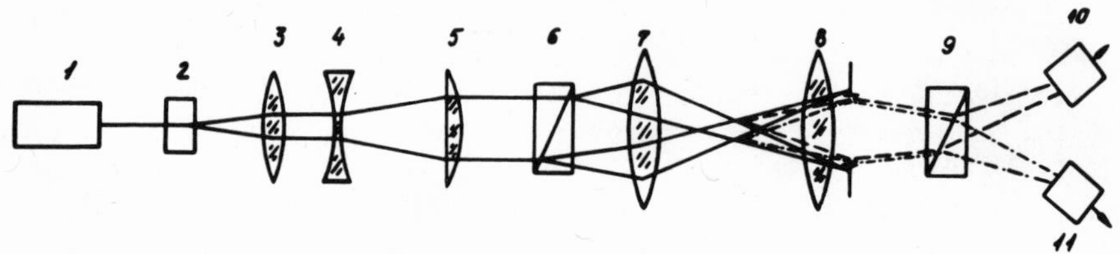

Fig. 14. The correlation LDV.

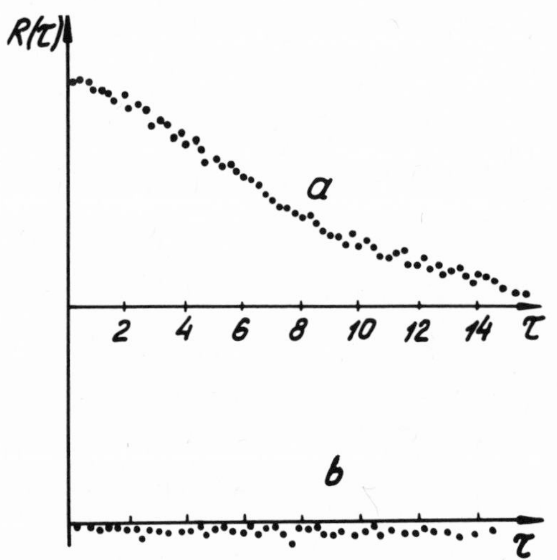

Fig. 15. a) signal autocorrelation function at the output of a single channel; b) cross-correlation function at the output of the 1 and 2 channels in the LDV.

of such a system is represented by Fig. 14 [23]. The device comprises the laser 1, acoustooptical beam splitter 2 functioning semultaneously as a one-band frequency modulator, afocal unit 3-4, cylindrical objective 5, Wollaston prism 6, objective 7, and two--channel photoreceiving system with polarizing filtration. The latter consists of the objective 8, polarizing prism-splitter 9, and two photoreceivers 10 and 11.

The optical device generates, in the flow area under investigation, two space-diversed orthogonally polarized interference fields with parallel oriented running fringes. The position of the afocal system defines localization of the interference fields within the flow. The cylindrical objective 5 integrally with the 7 serves to increase the number of interference fringes within scattering volumes. Interference field images in orthogonally polarized light beams are projected onto the photoreceivers. The electric signal frequency at the output of each receiver is proportional to the velocity of scatterers crossing the corresponding interference field. The optical scheme is oriented relative to the flow in such a way that one and the same particles do not pass probe fields. Signals from each of the scattering volumes are received and processed independently in the two-channel electronic measuring unit. The signals of each channel proportional to the instant Doppler frequency from the frequency demodulator output are fed to the correlator. Noises resulting from the superposition of the particles in the output signal of each channel and photoreceiver noises are statistically independent. Thus, they are eliminated from the resulting cross-correlation function. As the scattering volumes are spaced close enough the obtained mutual correlation function can be considered as an autocorrelative one. At sufficient averaging time it will practically coinside with the required autocorrelative function of turbulent pulses of the velocity of the flow under investigation.

Fig. 15 represents the autocorrelative function of the signal from the output of one channel obtained with changes within a laminar gradientless flow. The correlative function is of the exponent form and conditioned by "phase" noise. When a priori information on the laminar character of the flow is absent it would be erroneous to prescribe some equivalent turbulent fluctuations of velocity to such correlation function.

At the bottom of the same Figure the cross-correlation function of the same process is shown. The exponent is absent. Thus, noises in each of the channels are not correlated and they are defined not by the flow turbulence, but by statistics of scatterers space distribution.

REFERENCES

1. Izmerenie absolutnogo znatchenia gravitatzionnogo uskorenia. - Sbornik nautchnikh statei. Academia nauk SSSR, Novosibirsk, 1972, str. 77.

2. S.N. A t a u t o v, V.P. K o r o n k e v i t c h, A.I. L o k h m a t o v, V.V. S l a b k o, V.A. K h a n o v. "Autometria", N 1, 83, 1974.

3. A.Kh. K h i t r i n, E.M. S u n d u k o v a, B.A. O v s o v. Interferometritcheskie ustroistva kontrolya lineinikh peremeshenii i ikh primenenie v tekhnologitcheskom oborudovanii. Vipusk 1(77), TsNII "Electronika", Moskwa, 1973.

4. S.A. A l a k i s h e v, M.I. K a t a e v. Lasernie ismeriteli lineinikh peremeshenii dla tochnogo stankostroenia. Vipusk 6 (205), TsNII "Electronika", Moskwa, 1974.

5. J.N. K u k e s, G.B. G o r d o n. Hewlett-Packard Journal, 1970, august, p. 2-7.

6. M.J. R u d d. J. Phys. E., 1969, v.2, p. 55-58.

7. G.A. B a r i l l, Yu.G. V a s i l e n k o. Yu.N. D u b n i s t c h e v, V.P. K o r o n k e v i t c h. Autometria, 1973, str. 41-48.

8. Yu. V a s i l e n k o, Yu.N. D u b n i s t c h e v, V.P. K o r o n k e v i t c h, V.S. S o b o l e v, A.G. S e n i n, A.A. S t o l p o v s k i. E.N. U t k i n. Opto-Electronics, 1973, v.5, p.153-161.

9. Lazernie dopplerovskie izmeriteli skorosti. "Nauka", Novosibirsk, 1975, str. 170.

10. B. L e h m a n n. Geschwindig Keitsmessung mit Laser-Dopplerfahren. Wiss. Ber. AEG Telefunken, 1968, Bd. 41, N 3 ,S. 1340-1352.

11. C.P. W a n g. Applied Physics Letters. 1971, v.18, N 11, p. 522-524.

12. Yu.G. V a s i l e n k o, Yu.N. D u b n i s t c h e v. Autometria, 1972, N 5, str. 51-58.

13. H.H. B o s s e l, W.J. H i l l e r a G.E. M e i e r. "J. Phys.E.", 1972, v. 5, N 9,p.893-896.

14. Yu.G. V a s i l e n k o, Yu.N. D u b n i s t c h ev, and E.N. U t k i n. "Optika i spektroskopia",1973, t. 35, N 2, str. 306-308.

15. Yu.N. D u b n i s t c h e v, Yu.G. V a s i l e n - k o, Optics and Laser technology, 1974, N 10, p. 225-231.

16. A.I. V a t r u s h k i n, Yu.N. D u b n i s t c h ev, and T.Ya. P o p o v a. "Optika i spektroskopia", t. 34, N 6, 1974, str. 1184-1186.

17. R.J. A d r i a n. J. of Phys. E., v. 8, N 9, 1975, 723-726.

18. B u c h h a v e. "Optics and Laser Technology", v. 7, N 1, 1975, 11-16.

19. Yu.G. V a s i l e n k o, Yu.N.D u b n i s t c h e v, V.S. S o b o l e v, A.A. S t o l p o v s k i. Autometria, 1974, N 6.

20. J. C a m p b e l l, W. S t a i e r, IEEE, QE-7, N 9, p. 450-457.

21. Yu.N. D u b n i s t c h e v, Yu.K. K o v h o v. Autometria, 1971, N 3, 87-90.

22. Yu.N. D u b n i s t c h e v, Yu.G. V a s i l e n- k o. "Optics and Laser Technology", 1976, N 6,p.129--131.

23. Yu.N. D u b n i s t c h e v, V.P. K o r o n k e - v i t c h, V.S. S o b o l e v, A.A. S t o l p o v - s k i, E.N.E t k i n and Yu.G. V a s i l e n k o.

Applied Optics, N 1, 1975.

24. Yu.N. D u b n i s t c h e v, V.A. P a v l o v. A.I. S k u r l a t o v, V.S. S o b o l e v, A.A. S t o l p o v s k i, T.A. S h e l a p u t. Autometria, N 3, 1976, 53-60

OPTICAL PROCESSING IN FEEDBACK SYSTEMS

P.E. Kotljar, E.S. Nezhvenko, B.I. Spektor,
V.I. Feldbush

Institute of Automation and Electrometry
Novosibirsk, USSR

Up to now the major number of optical system for image processing is based on the open circuit principle. At the same time it is known that practically all electronic circuits more or less complicated consist of closed circuit, i.e. feedback circuits especially stricing it is demonstrated in analog computers which are based on the use of operation amplifiers, i.e. active system with feedback.

Two types passive and active feedback optical systems are considered in this report. The way of construction of such system is analised and the possibilities for image processing are enamined.

I. Passive feedback optical systems. Such systems were considered in the following works:

S.H. Lee. Mathematical Operation by Optical Processing. "Optical Engeneering" May/June, 1974, v.13,N3.

E.S. Hezhevenko, B.I. Spektor. Opticheskoe nelineinoe preobrazovaniya izobrazhenij, "Avtometriya", 1975, N 3.

N.C. Gallagner. Real-time Optical Signal Processors Emploing Optical Feedback: amplitude and phase control. "Applied Optics". April, 1976, v. 15, N 4.

E.S. Nezhevenko, B.I. Spektor. Afinnie preobrazovaniya izobrazhenij v opticheskikh sistemakh s obratnoj svyazy. "Avtometriya", 1976, N 6(to be published).

In the works of S.H. Lee, N.C. Gallagner the optical system was used which is represented on Fig.1.

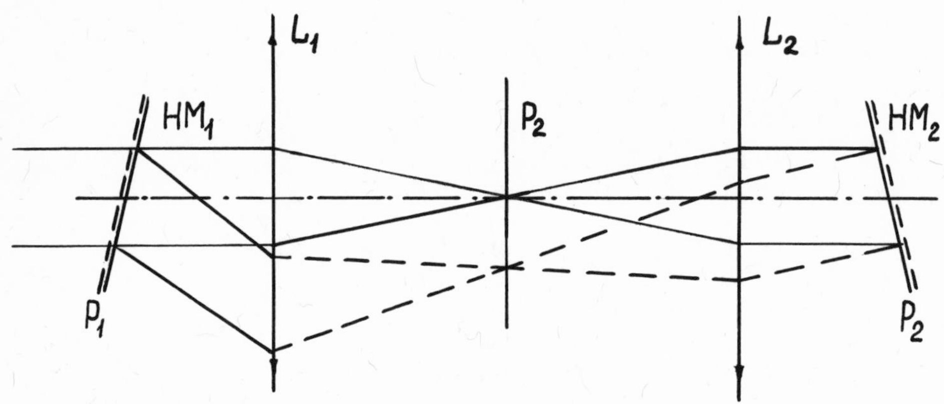

Fig. 1

In this system the image is in the plane P_1 and is introduced into the closed part through the half--mirror HM_1. The image is reflected many times from the plates HM_1 and HM_2 filtering at every add passage through the filter in the plane P_2. The light flows are taken out the half-mirror HM_2 and the processed image is formed in the plane P_3. The action of the system is based on the formal resemblanse of transfer function of closed system:

$$U_{out} = \frac{W_1}{1 \pm W_1 \cdot W_2} U_{in}$$

where U_{out}, U_{in} and W_1, W_2 are transfer functions of direct part and feedback circuit respectively, and the expression of sum of geometrical progression terms: $S = \frac{a}{1 - q}$, where a is the first term of geometrical progression and q is the denominator. In the optical system the processing image acts as first term of progression, and its production on transfer function of the filter in the plane P_2 is the donominator. Due to this resemblance it is possible to make such operations as the increase of image contrast solution of reverse problems, solution of partial derivative equations, etc.

Let us consider the system with open part and feedback circuit. We represent in a descriptive way because we shall analize its principle possibilitis although the constructive solution may be different from the given (Fig. 2).

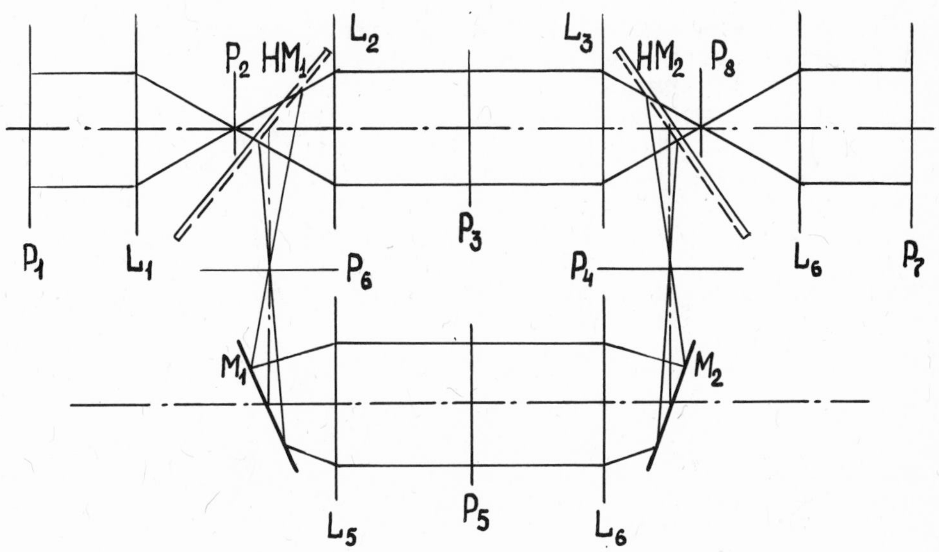

Fig. 2

In this system the planes P_1, P_3, P_5, P_7 are objective and P_2, P_4, P_6, P_8 are frequency planes. The planes P_1, P_2, P_7, P_8 are in the open part of system and planes P_3, P_4, P_5, P_6 are in the closed part.

Let us consider the ways of use of such system

1. Image in a closed cycle.

In this case the plane P_1 is free. The processing transparency $T(x,y)$ is placed in the plane P_3, the phase wedge with the transparency function $exp(j\alpha x)$ is in the plane P_5. The set of neutral filters is located in the plane P_7, which are in the row along the coordinate x and the distance between them is αf, where f is the focal distance of the objective L_3. Let us consider the operation of such system. We will not take into account the light relation in the closed system. If A is the amplitude of the illuminating beam then the amplitude distribution after transparant will be $A\sqrt{T(x,y)}$. One part of such flux after half-mirror HM_2 is weighted with the coefficient K_1 on the neutral filter in the plane P_8 and is focused in the plane P_7. Another part is reflected from the plate HM_2 and then in the plane P_2 it is added with the phase run and multiplied again by the amplitude transparency of the transparent. The amplitude of the wave after will be $(\sqrt{T(x,y)})^2 \cdot exp(j\alpha x)$. As a result of the phase run this part of the light beam after HM_2 will pass the neutral filter with transmittance K_2 and is focused in the plane P_7. All intensities of beams in the plane P_7 are added (the lenght of run in a optical delay line is choosen more then the lenght of laser coherency) and as a result we obtain the distribution

$$S(x,y) = K_1 T(x,y) + K_2[T(x,y)]^2 + \ldots + K_n[T(x,y)]^n$$

i.e. the power series expansion of some non-linear function . It is possible to change of the coefficients $K_1, K_2 \ldots, K_n$.

Special interest represents the case when in the plane P_3 instead of the amplitude transparent there is a phase operative one. Then besides the operating there another advantage appears. It is possibility to obtain the nonmonotonical non-linear relations (at amplitude transparant when $K_i \geqslant 0$ the function $\mathcal{F}(\cdot)$ may be only monotonical). In fact, at the definite install of the analizers after operative transparant it is possible to obtain the non-linear dependence

$$S(x,y) = \sum_{i=1}^{n} k_i [T(x,y)]^i + \sum_{j=1}^{n} k_j [T(x,y)]^j$$

Fig. 3

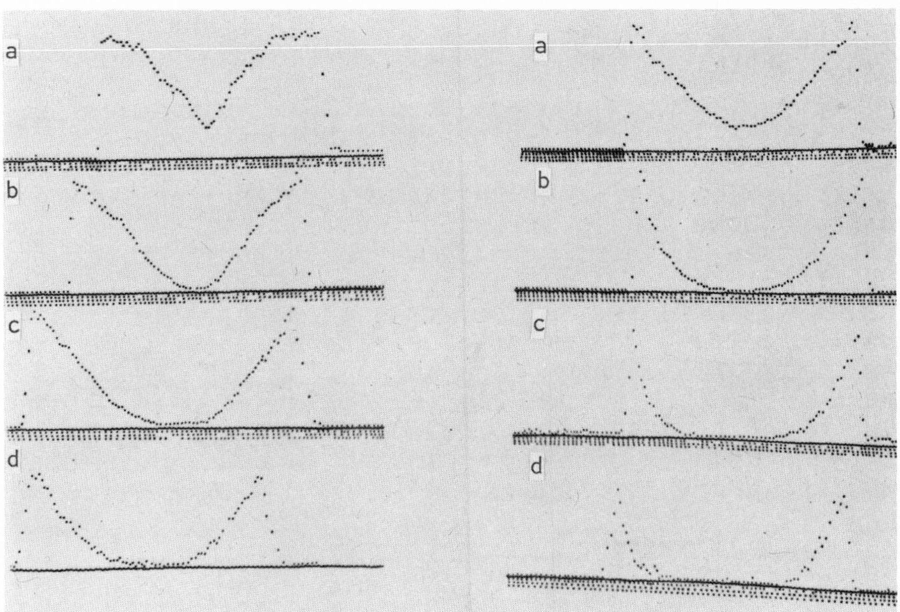

Fig. 4

The view of the system for non-linear transform is represented on the Fig. 3. On the Fig. 4 there are the results of scanning of linear and squared image,raised to the second, third and fourth power. Thus the install of the processing image inside the closed cycle gives the possibility to obtain the non-linear transform of the image, which may be used for correction of T-E characteristics of photomaterial, for increase of the image contrast, for increase the ratio of signal/noise at overlapping spectrum of signal and noise, ets.

2. Image in the open cycle.

In this case the image is placed in the plane P_1 and in the closed cycle a certain operator $R[T(x,y)]$ is acted. At first run the image is projected in the plane P_7, at second image transformed by the operator $R(\cdot)$, at third run - $R^2(\cdot)$, etc. For instance if the operator of scale change is acted then at the exit plain there will be a set of the different scale images, at the act of the operator of orientation change - a set of disoriented images, etc. At the illumination by the short laser pulse all transformed images will be presented on the output plane simultaneously.

The discribed above method allows to produce the affine image transformations at subnunoseconds. It gives the possibility to carry out the recognition with the reset on different parameters in the real time scale. On the Fig. 5 the experimental results are represented on image transformation (change of the scale, orientation) and also on correlation of transformed and standart images. As a standart object in the last experiment the image of a ring was taken, on the system input the image of a large ring was given. As it was expected the transformed images which didn't coincide with the standart correlation in the form of ring. If the diameters of rings coincide then the correlation function has a cone form.

Optical systems with active feedback

Operational amplifiers are their analogy in electronics. It is known how progressive jump was made after the appearance of this class of devices in the development of analog computers. We can expect the same effect in the optical computer systems if the optical

Fig. 5

operational amplifier will be created. There are several ways to design the optical operational amplifiers. We'll consider one of them based on the use real-time optical modulators of light (dynamic transparant). Let us enumerate the basic operations of dynamic transparants, which are necessary for creation of operational amplifier. Our considerations will be carried out for dynamic transparant of PROM type (Applied Optics, vol 11, No 12, December 1972, p. 2760-2767; Peter Nisenson and Sato Iwasa, "Real-time Optical Processing with $Bi_{12}SiO_{20}$ PROM").

1. Amplification of the light flux. It is based on the fact, that reading out light flux can 10^3 times be more than writing flux without destriction of information registered by PROM.

2. Composition of images. It is realized by series or parallel supply of two or more images in the record state.

3. Image deduction. Image - minuend $f(x,y)$ is given on PROM in the record state and image - subtrahend $\varphi(x,y)$ is supplied in the erase state.

The image - remainder is obtained in the form:

$$\mathcal{E}(x,y) = |f(x,y) - \varphi(x,y)| \cdot h(x,y)$$

where $$h(x,y) = \begin{cases} 1; & \text{if } [f(x,y) - \varphi(x,y)] \geqslant 0 \\ 0; & \text{if } [f(x,y) - \varphi(x,y)] < 0 \end{cases}$$

In order to obtain the alternating in sign $\mathcal{E}(x,y)$ we calculate at first

$$\mathcal{E}_+(x,y) = |f(x,y) - \varphi(x,y)| \cdot h_+(x,y)$$

then $$\mathcal{E}_-(x,y) = |f(x,y) - \varphi(x,y)| \cdot h_-(x,y)$$

It is clear, that $h_+(x,y) + h_-(x,y) = 1$.

4. Image integration. This operation is performed if the dynamic transparant pasess by large enough memory and the time of image erasing must be more than the time of the problem solving.

5. Image multiplication. It is carried out by the recording of one image-factor on the dynamic transparent and passing through it another image-factor with account of discribed above properties.

Let us consider the optical feedback processing information system which posses by two dynamic transparants. One of then is used as an amplifier - integrator, another as a summator, first is made on the base of PROM and second on the base of PHOTOTITUS and are discribed in Acta electronica 18,3,1975. J. Hazan "Les applications de Titus et de Phototitus traitment dimages et de donneas".

The diagram of the system is similar to that used in electronics (Fig. 6).

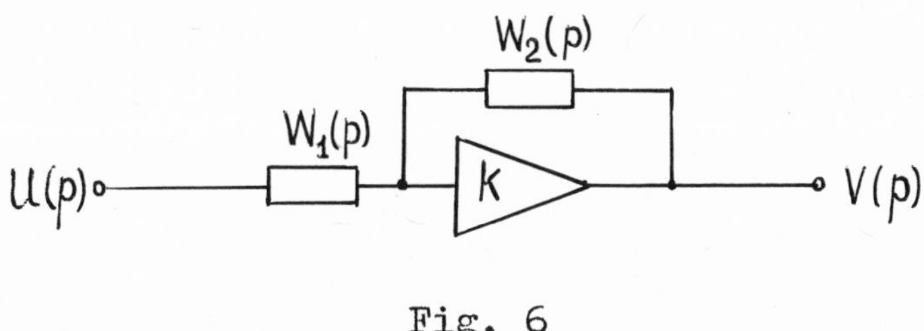

Fig. 6

Where $W_1(p)$, $W_2(p)$ are the transform functions of direct and back circuits, $U(p)$, $V(p)$ - Laplash images of input and output signals. At sufficient enough K

$$V(p) = \frac{W_1(p)}{W_2(p)} U(p)$$

The optical diagram of operational amplifier is represented on the Fig. 7.

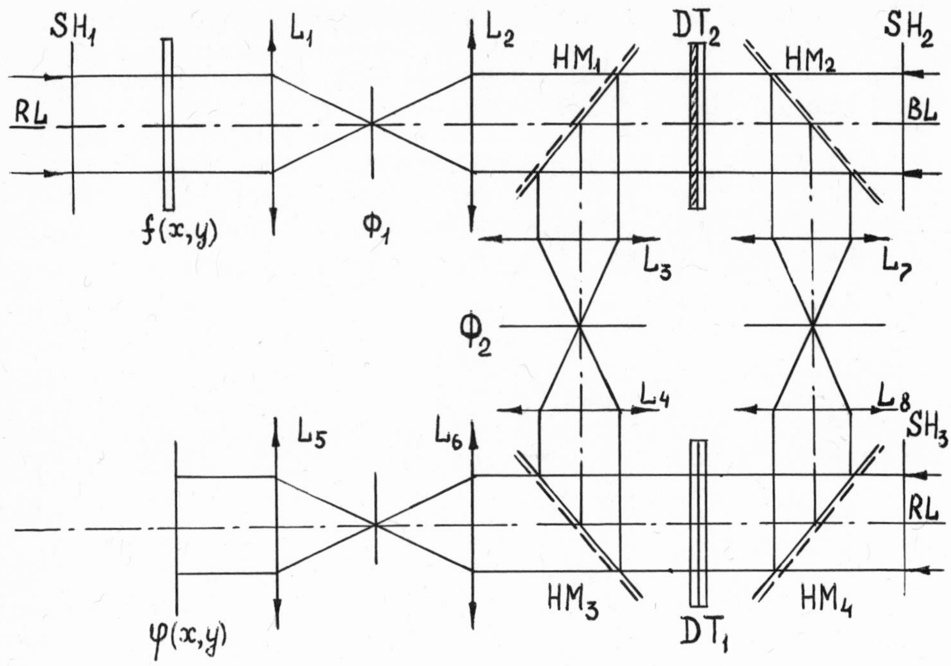

Fig. 7

Besides the dynamic transparants DT_1 , DT_2 the system constains: 4 half mirrors HM1-HM4, plaing at the same time the role of analizers for dynamic transparants; 8 objectives L1 - L8, providing the frequency and projective image transformations; input image $f(x,y)$; shutters SH1 - SH3 to shut the light beam ("RL" is the light beam of $\lambda = 633$ HM ,"BL" - $\lambda = 440$ HM); the filters Φ_1 and Φ_2 with transform functions $W_1(\omega_x,\omega_y)$ and $W_2(\omega_x,\omega_y)$.

For the record of alternating the sign signal on the dynamic transparant the positive and negative components of the signal are recorded at different time intervals thus one cycle of the system change is realized in some tacts. That cycle may be described by the state table of individual system elements (Table).

First column are the working tacts of system, 2nd, 3d and 4th columns show the shutter states: "I" - the shutter is open, "O" - the shutter is closed. 5 and 6 columns show the dynamic transparant states: "+" - the transparant is in the state of image recording, "-" - erase state, " " - reading out from the transparant. In 7 and 8 columns there are the functions recorded on the dynamic transparants after the end of corresponding tact. "O" in these columns means that at the end of the tact the erase of information recorded on corresponding dynamic transparant. The index " " means the cycle number, "*" - is a convolution symbol.

Table

	3_1	3_2	3_3	DT_1	DT_2	DT_1	DT_2
1	1	0	0		+	φ_n^2	$(f * W_1)^2$
2	0	0	1	c	-	φ_n^2	$\varepsilon_+^2 = \|(f * W_1)^2 - (\varphi_n * W_2)^2\| \cdot h$
3	0	1	0	+	c	$\varphi_n^2 + K\beta\varepsilon_+^2$	0
4	0	0	1		+	$\varphi_n^2 + K\beta\varepsilon_+^2$	$(\varphi_n^2 + K\varepsilon_+^2) * W_2 \approx (\varphi_n * W_2)^2$
5	1	0	0		-	$\varphi_n^2 + K\beta\varepsilon_+^2$	$\varepsilon_-^2 = \|(\varphi_n * W_2)^2 - (f * W_1)^2\| \cdot h(\cdot)$
6	0	1	0	-	c	$\varphi_{n+1}^2 = \varphi_n^2 + K\beta\varepsilon_+^2 - K\beta\varepsilon_-^2$	0

At the first tact the convolution of the processing image $f(\cdot)$ with pulse reply of open system is recorded in DT_2. At the second tact the signal of feedback is subsracted from the recorded signal, and the positive part of discrepancy is calculated. At third tact that part is added (or substracted if the feedback is negative) with the recorded an DT_1 image and the information recorded on DT_2 is erased. At 4th and 5th tacts as at 1st and 2d tacts the negative discrepancy part is on DT_2. At 6th tact is substracted (or added) from the image recorded on DT_1

At analysis for simplicity of calculation it wasn't taken into account that filtration of the image recorded on DT_1 is carried out at 4th tact with the add of $K\beta\mathcal{E}_+$.

After 6th tact on DT_1 there will be recorded the function:

$$[\varphi_{n+1}(x,y)]^2 = [\varphi_n(x,y)]^2 - [(\varphi_n * W_2)^2 - (f * W_1)^2]\,k\frac{\Delta t}{T}$$

where k is the amplification factor of operative transparent, $\Delta t/T$ is time constant, which is determined by the exposition time of transparents.

In the operative form the equation (1) will be:

$$(\Phi_{n+1})^{②} = \Phi_n^{②} - [(\Phi_n W_2)^{②} - (FW_1)^{②}]\cdot k\frac{\Delta t}{T}$$

or

$$[(\Phi_{n+1})^{②} - (\Phi_n)^{②}] : \frac{k\Delta t}{T} = (FW_1)^{②} - (\Phi_n W_2)^{②}$$

where $A^{②} = A * A$.

At sufficiently big n and satisfaction of equaation (1) stability conditions the value $(\Phi_{n+1})^{②} - (\Phi_n)^{②} \to 0$ then $\Phi = F\frac{W_1}{W_2}$, what is equivalent to the expression discribing the amplifier with feedback, if the time coordinate and transfer function to subsitute on corresponding space parameters.

Let us consider typical uses of optical operational amplifiers.

1. Let be $W_1 = \beta_1$, $W_2 = \beta_2$. Then

$\varphi(x,y) = \frac{\beta_1}{\beta_2} f(x,y)$ i.e. transmittance function of operative transparant is proportional to the supplied on it the light distribution with β_1/β_2 amplification. In the contrary to the ordinary system of information input in the coherent optical system which does the same function, in that case the change of the dynamic transparant parameters on the space (electro-optical coefficient, photosensitivity) comparativly less influences on the proportion between recorded and reading out the image, because in the scheme of optical operational amplifier such change gives only the change of the discrepancy $[\varphi_{n+1} - \varphi_n]/[\frac{k\Delta t}{T}]$ and the value of β_1/β_2 is determined by the properties of optical system only. For this reason the linearization of dynamic transparant characteristics is provided: in spite of the non-linearity characteristics each of them the system as a whole will be nearly linear.

2. Let the function $f(x,y)$ be the result of convolution of function $f_0(x,y)$ with the nucleus $S(x,y)$:

If in the feddback circuit to set a filter with the transform function $W_2(\omega_x, \omega_y) = S(\omega_x, \omega_y)$, than the optical operational amplifier solves the reverse problem

$$\Phi = F \cdot \frac{1}{S} = F_0$$

There is no need of tedious calculations and laborious fabrication of the inversion filter. It should be noted, that if there is an optical system introdusint the distortion or, for instance, the distortion is a result of defocusing so there is no need to search the impulse response of the filter and to fabricate it: it is sufficient to set the distortion system in the feedback circuit or introduce corresponding defocusing in it.

Actually the discribed procedure is not different from the iteration method of integral equation solution. It is obvious that for the solution existance and it's stability the nucleus $S(x,y)$ must match the conditions providing the condergence of the solution.

3. Nonlinear image transformations.

The characteristics of dynamic transparant are similar to diodes characteristics in electronics, so the inclusion of it in the feedback circuit permit to obtain the relay characteristic of the system.

In fact, if the positive feedback is provided (by the change of the order of the operation addition--substraction on DT_2) an substrate at each cycle from the image $f(\cdot)$, recorded on DT_2, some constant h (it is realized by the uniform lighting of DT_2 in the state of substraction) thus the equation (2) will be:

$$\frac{(\Phi_{n+1} - \Phi_n) * (\Phi_{n+1} - \Phi_n)}{k \frac{\Delta t}{T}} = \Phi_n * \Phi_n - (F * F - H)$$

This equation has solution:

$$\varphi = +\infty, \text{ if } f > h, \text{ or } \varphi = -\infty, \text{ if } f < h$$

Sinse the upper limit of transparency of the dynamic transparant DT_1 is limited by the value 1 and lower - 0 thus after sufficient number of cycles on the DT_1 there will be recorded the image:

$$\varphi = \begin{cases} 1, & \text{if } (f-h) \geq 0; \\ 0, & \text{if } (f-h) < 0; \end{cases} \quad \text{or} \quad \varphi = \begin{cases} 1, & \text{if } (h-f) \geq 0; \\ 0, & \text{if } (h-f) < 0; \end{cases}$$

It is possible to show that at some complication of the procedure it is real to blain the equidensities of the image f, i.e. the function

$$\varphi = \begin{cases} 1, & \text{if } |\varphi - h| < \Delta/2; \\ 0, & \text{if } |\varphi - h| > \Delta/2; \end{cases}$$

where Δ determes the transmitancy range in side of which the equidensity is detected.

4. Solution of partial differential equations.

Up to now the static (established) state of working system was considered. The dynamic state is of special interest. As far as in this state there are both space coordinate and time so the system work will be discribed by the transparensy function of the transparant DT_1 will be in agreement with its solution. The problem is complicated due to fact that the space filtration is carried out on amplitude and the record on the operative transparants is made on the intensities. Probably this difficulty will be overcome with time and then the optical operational amplifier will permit to solve a wide class of partial differential equations. Now let us consider one special case.

Let us write equation (1) in the form:

$$\frac{(\varphi_{n+1})^2 - (\varphi_n)^2}{k\,\frac{\Delta t}{T}} + (\varphi_n * W_2)^2 \;;$$

if $\quad W_2 * \varphi_n = \dfrac{\partial^2 \varphi}{\partial x \partial y} \quad ; \quad W_1 = 1$

then taking into account that the function

$\left[(\varphi_{n+1})^2 - (\varphi_n)^2\right] : \dfrac{k\Delta t}{T}$ is the finite-difference representation of partial derivative on time of function $[\varphi(x,y,t)]^2$ we obtaine

$$\frac{1}{k\,\frac{\Delta t}{T}} \cdot \frac{\partial[\varphi(x,y,t)]^2}{\partial t} + (\nabla\varphi)^2 = f^2$$

The problem of the unitial and boundary conditions input into the optical system are simple: by the suppling at the definite tacts of light distributions corresponding to this conditions.

In the present paper it is impossible to consider all problems solving by the optical feedback systems. Let us remined the most important of them.

By the use of parametric feedback (in optics it

is realized more simple then additive feedback) it is possible to create the optical systems similar to the systems of automatic tuning of amplification in electronics. If on the optical system input to give not a single image but a series of images, which represents the teaching sampling, then at a definite organisation of memory on the dynamic transparant the optical system acquires the adaptive properties. The combination of the system for nonlinear image transformation and optical operational amplifier allows to solve nonlinear partial differential equations, etc.

All above considerations proves that the development of the optical system with the feedback permit to carry out the optical processing of information on more high quality level.

RECENT DEVELOPMENTS IN OPTICAL INFORMATION PROCESSING USING NONLINEARITY AND FEEDBACK

Sing H. Lee

Department of Applied Physics and Information Science
University of California, San Diego
La Jolla, California 92093

I. INTRODUCTION

By using nonlinearity and optical feedback the flexibility of optical processing can be increased. Logarithm, equi-density contour mapping and A/D conversion are nonlinear operations which have already been successfully implemented with halftone screen processes [1-3]. These and other nonlinear operations as well as optical digital processing can be performed by suitable image thresholding and planar logic devices [4,5]. To accomplish optical feedback, partially reflecting mirrors have been utilized [5-7]. Optical processing systems with coherent feedback allow for the realization of the well-known feedback transfer function in two dimensions. In this paper some progress during the last year in the studies at UCSD involving nonlinear optical devices and optical feedback is reported, together with brief reviews of relevant background information.

II. OPTICAL PROCESSING WITH NONLINEARITY

Image thresholding and planar logic devices are two of the most important types of nonlinear optical devices. Optical thresholding devices with real-time capability are essential to real-time analog-to-digital conversions of images (or two dimensional data). Planar logic devices are crucial for processing the converted images in digital mode.

II.A Image Thresholding Device Using Saturable Absorbers [8]

Image thresholding can be achieved with the plane parallel, partially reflecting mirror system of Fig. 1 with a saturable absorber between the mirrors. Multireflections between the mirrors enhance the nonlinear transmission characteristics of the saturable absorber. As a result, the overall transmission is reduced to a very low level when the image intensity is below some threshold value, and increased to a very high level when the image intensity is above threshold.

The threshold value is determined by the mirror reflectance and the saturable absorber. For real-time image thresholding, it would be desirable to have a saturable absorber with a low saturation intensity and fast response time. In the past year, the saturable transmission characteristics of mercurybis(dithizonate) have been studied. Figure 2 shows the measured transmission of mercurybis(dithizonate) dissolved in 1, 1, 2-trichloroethane. It has a strong absorption band in the blue region of the spectrum (4900 Å), which matches the blue line of the argon laser (4880 Å). The saturation intensity is reasonably low ($\sim$1 mw/cm^2) and the response time is about 1 second.

To control the response time, which is dependent upon the lifetime of the excited state, different solvents can be used [9-12], which is an interesting property of mercurybis(dithizonate). However, since the saturation intensity varies inversely as the excited state lifetime, a compromise must be made between keeping the response time of the system as short as possible and keeping the saturation intensity at a level compatible with the available laser power. New solvents are being investigated and the saturable resonator of Fig. 1 is being assembled with mercurybis(dithizonate) to measure the image thresholding and response time of the system.

When thresholding operations can be performed on images in real-time, analog-to-digital conversions can be accomplished in real-time either with the optical system of Fig. 3, in which a halftone screen and spatial filtering technique are utilized [13], or with the optical system of Fig. 4, which works on a principle similar to that used in an electronic digital computer [14]. Figure 5 shows an example of an analog image (Fig. 5a) converted into three binary images, each containing bits of the same significance (Fig. 5b-d)[15].

II.B Planar Logic Devices [16]

Logic operations on two binary images of the same significance can be performed with planar logic devices. To address the planar logic devices properly, the two binary images need be combined either with interwoven fiber bundles (Fig. 6a) or image combiner (Fig. 6b,c). In either case, the combined image consists of alternating columns from the two original binary images as illustrated in Fig. 7. Combining images with fiber optics probably offers the

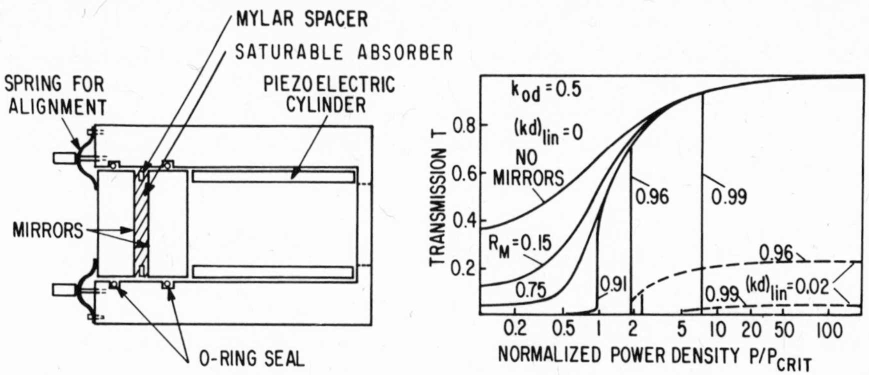

Figure 1. (a) Construction of the saturable resonator. (b) Transmission of a saturable resonator versus the incident power density for different mirror reflectivities. The full curves are for the case that all losses are saturable; the dashed curves are for additional linear losses $(kd)_{lin}$ = 0.02. (E. Spiller, J. Appl. Phys. 43, 1673 (1972)).

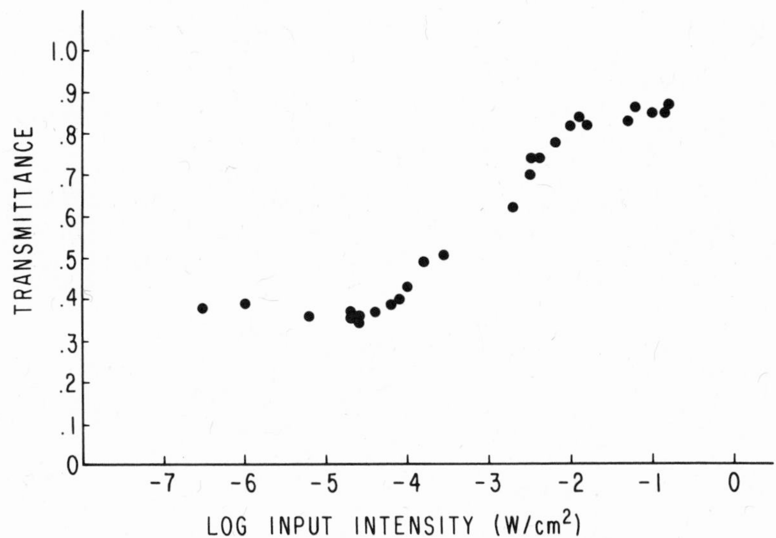

Figure 2. Measured values of transmittance of mercurybis(dithizonate) dissolved in 1,1,2-trichloroethane as a function of incident intensity at 4880 Å of Argon laser.

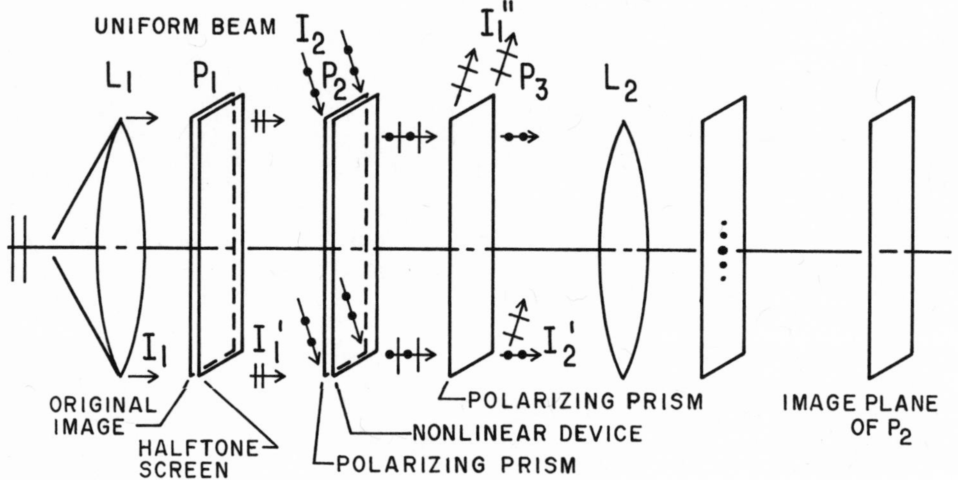

Figure 3. A/D conversion employing halftone screen, nonlinear (threshold) device and spatial filtering. I_1 and I_2 have orthogonal polarizations. The intensity of the uniform beam I_2 is set such that its sum with the desired threshold intensity in image $I_1'(x,y)$ is equal to the threshold level of the nonlinear device (TD). With the uniform beam I_2 properly biasing TD, I_1' will control the transmission of TD at two binary states. Then, as I_2 passes through TD, and experiences the binary transmission states controlled by I_1', the halftone image I_2' results. Different bit planes can be obtained by filtering different diffraction orders in the Fourier plane as described in Ref. 3.

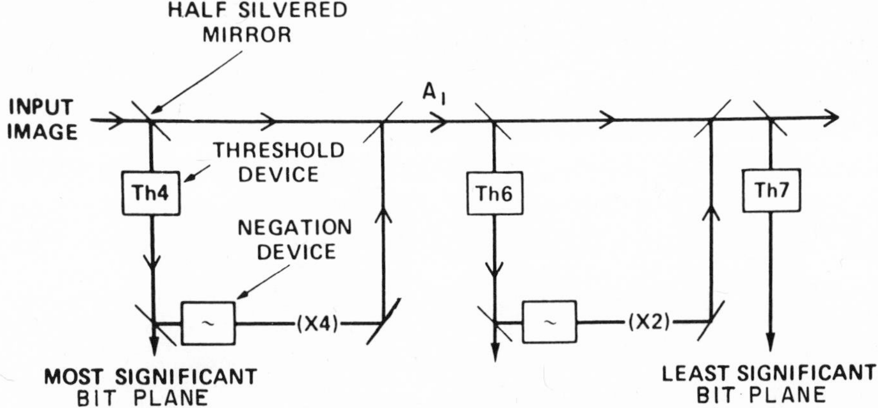

Figure 4a. A/D conversion of an eight level image. The most significant bit plane is obtained by thresholding the input at intensity level 4. It is white above threshold and black otherwise. The *negative* of this bit plane is then taken (i.e. white areas made black and black areas made white), with the white areas given an intensity value of 4, and added to the original input. The resulting image (A_1) is thresholded at intensity level 6 to produce the second bit plane. It is white in areas above the threshold level 6 and black otherwise. Next, the *negative* of the second bit plane is taken, with the white areas given an intensity level of 2. This *negative* is added to image A_1; the resulting image is thresholded at intensity level 7 to give the least significant bit plane. It is white in areas above threshold level 7 and black otherwise (see Ref. 14).

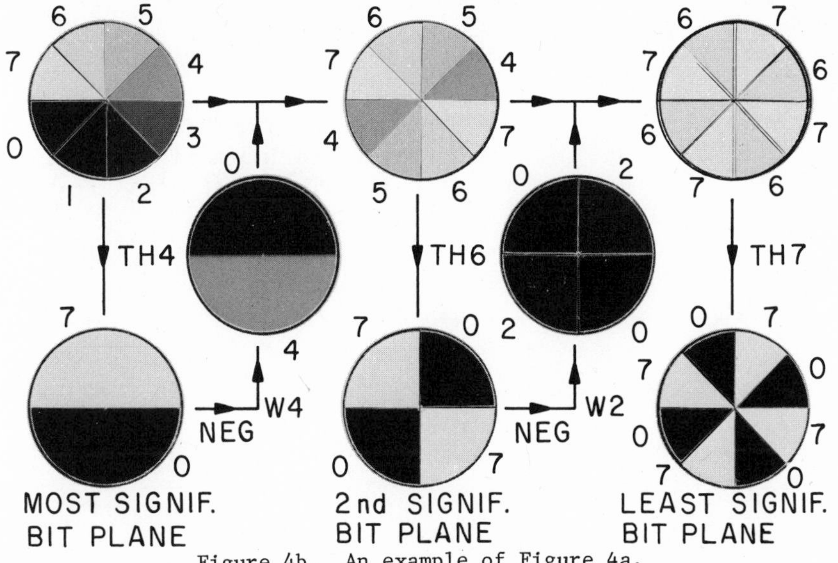

Figure 4b. An example of Figure 4a.

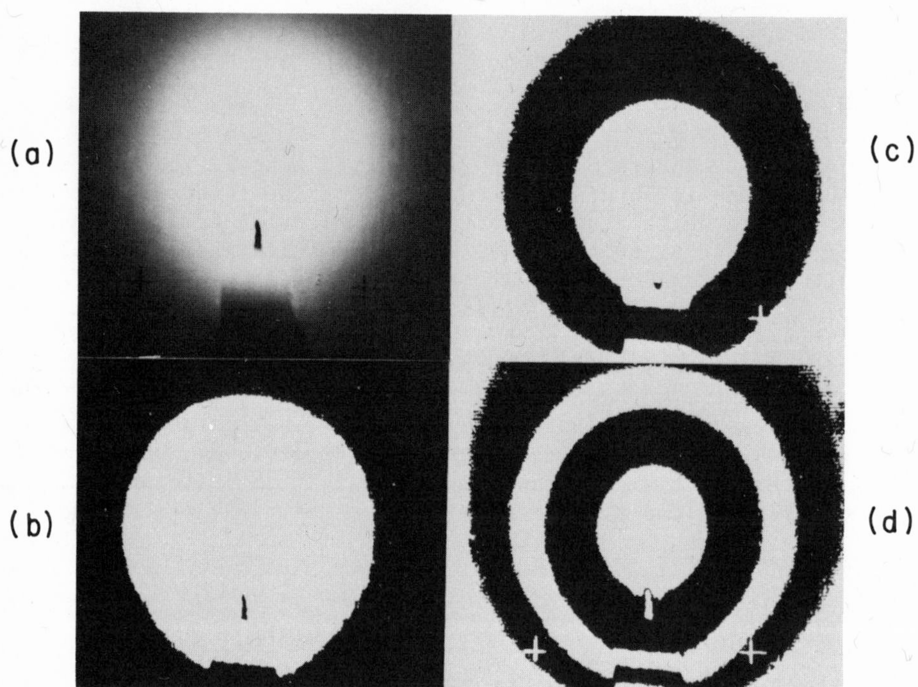

Figure 5. Results of bit plane generation. (Reference 15). (a) Original image, (b) First bit plane (most significant bit), (c) Second bit plane, and (d) Third bit plane (least significant bit).

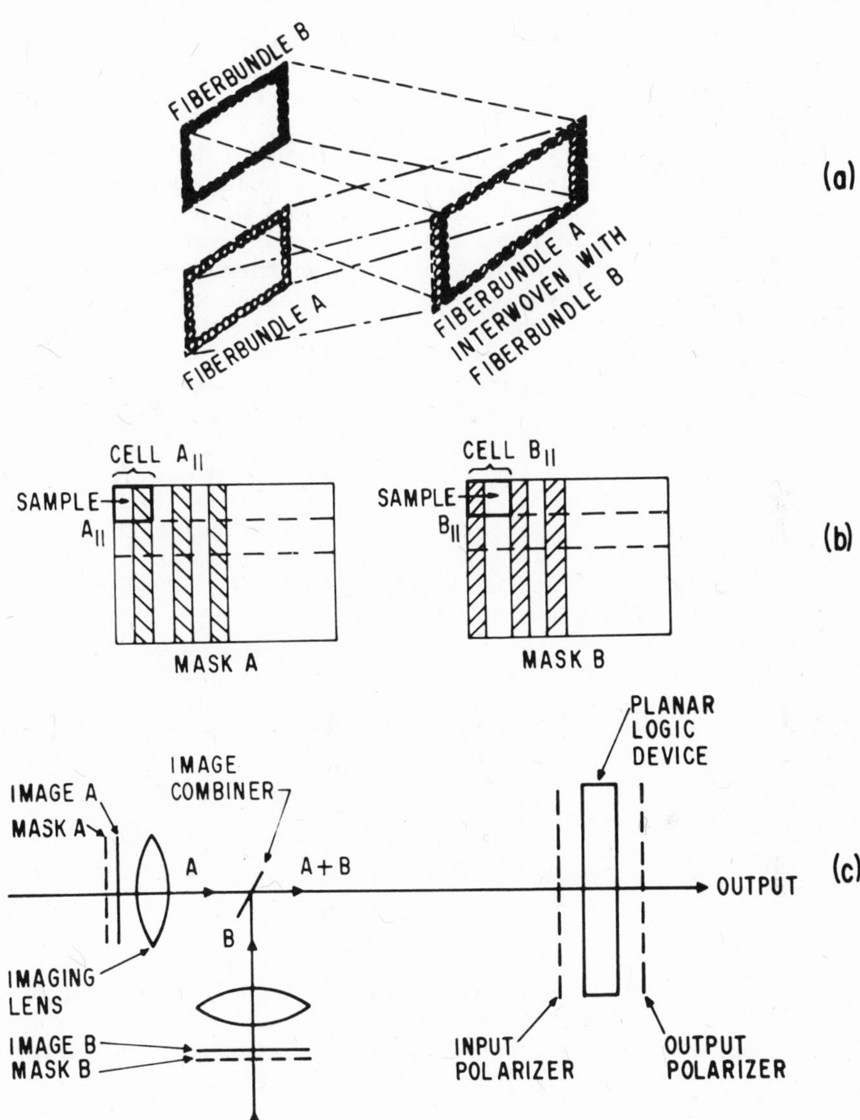

Figure 6. (a) Image combining using optical fiber interweaving; (b) Masking or sampling of input images; (c) Planar logic device with associated optical system.

advantage of greater compactness, especially for signals of high data rates. This permits the many devices required for large scale data processing to be assembled within a reasonable volume. At Galileo Electro-Optics Corporation, Sturbridge, Mass., there is a program currently devoted to developing fiber optic image combiners (and other image processing fiber optics devices [17]).

Device Description and Operation. Figure 7 illustrates the schematic diagram of a planar logic AND gate under investigation at UCSD. It consists of a glass substrate coated with a layer of transparent electrode, followed by a patterned layer of photoconductor, a patterned layer of electrode, a uniform layer of electro-optic material and another glass substrate coated with a transparent electrode. Incident upon the planar AND gate is the combined image just after passing through a polarizer. The light incident on cell A_{ij} causes the potential across the photoconductive material in the cell to drop to zero. This drop in potential will then be transformed into an increase in voltage across the electro-optic material in cell B_{ij}, since the electrode pattern above the electro-optic layer links cells A_{ij} and B_{ij} together. The voltage increase across the electro-optic material rotates the polarization of the light through cell B_{ij} by 90°. In other words, had it not been for light incident on cell A_{ij}, the potential across the photoconductive region in cell A_{ij} would be high and that across the electro-optic material in cell B_{ij} low. Then the light passing through cell B_{ij} would not be rotated in polarization. An output polarizer orthogonal to the input polarizer can easily convert the output polarization information from cell B_{ij} into binary intensity information. Hence, the planar AND gate performs an AND operation, i.e. there will be light passing through cell B_{ij} only when there is light incident on both cells A_{ij} and B_{ij}.

The scheme of Fig. 7 can also be used for several other important processing operations. (a) Pseudo-amplification: Pseudo-amplification will be obtained when A_{ij} is binary, the intensity of the "on-state" of A_{ij} is at a level still strong enough to activate the photoconductor and the intensity of all B_{ij} is at a level higher than that of the on-state of A_{ij}. Amplification becomes necessary when the image signal is attenuated after passing through a number of transmission channels or logic gates. (b) Standardization and noise suppression: When A_{ij} represents the intensity pattern, which contains noise and is to be standardized, and the intensity of all B_{ij} coming into the device is at a desired standardized level, the outgoing C_{ij} from the device will be the standardization of A_{ij} and the noise contained in A_{ij} eliminated, provided that the photoconductor is operated in its nonlinear region. (c) Negation Operation: When the output polarizer is oriented in parallel to the input polarizer and all B_{ij} are in their "on-state", the output C_{ij} will be the NEGATION of the input A_{ij}. The reason is that an A_{ij} element

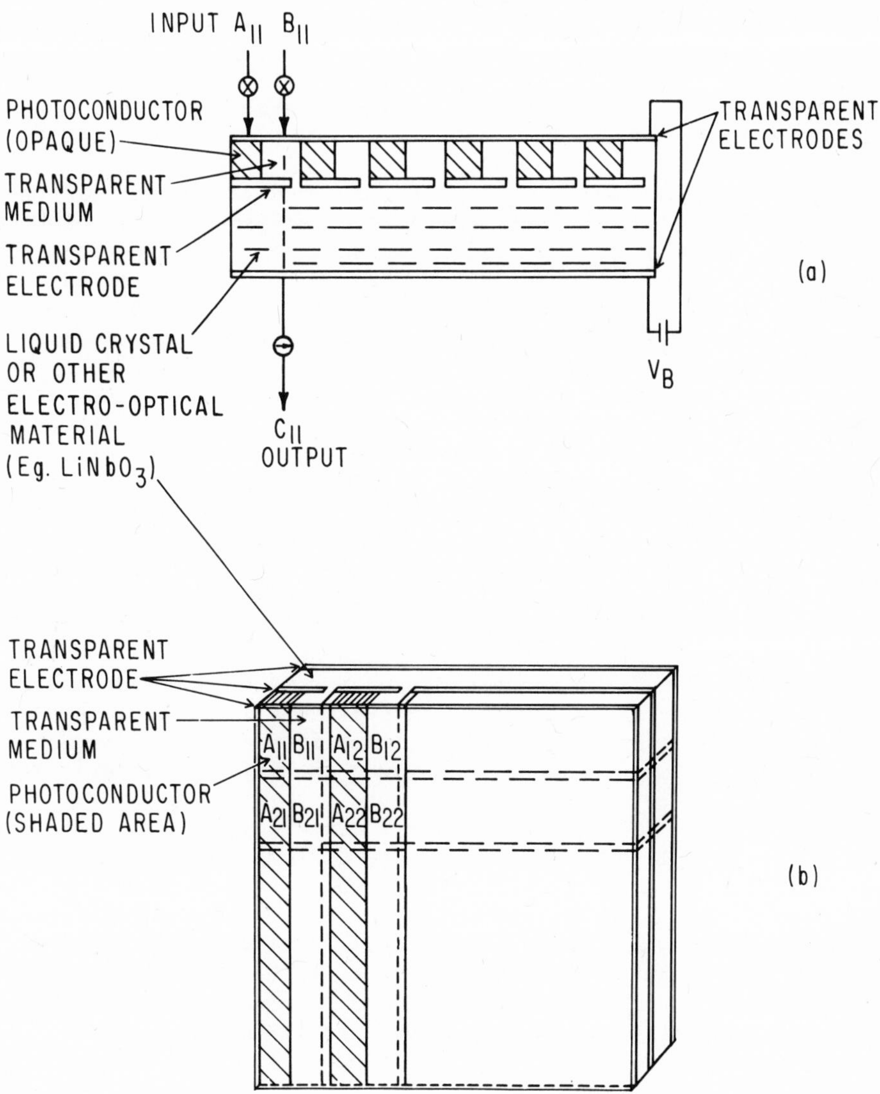

Figure 7. Planar logic AND gate. (a) Side view. (b) Perspective view.

in its "on-state" will rotate the polarization of light going through cell B_{ij} by 90° and be blocked by the output polarizer. (d) OR operation: If there are two input images addressing the A_{ij} elements in similar ways as shown in Figs. 6 and 7, either input image will be able to control B_{ij}, which is in its "on-state", to provide an output image C_{ij} through an analyzer orthogonal to the input polarizer satisfying the logic OR conditions. The underlying principle involved here is similar to those for other logic operations discussed above.

The attractive features of planar logic devices are: (a) they are reasonably simple to fabricate by the well-known semiconductor technology (the photolithographic process), (b) they are passive devices whose operations can rely on the most efficient optical source available external to the devices (e.g. heterostructure injection lasers), (c) they promise to consume reasonably small power (see power consumption calculation below). In the past year, the necessary facility to fabricate planar logic devices has been developed. We have also estimated the power consumption requirements of several important computing circuits built with planar logic devices. Power consumption often concerns those who contemplate using the optical approach for computing [18].

Power Consumption Estimate for a Half-Adder Circuit. The binary addition of two bit planes of the same significance can usually be accomplished with three AND gates, two NOT gates and one OR gate (Fig. 8). Assuming that the logic device has an area of 1 square inch (6.25 cm^2) and that the photoconductor used in the device is PVK:TNF (1:1 molar ratio), which can provide a light to dark current ratio of 40:1 in response to a light intensity as low as 5μw/cm^2, the minimum input optical power to activate an logic gate is 31.25μw.

$$\text{Min. input optical power per logic gate} = 5\mu w/cm^2 \times 6.25\ cm^2/\text{gate} = 31.25\mu w/\text{gate} \quad (1)$$

Each of the inputs A and B in Fig. 8 are required to drive three gates and will require the optical power of 187.50μw.

$$P_1(\text{optical power for inputs A and B}) = 6\ \text{gates} \times 31.25\mu w/\text{gate} = 187.50\mu w \quad (2)$$

The two NOT gates and the OR gate in Figure 8 will require an additional optical power of 93.75μw.

$$P_2(\text{optical power for two NOT and one OR gates}) = 3\ \text{gates} \times 31.25\mu w/\text{gate} = 93.75\mu w \quad (3)$$

In the operations of planar logic gates (Fig. 7) it is observed that one of their inputs is transmitted directly through to the output, which can be used to drive one subsequent gate. Hence, the total optical power needed to operate the half-adder circuit will be 281.25μw.

$$P_{HA}\text{(total optical power for half-adder circuit)} = P_1 + P_2 = 187.50\mu w + 93.75\mu w = 281.25\mu w \quad (4)$$

At 5% conversion efficiency [19], the electrical power required to generate the needed optical power will be 5.625 mw.

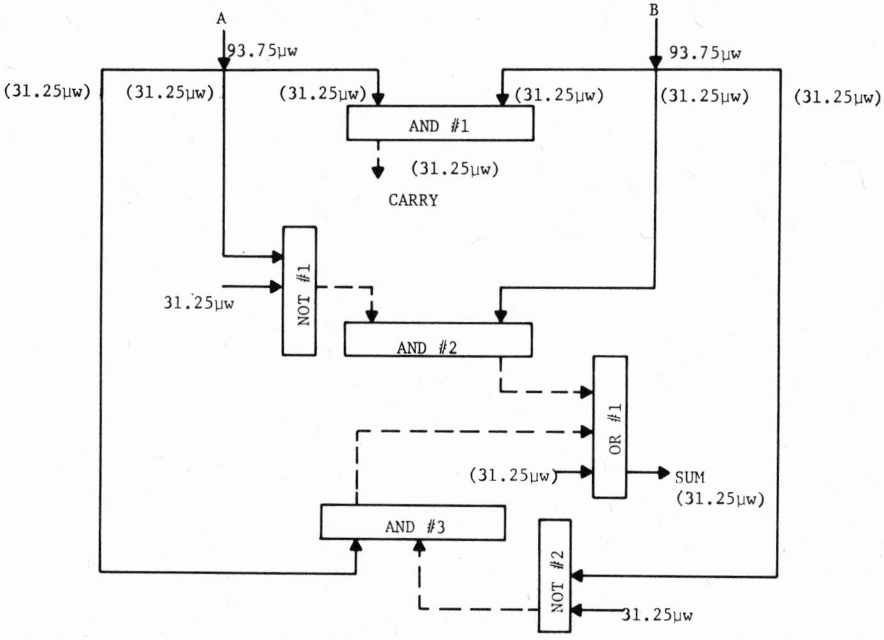

Figure 8. Half-adder circuit. The binary addition of two bit planes of the same significance can be accomplished with three AND gates, two NOT gates, and one OR gate: A exclusive OR B = Sum, A AND B = Carry. Solid lines indicate power supplied from external source to the half-adder, dashed lines indicate power transfer between logic gates within the half-adder.

<u>Power Consumption Estimate for a Full-Adder Circuit</u>. A full-adder circuit can be built by combining two half-adder circuits and an OR gate as illustrated in Fig. 9. The power consumption calculation for half-adder I is similar to that given above. The calculation for half-adder II is also the same except that additional power needs be provided to the OR gate in HA-II. The additional power is needed because the SUM output from the OR gate in HA-II is used to drive one of the two inputs to HA-I (93.75μW vs 31.25μW). Therefore, the power consumption for the two-half adders consists of

C_{in} input to HA-I	93.75μw	
Sum input to HA-I is derived from HA-II	0	
Power supplied to 2 NOT, 1 OR gates in HA-I	93.75μw	
Input 1 to HA-II	93.75μw	
Input 2 to HA-II	93.75μw	
Power supplied to 2 NOT, 1 OR gates in HA-II	156.25μw	
Total optical power input to HA-I and HA-II	531.25μW	(5)

The OR gate in Fig. 9 provides the CARRY output from one full-adder stage, which is normally used to drive the C_{in} input to HA-I in the subsequent stage. So, the OR gate requires 93.75μw input power in order to provide the desired optical power at its output. The total optical power required to operate one full-adder stage is then 625μw (=531.25μw+93.75μw). At 5% conversion efficiency, the electrical power required to generate 625μw is 12.5 mw.

One more example of power consumption calculation is given in Appendix A. The calculation is for a single-stage of the adder-subtractor combined with accumulator, which is a very common unit in computers for performing arithmetic operations. The important conclusion derived from this example as well as those above is that planar logic devices can be employed to perform optical digital processing at reasonable power levels. Furthermore, assuming a moderate resolution of 20 cells/mm on 3" devices and a switching speed of 100μsec, logic operations can be performed at the rate of 2.25×10^6 bits/frame and 22.5 gigabits/sec., which is quite attractive. Faster rates can be expected if electro-optic materials of faster switching speed are available.

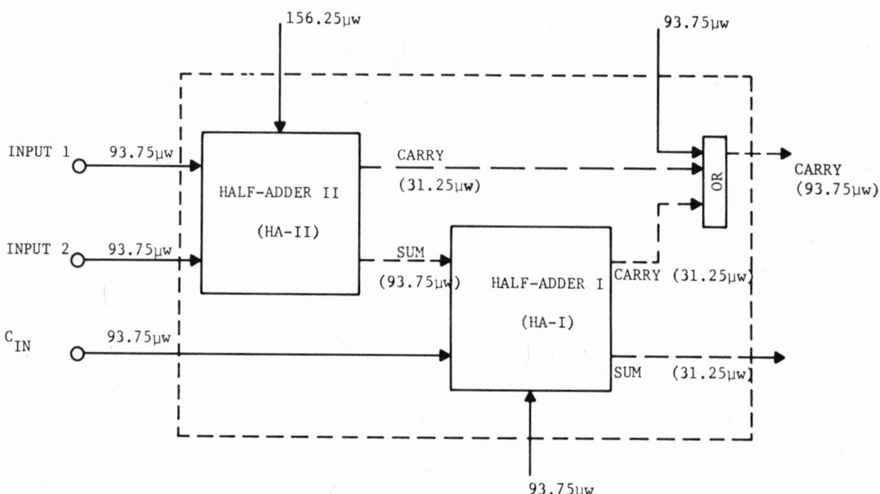

Figure 9. Full-adder circuit. A full-adder circuit can be built by combining two half-adder circuits and an OR gate. (Solid lines indicate power supplied from external source to the full-adder; dashed lines indicate power transfer between logic gates within the full-adder).

III. OPTICAL PROCESSING WITH COHERENT FEEDBACK [20]

In coherent optical processing, part of the output signal can be fedback to the input using plane mirrors together with or without lenses. The threshold device uses plane mirrors to feedback light for increased nonlinearity (section II.A). If lenses are used and the mirrors are symmetrically tilted at a small angle with respect to the optical axis as shown in Fig. 10(a), two physically separated Fourier transforms become available for spatial filtering, one transform appearing in the forward path and the other in the feedback path [6]. The transfer function for the coherent optical feedback system is

$$T(u,v) = \frac{A_o(u,v)}{A_i(u,v)} = \frac{t_1 t_2 G(u,v)}{1-[r_1 r_2 t_0]G(u,v)H(u,v)e^{i\phi}} \tag{6}$$

where

$A_i(u,v)$ and $A_o(u,v)$ are the Fourier transforms of the input and output images respectively,
$G(u,v)$ and $H(u,v)$ are the spatial filtering functions in the forward and feedback path,
r_1 and r_2 are the amplitude reflectances of M_1 and M_2,
t_0, t_1, and t_2 are the amplitude transmittances of the optics between M_1 and M_2, M_1, and M_2, respectively,
ϕ is the phase of the feedback signal.

ϕ can be controlled to give positive, negative or complex feedback by mounting one of the feedback mirrors on a piezoelectric translator. r_1, r_2 and t_0 are typically 0.98 while t_1, and t_2 are 0.2. Aside from the constant multipliers of the "r"s and "t"s and the fact that it is two-dimensional, $T(u,v)$ is similar to the well-known transfer functions of other feedback systems.

Several applications which lend themselves to the use of coherent optical feedback have been studied. These include image restoration, contrast control and solving partial differential equations [6].

Confocal Feedback System. In the past year we have been studying a method to simplify the optical system of Fig. 10(a) by using spherical mirrors both as reflecting and Fourier transforming elements in the confocal Fabry-Perot arrangement shown in Fig. 10(b). The operation of the confocal system is as follows: An input a_i is imaged telecentrically to the mid-plane, P. The mirror M_2 produces the Fourier transform, A_i, of the input centered at g in plane P. Because the plane wave illuminating the input image is at an angle Θ

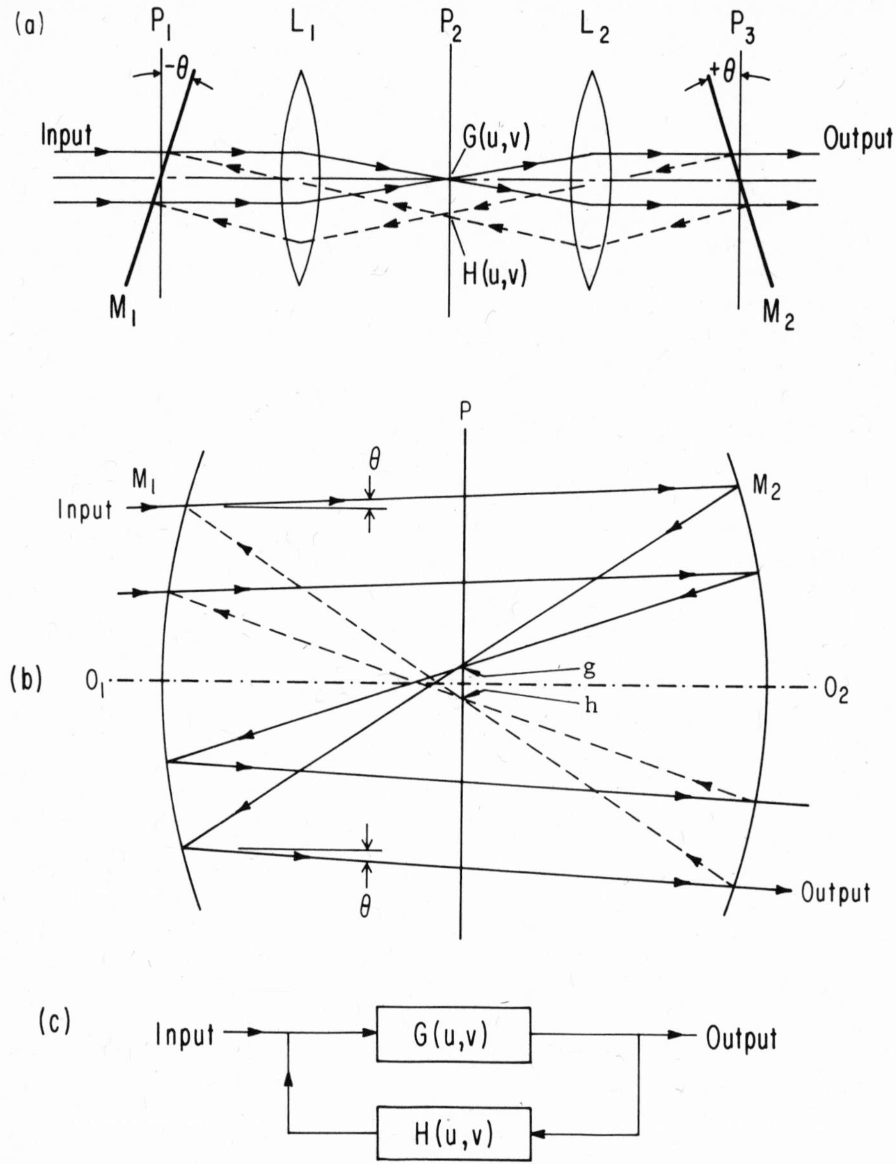

Figure 10. The optical analog computer with coherent optical feedback. (a) The previous system using two Fourier transform lenses, L_1 and L_2, and two plane mirrors, M_1 and M_2. (b) The simplified system consisting of two spherical mirrors, M_1 and M_2, in a confocal arrangement. (c) The block diagram of the optical computer with Fourier transform filters, G(in the forward path) and H (in the feedback path).

to the optical axis, the point g is located at a distance y above the optical axis:

$$y=(r \sin\theta)/2 \tag{7}$$

where r is the radius of the mirror. A_i is spatially filtered by G at the point g and the result transformed again by mirror M_1. The part of this light which is transmitted by M_2 is the output, a_o, of the system. The rest of the light is reflected into the feedback path and Fourier transformed by M_2. This transform appears at point h and is spatially filtered by H. Finally, the feedback beam is transformed by M_1 and rejoins the original beam with a phase shift which depends on the mirror separation.

Comparison to Original Feedback System. The confocal feedback system has a number of advantages over earlier feedback systems. The spherical mirrors produce good Fourier transforms, yet are considerably less expensive than high quality Fourier transforming lenses. Because of the confocal Fabry-Perot configuration, the adjustment of the mirrors is less critical and the stability of the two physically separated transforms is greater than for systems using plane mirrors. It is interesting to note also that the input and output planes of the device are located in the midplane P (Fig. 10b) rather than on the surfaces of the mirrors as in the original system (Fig. 10a). These planes are therefore easily accessible.

The confocal system of Fig. 10b has been constructed and successfully tested. Experiments are now being conducted to evaluate the extent of its ability to solve partial differential equations (e.g. Helmholtz, LaPlace, and wave equations of two variables).

IV. SUMMARY

Some progress in optical processing with nonlinearity and feedback has been reported. In our study on saturable absorbing materials for image thresholding, mercurybis(dithizonate) has been found to have an unusually low saturation intensity and controllable response time. In the planar logic device study, educated estimates of power consumption of several important computing circuits have been made. The estimated power consumptions are all at reasonable levels. In addition, based on the present electro-optic technology, the processing speed of planar logic devices is attractive enough to justify further research in optical digital processing. In the study of coherent optical feedback, a simplified system using a confocal Fabry-Perot interferometer has been developed with improved performance characteristics.

APPENDIX A

A Single Stage of the Adder-Subtractor with Combined Accumulator. The combined adder-subtractor and accumulator unit is one of the most common units in a computer for performing arithmetic operations. This unit contains a full-adder circuit to perform the arithmetic operation of addition or subtraction and two set-reset flip-flop memories, one of which is used as a memory data register and the other one as an accumulator. The memory data register is for storing one of the two binary numbers on which the operation is to be performed. The accumulator stores the other binary number and the result of the operation. There are also five AND, two OR and three NOT gates in this unit controlling the operation of these circuits and eight input terminals at which control signals are applied. The schematic diagram for a single stage of the combined adder-subtractor and accumulator unit is given in Figure A.1 [21].

Circuit Operation. The memory bus #1 carries one binary digit from the core memory which is stored in the memory data register. A signal at terminal 2 (Transfer) will cause the transfer of the digit stored in the memory data register to the INPUT 1 of the full-adder circuit, if a signal is simultaneously applied to terminal 4 (Add). But if a signal is simultaneously applied to terminal 3 (Subtract) instead of terminal 4 (Add), the complement of the digit stored in memory data register will be transferred to the INPUT 1 of the full-adder circuit. One digit of the other binary number on which an operation is to be performed is transferred from memory bus #2 to the accumulator, when a signal is applied to the terminal 6 (Load Accumulator From Memory Bus). The output of the accumulator is connected to the INPUT 2 of the full-adder circuit. The CARRY input to the full-adder in the least significant stage is set at 1 when subtraction is to be performed. It is set at 0 when addition is to be performed. To reset the accumulator for storing the result of the operation, a signal is now applied at terminal 7 (Clear Accumulator). Then the full-adder carries out the desired operation. A signal at terminal 5 (Load Accumulator From the Full-Adder) transfers the SUM output from the full-adder to the accumulator. A signal at terminal 8 (Transfer Accumulator To Memory Bus) transfers the results of the operation to core memory. The memory data register and accumulator are reset for the next operation by applying signals to terminal 1 (Clear Memory Data Register) and terminal 7 (Clear Accumulator) respectively.

Power Consumption Calculations. The circuit in Figure A.1 can be broken up into four parts for power consumption considerations. They are:

(i) the input terminals where the control signals are applied,

(ii) the logic gates controlling the operation of the circuit,

(iii) the full-adder circuit and
(iv) the two flip-flops (Memory data register and accumulator).

There are eight terminals to the unit where control signals are applied. But terminal 2 (Transfer) is required to drive two gates (AND #1 and AND #2). The optical power that needs to be provided to the terminals thus becomes 281.25μW [=(8+1)×31.25μW].

There are five AND gates, two OR gates and three NOT gates in the circuit. These gates control the flow of information in the unit in accordance with the control signals applied to various terminals. One of the OR gates (OR #1) supplies optical power to drive INPUT 1 of the full-adder and requires at its output three times the power needed for driving a single gate (Section II.B). The remaining OR gate (OR #2) and the three NOT gates (NOT #1, #2, #3) drive only one subsequent gate each. Since the optical power required to drive one logic gate is 31.25μW (Section II.B), the external optical power required by the logic gates is 218.75 W [= 3×31.25μW (OR #1)+31.25μW (OR #2)+3×31.25μW (3 NOT gates)].

The full-adder circuit consists of two half-adder circuits and an OR gate (Section II.B). Since the power supplied to INPUT 1 and INPUT 2 comes from parts within the same stage, the external power required to operate the full-adder will be 437.5μW [= 93.75μW (C_{in}) + 93.75μW (2 NOT, 1 OR in HAI) + 156.25μW (2 NOT, 1 OR in HA II) + 93.75μW (OR in FA)].

The accumulator flip-flop output drives one AND gate (AND #5) and provides power to INPUT 2 of the full-adder. Thus the accumulator flip-flop must have at its output 4 [= 1 + 3] times the power needed for driving a single gate. The optical power required by this flip-flop is calculated to be 156.25μW (Appendix B).

The memory data register flip-flop output drives two AND gates (AND #1 and AND #2). The optical power required by the memory data register is calculated to be 93.75μW (Appendix B).

The results of these calculations (assuming 5% efficiency in generating the external optical power) are summarized in the following table:

Part	Optical Power Required	Elec. Power Consumed
The input terminals	281.25μW	4.375mW
The logic gates	218.75μW	4.625mW
The full-adder	437.50μW	8.750mW
Accumulator	156.25μW	3.125mW
Memory data register	93.75μW	1.875mW
TOTAL	1.1875mW	23.750mW

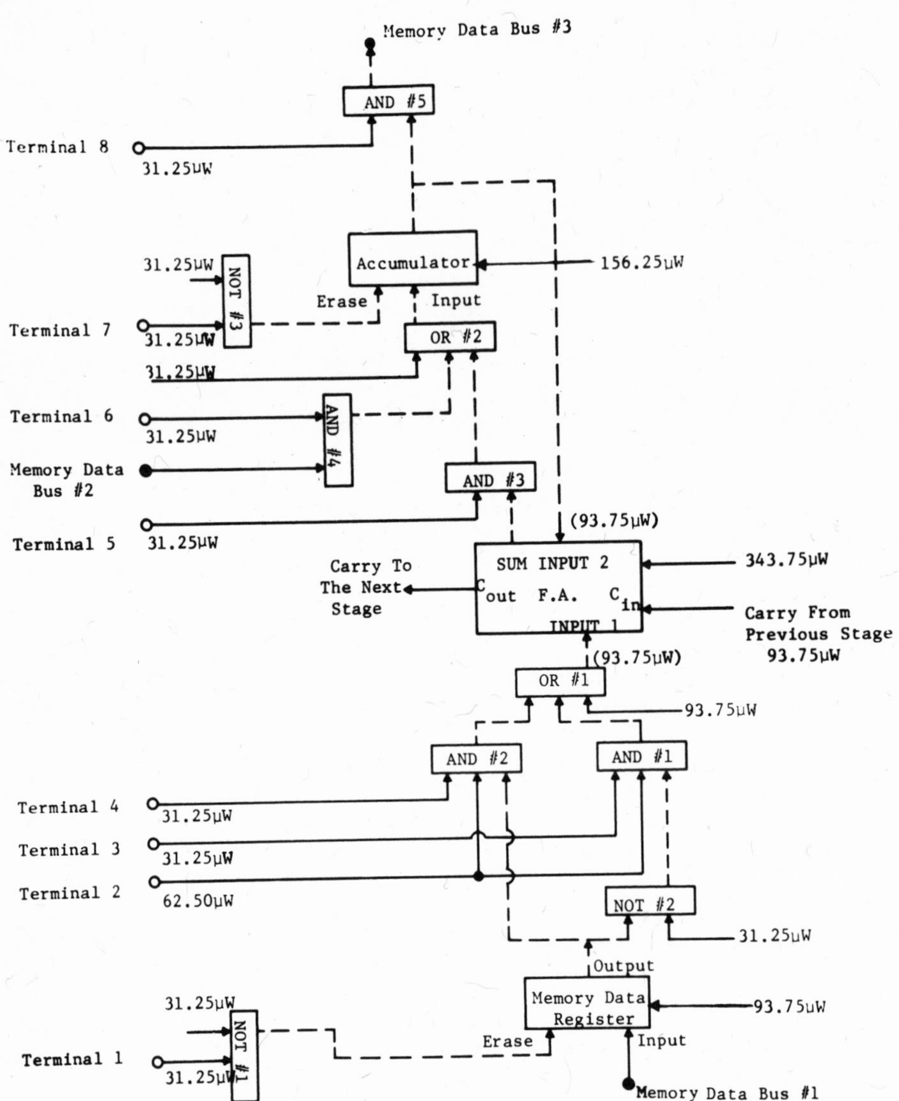

Figure A.1. Schematic diagram for a single stage of the adder-subtractor combined with accumulator. Terminal 1-Clear memory data register; Terminal 2-Transfer; Terminal 3-Subtractor; Terminal 4-Add; Terminal 5-Load accumulator from full-adder; Terminal 6-Load Accumulator from memory bus; Terminal 7-Clear accumulator; Terminal 8-Transfer Accumulator to memory data bus; F.A.-Full Adder.

APPENDIX B

Set-Reset Flip-Flop Memory. A set-reset flip-flop memory is essential to a computer for temporary storage of data. One way to construct a flip-flop is by use of optical feedback between logic gates as illustrated in Figure B.1 [14].

The AND gate in the circuit controls the feedback to the OR gate. Thus when the ERASE signal is in the 0 state, the feedback cannot reach the input 2 of the OR gate. At this stage, if there is no signal at the INPUT port (input 1 to the OR gate), the OUTPUT of the flip-flop becomes 0 and the flip-flop is said to be reset. When information is to be stored in the flip-flop, the ERASE signal is switched to 1 state which now allows the OUTPUT of the OR gate to pass through the AND gate to the input 2 of the OR gate via the feedback loop. Since the input 2 to the OR gate is in 0 state to begin with, the OUTPUT of the flip-flop assumes the same state as the INPUT signal when the signal to be stored is applied to the INPUT port. This signal now appears at the input 2 via feedback and through the AND gate and keeps the OUTPUT of the flip-flop in the same state even when the INPUT signal is removed. To reset the flip-flop, the ERASE signal 0 is applied to break the feedback loop as described before.

To operate the flip-flop, enough optical power needs be supplied externally to the OR gate (Fig. B.1) in order to provide the feedback signal and an output at a proper power level. (The power associated with the ERASE signal has already been accounted for in Appendix A). Thus, for the accumulator flip-flop whose output drives one AND gate (#5) and INPUT 2 to the full-adder circuit (Figure A.1), the proper output power level is four times the power needed to drive one gate. The optical power to be supplied externally to the OR gate of the accumulator flip-flop is 156.25μW [= 31.25μW (FEEDBACK) + 4 × 31.25μW (OUTPUT)]. For the memory data register flip-flop whose output drives two AND gates (#1 and #2 in Figure A.1), the proper output power level is two times the power needed to drive one gate. The optical power to be supplied externally to the OR gate of the memory data register flip-flop is 93.75μW [= 31.25μW (FEEDBACK) + 2 × 31.25μW (OUTPUT)].

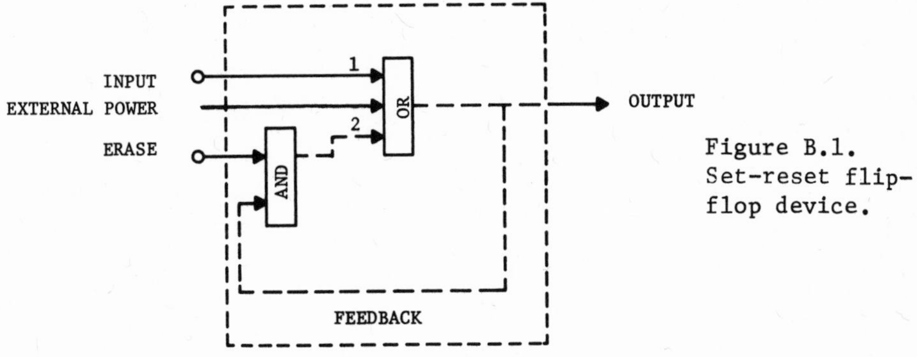

Figure B.1. Set-reset flip-flop device.

ACKNOWLEDGEMENT

Much of the recent progress reported in this paper is obtained jointly with R. Athale, B. Bartholomew and J. Cederquist of UCSD. The planar logic device concepts described in Section II.B are extensions of those concepts initially described by D. Schaefer and J. Strong of NASA GSFC in Ref. 14. The research is supported in part by the National Aeronautics and Space Administration at Goddard Space Flight Center and by the National Science Foundation.

REFERENCES

1. H. Kato and J.W. Goodman, "Nonlinear Filtering in Coherent Optical Systems through Half-tone Screen Process", Appl. Opt. 14, 1818 (1975).

2. S.R. Dashiell and A.A. Sawchuk, "Analysis and Synthesis of Nonlinear Optical Processing", Proc. 1975 Electro-Optical Syst. Design Conf., pp. 9-15. Industrial and Scientific Conference Management, Inc., Chicago, Illinois (1975).

3. A.W. Lohmann and T.C. Strand, "Analog-to-Digital Conversion of Pictures with Optical Means", Proc. 1975 Electro-Optical Syst. Design Conf., pp. 16-21. Industrial Scientific Conference Management, Inc. Chicago, Illinois (1975).

4. S.H. Lee, "Nonlinear Optical Processing" in Optical Information Processing (Yu E. Nesterikhin, G.W. Stroke and W.E. Kock, ed.) Plenum Press, N.Y. (1976).

5. S.H. Lee, B. Bartholomew and J. Cederquist, "Two Improved Coherent Optical Feedback Systems for Optical Information Processing", To be presented in the 20th Anniversary Technical Symposium of SPIE at San Diego, August 25, 1976.

6. D.P. Jablonowski and S.H. Lee, "A Coherent Optical Feedback System for Optical Information Processing", Appl. Phys. 8, 51 (1975).

7. J.C. Wyant and P.N. Tamura, "On-Axis Coherent Optical Feedback System for Image Processing", J. Opt. Soc. Am. 66, 169A (1976).

8. The research on image thresholding devices is jointly conducted currently with Bruce Bartholomew of UCSD.

9. L.S. Meriwether, E.C. Breitner and C.L. Sloan, "The Photochromism of Metal Dithizonates", J. Am. Chem. Soc. 87, 20, 4441-4448 (October 20, 1965).

10. L.S. Meriwether, E.C. Breitner and N.B. Colthup, "Kinetic and Infrared Study of Photochromism of Metal Dithizonates", J. Am. Chem. Soc. 87, 20, 4448-4454 (October 20, 1965).

11. L.S. Meriwether, E.C. Breitner and C.L. Sloan, U.S. Patent No. 3,361,706 assigned to American Cyanmid Company (January 2, 1968).

12. British Patent Nos. 1,146,309; 1,146,310 assigned to Nuclear Research Associates, Inc., (March 26, 1969).

13. S.H. Lee, "Optical Processing with Nonlinearity and Feedback", Proc. 1975 Electro-Optical Syst. Design Conf., p. 29, Industrial Scientific Conference Management, Inc., Chicago, Illinois (1975).

14. D.H. Schaefer and J.P. Strong, "Tse Computer". NASA Technical Report X-943-75-14 (January 1975).

15. T.C. Strand, "A Method of Nonlinear Optical Information Processing". Ph.D. dissertation, University of California, San Diego (1976).

16. The research on planar logic devices is jointly conducted currently with Ravi Athale of UCSD. The planar logic device concepts described in Section II.B are extensions of those concepts initially described by D. Schaefer and J. Strong of NASA GSFC in Ref. 14.

17. C.H. Tosswill, "Image Processing by Fiber Optics Array", to be published in Applied Optics.

18. R.W. Keyes and J.A. Armstrong, "Thermal Limitations in Optical Logic", Appl. Opt. 8, 2549 (1969).

19. H. Kressel, H.F. Lockwood, I. Ladany and M. Ettenberg, "Heterojunction Laser Diodes for Room Temperature Operation", Opt. Eng. 13, 416 (1974).

20. The research on confocal feedback systems described in Section III is jointly conducted currently with Jack Cederquist of UCSD.

21. D.J. Woollons, Introduction to Digital Computer Design, Chapter 7, McGraw-Hill, New York (1972).

A SURVEY OF MICROIMAGE TECHNOLOGY AND PRACTICE

MICHAEL P. ZAMPINO

ASSISTANT VICE PRESIDENT

CITIBANK, NEW YORK

1. Introduction - Citibank, The Paper Machine

Citibank is one of the world's leading private banking institutions. With $60 billion in total assets, offices in every major city of the world, including Moscow, 20,000 employees in the New York City area and over 20,000 additional employees throughout the world, Citibank is a major business undertaking in its own right. More significant is Citibank's scope of customers: nearly one million personal banking accounts, banking relationships with nearly every major American and many non-U.S. business corporations, as well as many national governments, including the Soviet Union's.

The most striking aspect of Citibank's business profile is that its product is, for the most part, completely intangible. That product is service. It is Citibank's business to accept funds and other assets from its depositors, transfer them in accordance with their instructions and offer credit. In industrial economies, these transactions are conducted mainly by paper transference or ledger entries. Although these processes conjure images of the 19th century clerk with green eyeshades and endless ledgers, the very size of Citibank's business and the need to execute our customers' transactions promptly and accurately has rooted the bank firmly in the 20th century.

2. Need

By 1976, the paper and ledger accounting orientation of Citibank's service business and its coupling to our enormous customer base had led to an incredible paper explosion. In our New York City operations alone, several million checks, nearly one hundred thousand stock certificates, tens of thousands and non-check funds transfers, letters of credit, items for collections and loans are processed daily. Added to these documents are countless internal work papers, internal and external correspondence, advices to our customers and computer printout. Computer printout, for example, averaged in excess of three tons per day.

Processing the daily flood of transactions has long been the task of sophisticated digital computing equipment. However, paper input and output from our accounting systems has resisted the advance of technology. There has been many reasons for this. Paper-based systems offer:

- simplicity
- low capital investment
- familiarity with the medium

Furthermore, management, and the scientific and engineering communities have focused their attention on the more glamorous aspects of data processing. In the face of literally millions of transactions, the advantages of paper quickly evaporate.

Our need was where to put our incredible volumes of paper and how to find specific items on short notice. Historically, solutions to these problems have been based on the file clerk, te file cabinet and file room, and the records warehouse. In Citibank's case this represented hundreds of clerical personnel, thousands of square feet of high-rent office space used as storage facilities and a 300,000 cubic foot capacity warehouse. The system's chief drawback was that customers and executives waited days for documents they needed. Moreover, with disturbing frequency, desired documents could not be found.

In 1975, Mr. Robert B. White, Citibank Executive Vice President, commissioned a research project to discover technologies which would solve our paper explosion. This paper represents the findings and some tangible results of that study.

3. The Goal - The Ideal and The Real

Before setting out on our quest, we had to better understand our needs and, then, to envision the instrumentality of solution.

Citibank's "transaction machine" was driven by our customers' demands upon it. These demands were in the form of checks or loans to be posted, funds or stock to be transferred in ownership, or other paper documents which our customers sent to the bank. Processing these documents in our "transaction machine" created, in turn, an even larger bulk of output reports and other supporting papers.

The ideal solution presented itself very quickly. That solution was to require all of our customers to communicate their transactions to the bank in formatted electronic code. This approach has received great coverage in banking literature under the heading of Electronic Funds Transfer System (EFTS). Citibank has been a pioneer in EFTS with its introduction of Citicard as a substitute for certain checking transactions and with formatted telegraphic money transfer systems. Schematically, EFTS would appear as in Figure 1.

However, owing to custom, legal requirements, as well as the design and introduction of new systems, EFTS is not expected to become a significant factor for many years. As paper initiated transactions appeared to be given for the bulk of our processing for the near to intermediate future (5-10 years), with certain types of transactions sure to be paper bound for even longer periods, we turned to the concept of capturing our customers paper-initiated transactions on microimage as well as data files and directing all subsequent referral to paper-bound information to microimage files. The nature of our desired instrumentality, then, was a device which could readily capture document images in conjunction with coded document descriptions and assemble those images into logical files which could be readily accessed. Moreover, the system had to be:

- cost effective
- reliable
- easy to operate
- have the file management facility of a computer
- available in a reasonable time period.

Schematically, a system of this type would appear as in Figure 2.

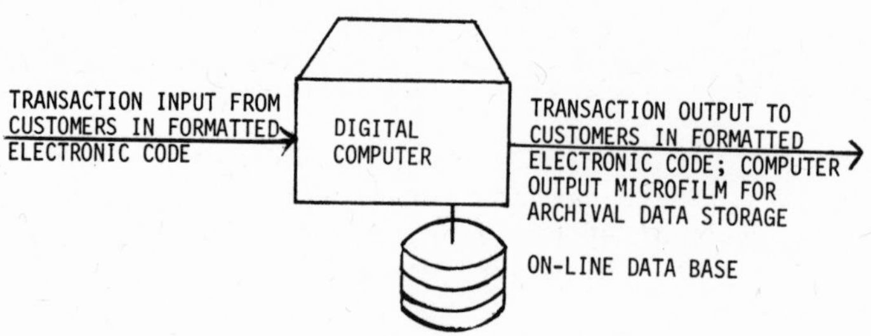

Figure 1

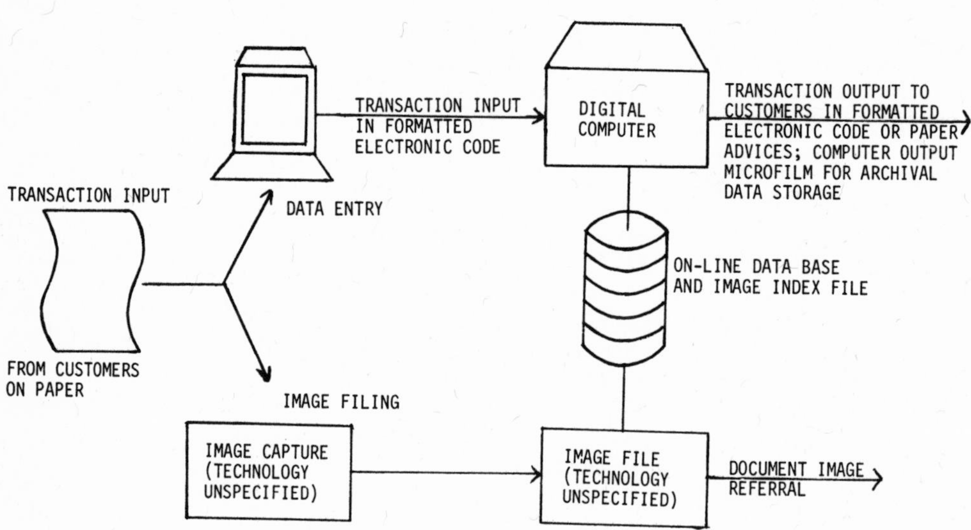

Figure 2

4. Theory and Technology

The four requirements of the device which we evisioned for our needs were:

- o Filing
- o Storage
- o Indexing
- o Retrieval and Display.

At the outset of this discussion, I would like to pose the children's trick question: Which is heavier, a pound of iron or a pound of feathers? The answer of course is that they are the same. The same type of question arises in the analysis of microimage technologies. Which is better: digital or analog, magnetic or photographic? The answer is that only in context can one make the right selection with assurance. Likewise, for the remainder of this section, we will examine alternatives, but not until the final section, "Some Solutions", will choices become apparent.

4.1 Filing

For the purpose of this discussion, filing is the process of capturing document images in conjunction with creation of an organized image index. Since indexing methods will be discussed later, this section will concern itself with image capture.

Cameras are perhaps the oldest of imaging devices. Because I am sure that the reader is familiar with lenses, it is sufficient to report that microimage photographic camera systems were readily available for commercial purposes at various reduction ratios of from 18X to 48X and 50X.

One process which will be examined in "Some Solutions" employed a two-stage process with a first-stage 24X reduction and a subsequent 10X photographic reduction of the 24X negative to 240X.

Vidicon tubes in conjunction with lenses were also available for producing an analog electronic signal. Empirical evidence as well as theoretical analysis demonstrated that a minimum of 1,250 scan lines over the 11 inch height of a typical business document were necessary for adequate resolution of 10 point type.

A wide variety of techniques are available for digital image scanning. Empirical evidence coupled with theoretical analysis yielded a minimum binary sample density of 120 points to the inch in a square pattern for adequate resolution of business documents. Various document scanning systems have been commercially available which use Reticon arrays or flying spot laser scanners. Transfer rates were typical of these devices for scan time of a 8-1/2 by 11 inch business document in approximately 3 seconds. Components to build such systems remain readily available.

4.2 Storage

A wide range of storage devices present themselves when one considers the possible combinations of digital or analog format and photographic, mechanical or electronic media. Research uncovered at least one class of executed device for each possibility. Following is a summary of systems and devices with reference to specific vendors for unique devices or technologies.

Magnetic/Digital

Magnetic tape and disc are well established as the prime storage media for use with the digital computer. Greater bit densities on both tape and disc have become standard and will probably continue to increase. Discs drives of 300 megabytes are commercially available. The IBM 3850 with maximum storage capacity of almost 400 billion bytes represents the largest self-contained on-line tape library.

Magnetic/Analog

Ampex has been active in the field of very large magnetic analog storage devices. They have built large video disc buffers and video tape drives for use with their Video File System.

They have also converted standard digital discs to analog for the television broadcast industry. These discs store TV frames for instant display at the command of TV directors broadcasting live events.

Digital/Optical

Various attempts at digital/optical memories using photographic film as a medium have been undertaken. Digital bit patterns exposed on film were subsequently read by optically replaying the bit patterns. A commercially available optical/digital memory by Precision Instruments was examined. This device employs a flying spot laser on a rhodium coated mylar substrate. When writing, the laser is given sufficient power to deform the rhodium, thereby creating a bit 3 mil in size. For reading, a lesser power laser scans the rhodium surface, determining bit patterns by differing reflection from the deformed and clean surface areas. The Precision Instrument memory has 16,000 megabytes on line per device woth up to 8 devices on line for a total memory of 128,000 megabytes on line. Any number of data cartridges may be stored offline.

Optical/Analog

Photographic film is, of course, an analog memory medium. A wide range of film sensitivities and resolving capacities are available. Since film product specifications are readily available and the process is familiar, I will conclude this section with the observation that of all the media examined, photographic film offered the highest storage capability per unit of surface area.

Exotic Memories

Various exotic memories were considered and are presented briefly as follows:

Electron Beam/Silicon Dioxide: A technology developed by CBS Laboratories which used the secondary effect principal of electron beams was bought to detailed specification stage but was dismissed as a higher technological risk than we were prepared to take. The specification called for 60 billion bits of on line storage.

Electron Beam/MOS: A General Electric device called BEAMOS, an electron beam on MOS, was examined. This device had storage capacity of only 32 megabytes and was dismissed as having insufficient capacity.

Bubble Memory: Bubble memory devices employ the localized magnetic field reversal effects which may be generated in wafers of garnet crystals to store digital bit patterns. Although this technology offered great promise, there were few commercially available devices and these were under one megabyte.

Holographic Memory: A Harris Radiation device in the prototype stage which employs holographic bits on photographic film was examined. Any possible advantages of holographic bits over ordinary optical bits was not perceived to outweigh the complexity of the process for the specific application.

4.3 Indexing

The ultimate aim of commercial applications of microimage technology is very often lost amid the analysis of bit densities exotic memories, transfer rates and the like. A commercial users interest transcends technology and simply asks for a system that will permit organization of very large image files (hundreds of thousands to millions of pages) with ready access to any one page by calling for it by descriptive characteristics. This is the function of the index.

There are two conceptical index classifications to which all index schemes must belong. The first is physical organization of the file material by alpha, numeric, chronological or other logical key. This procedure is the one most usually used by secretaries and file clerks. It is quite manual and quickly becomes unwiedly for very large files or in situations where quick retrieval is necessary.

The second index scheme is creation of a supporting index of positional coordinates of the file's physical array in unsorted state. Indexes again describe individual documents by alpha, numeric, chronological or other logical key.

Given the ready availability of sophisticated data base management software (DBMS) for both IBM 370 and a number of mini-computers, indexing of random arrays of large document or microimage files is becoming more attractive.

A data base management system (DBMS) is a software package designed to operate interactively upon a collection of computer-stored files, or what is called a data base. Its primary operations are to count data base records that have in common whatever characteristics may be specified by a user and to retrieve those records for further processing and display. A typical DBMS includes a:

- Schema - a high-level, compiler-like language that can be used to describe the location, contents, relationships, and security level of data that are stored.

- o Data Manipulation Subroutines - callable from high level languages such as Cobol or Fortran, that are used to manipulate and extract data from the data base.
- o Inquiry Language - an English-like language, incorporated into some DBMS, that acts as an impromptu report generator. The most sophisticated inquiry languages enable searches on prime keys, sub-keys, specific records or ranges and concurrent searches on multiple keys.
- o Utility Programs - off-line routines that are used to load the data base on to disk from other media and to duplicate the data base as a backup.

Among the most attractive DBMS systems for this purpose were:

- o Ragen Precision Industries Retrieval System for use with their special mini controller and retrieval terminal.
- o Microdata REALITY, currently an OEM item in the Microform Data Retrieval System.
- o Datapoint DART, currently marketed in conjunction with Kodak's Computer Assisted Microfilm Retrieval System.
- o Hewlett Packard IMAGE 1,000 and 3,000, not currently affiliated with any microimage system for use on the Hewlett Packard 1,000 and 3,000 minicomputers.
- o Cullinane IDMS: not currently affiliated with any microimage system, for use with IBM 370 systems.

4.4 Retrieval and Display

The ultimate object of the filing. storage and indexing techniques discussed is the eventual retrieval and display of specific document images on demand. Except for the few solid state memories examined, we found that all required mechanical transport of a physical medium to retrieve an image. Furthermore, those that were advanced to the point of being considered memory systems were controlled by a computer index addressing physical locations.

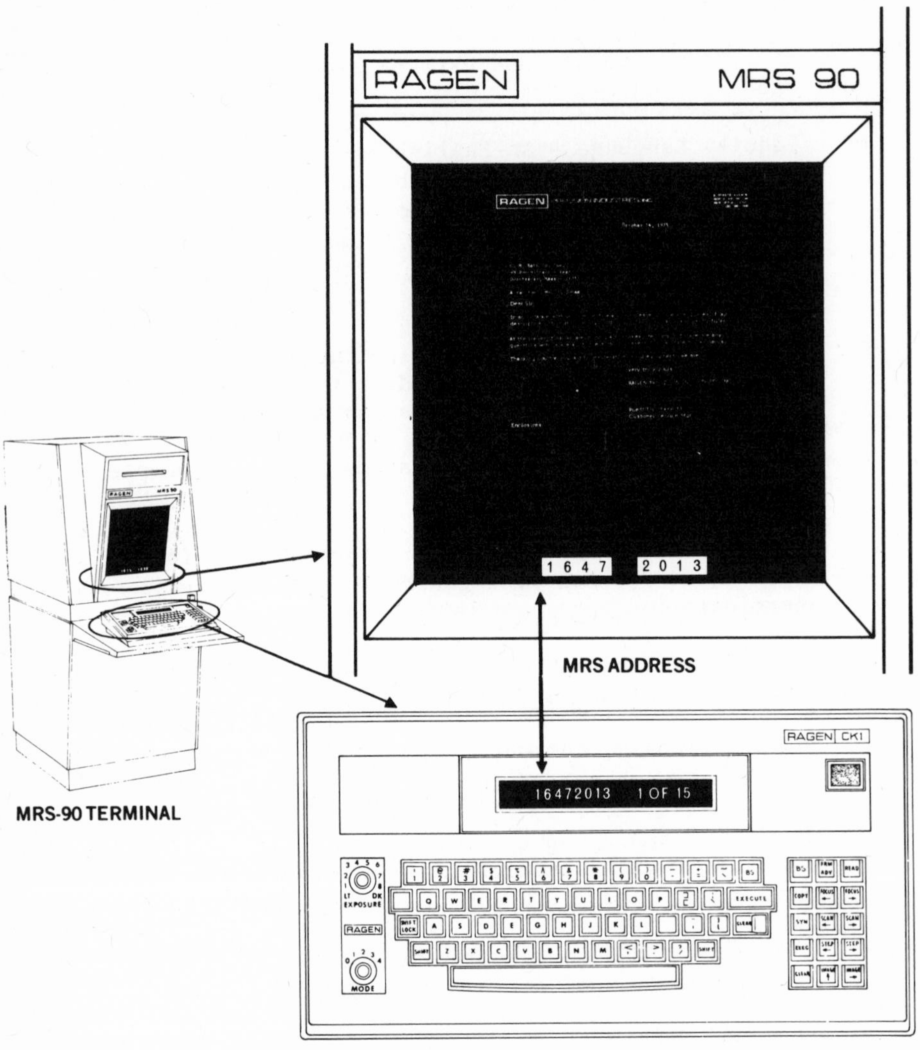

Figure 3. Ragen MRS ADDRESS interrogation, retrieval.

Striking conceptual similarities in technique among memories employing magnetic, photographic, or other mediums were apparent. For example, the IBM 3850 mass memory consisted of numerous cartridges of magnetic tape. Upon request for a particular record, a disc resident index was searched and a jukebox like mechanism fetched the specified magnetic tape cartridge which was unreeled under a magnetic read/write head to the requested record.

A microfilm retrieval device by Ragen Precision Industries used essentially the same techniques to retrieve photographic records. In this case storage was on cartridges of 16mm photographic film. Upon request for a particular image, a disc resident index was again searched, and a jukebox-like mechanism fetched the specified film cartridge, which was unreeled under a projection lens to the requested image (Figure 3). Of course, photographic projection is a human readable memory and may not be machine read or updated. However, if the object was to store mass document image files, the Ragen system provided the equivalent of 62,500 megabytes of on-line compressed digitized document images.

Display of an image is dependent on the form in which it is stored. Photographic memory calls for image projection. A wide variety of vendors, devices, and magnification ratios is commercially available. Electronic memories may be displayed on high resolution (at least 1240 scan lines over the 11 inch height of a business document) CRT monitors. Again, a wide variety of vendors and systems is available.

A hybrid technique of facsimile image transmission to remote CRT monitors from a central photographic memory was also explored.

5. Some Solutions

Science and the study of technology are driven by man's need to explore the unknown. However, application of science and technology to commercial situations is bound by additional constraints.

- Does a new technology offer the opportunity to perform an existing process less expensively or with greater quality at an equivalent expense?

- Does a new technology producing a new product (or an enhanced version of an existing product) offer sufficient added value to offset its cost?

Only when viewed within the foregoing bounds may one technology be said to be "better" than another for a given time and situation.

Assuming that Citibank's investment decisions, based upon aggressive research into available technology, were true indicators of the best technology for the given time and situations under the constraints of commercial enterprise, some broad conclusions may be drawn.

These are:

- Photographic film is the most cost effective storage medium for large files of document images.
- The facility of computer systems to index and address large random files of document images has made storage and retrieval of large document image files in random arrays more effective than physical reorganization of document files into logical sequence.
- The physical apparatus to capture, store and retrieve large document image files at an equivalent level of image quality is of significantly greater complexity for digital systems than for photographic systems.

Support for these conclusions may be drawn from the example of solutions by Citibank to the following document filing, storage and retrieval situations.

These solutions do not imply Citibank endorsement of any vendor, but are provided only for illustration.

5.1 TRACE

Citibank processed an average of 2.5 million checks a day in 1976. Each of these checks was microfilmed in random sequence. The microfilm process was integrated into our check readers in 1973, thereby permitting simultaneous capture of images and an index file. With the use of the index, the image of any one of the 625 million checks filmed in 1976, may be located in about ten minutes with greater than 98% probability.

Until 1973, Citibank relied upon physical reorganization of its check files into numeric account sequence before microfilming to enable retrirval. This process resulted in numerous lost checks prior to capture on microfilm. Paradoxically, lost checks were precisely the ones for which customers were likely to request evidence. Not suprisingly, the probability of finding a check requested by a customer was less than 50% under this system.

Citibank's file of random check images has been growing by several million each business day since 1973. It is undoubtedly one of the largest random film files in existence.

Intellectually, the notion of indexed random files was simple. But, a $7.3 million capital decision based on theory (Citibank pioneered the concept) was not quite so obvious. Since the system performed many more functions than microfilm and indexing, isolation of these components in a cost sense was difficult; the system consumed in excess of $300,000 of film and directly related photographic supplies annually. The reward has been in control of lost items. Before the system was implemented, dollar losses to missing checks had been over six figures. Losses now are counted in tens of thousands. Very large indexed random document image files are entirely manageable in a commercial environment.

Citibank pioneered the TRACE system by Recognition Equipment in 1973. Since then, systems with similar capability have been made available by Burroughs and IBM.

A schematic diagram of TRACE appears in Figure 4. A schematic diagram of the camera system appears in Figure 5.

5.2 CITISEARCH

After being computer sorted, the index to the TRACE file was maintained off-line on computer output microfilm. This method was chosen because of the enormity of the supporting index and the infrequent access relative to overall file size.

However, since 1975 Citibank operations have been organizing into divisions serving market oriented customer sets. Certain of these sets serving large commercial and industrial customers are characterized by relatively low transaction volume (tens of thousands daily rather than millions) and high average unit value per transaction.

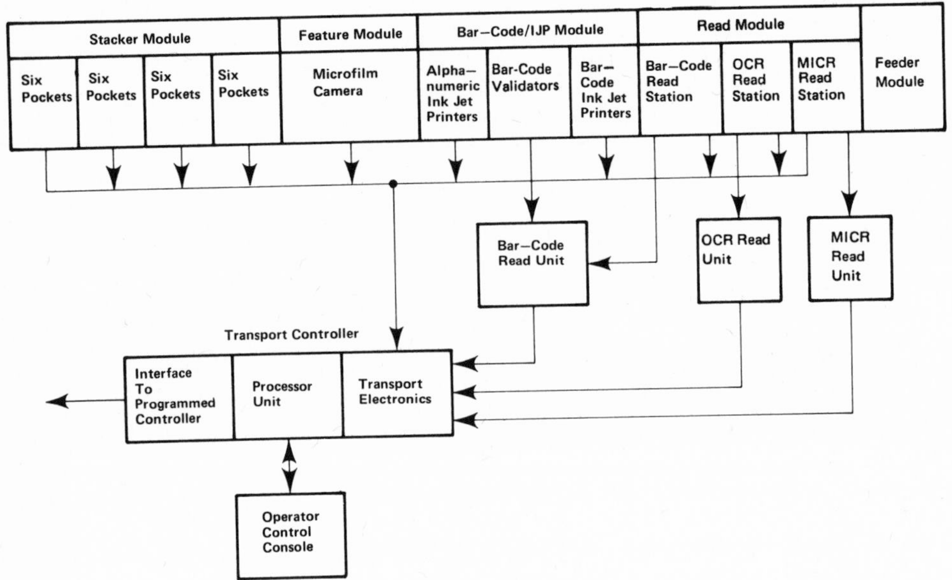

Figure 4. Typical TRACE capture system.

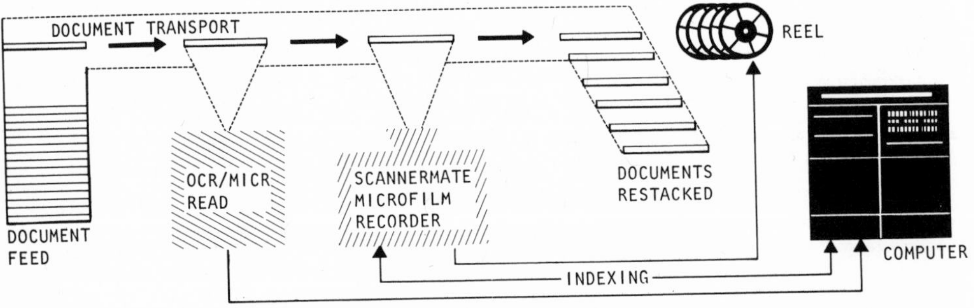

Figure 5. Terminal Data Corporation Scannermate microfilm recorder.

It has been possible to develop economic computer indexed microimage filing, storage and retrieval systems for these situations.

These situations also served as the analytic field for comparing photographic and electronic systems.

The first of these systems adopted was a small application whose function was to produce customer signature cards quickly for use in verifying the authenticity of certain transaction documents. The system employed the Microform Data Ultrastrip Terminal.

The MDS Terminal provided a high-speed microfilm retrieval unit designed for applications requiring rapid access to image files of several hundred thousand to over a million images. The unit automatically retrieved and displayed any individual page in a 100,000 page file within one-half to four seconds after receiving a command through entry on a keyboard.

The MDS Terminal was designed to accept cartridge-loaded Ultrastrips. Each Ultrastrip contained as many as 2,000 ultra-high-reduction images, each typically one two hundredth (200X reduction) its original size. Figure 6 illustrates the dimensions and format of an Ultrastrip. The MDS terminal accommodated up to 50 Ultrastrips providing access to as many as 100,000 pages.

Image capture involved a two-step filming process. The first step reduced the source document or computer generated data to images onto a roll of 16mm film (20X reduction ratio). The second step further reduced the film image onto a sheet of high resolution film called an Ultrastrip master. The reduction at this step was 10X, resulting in a total reduction ratio of 200X referenced to the original. Multiple copies for distributing were contact printed from the Ultrastrip master.

Index keys were manually key-entered during source document filming or were stripped from the magnetic tape file used to produce computer output microfilm.

The index system consisted of a minicomputer, disc storage and associated peripherals. Computer storage contained the index to the file. The MDS system usually employed the Microdata mini-computer and the associated REALITY operating system and data base manager.

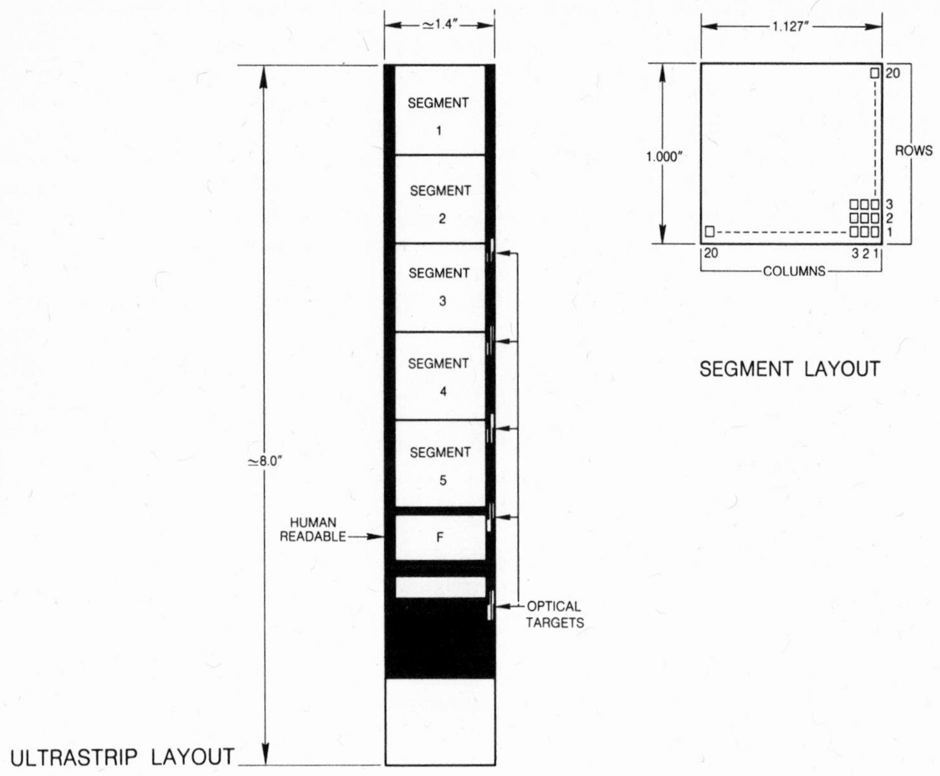

Figure 6. Ultrastrip dimensions and format.

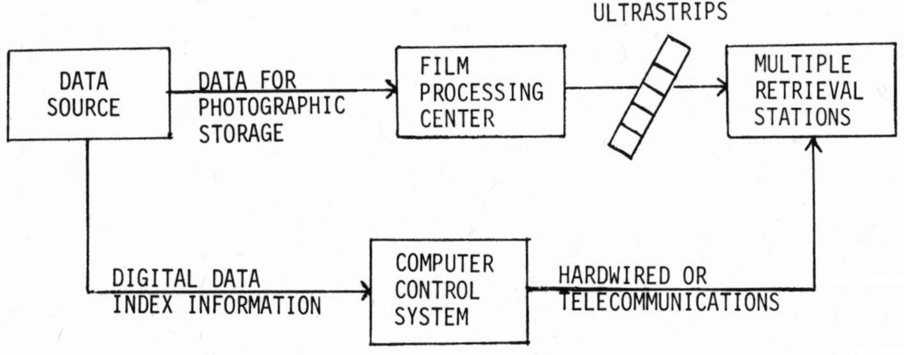

Figure 7. Indexing system.

Overall the system appeared schematically as in Figure 7. The first opportunity for head-on competition between a photographic system as described above and an electronic image system presented itself in the form of a business requirement in Citibank's Funds Transfer operations. Funds Transfer, an early form of EFTS, employed telegraphic links to effect transfer of funds in the account of a customer to another account at another bank. The dollar value of transactions are typically very large and a customer used the service to assure transfer during the same business.

The process in 1975 was not true EFTS since human intervention was required to receive incoming cables and route outgoing messages. In the few cases where the process failed to accurately execute our customers' instructions, it was incumbent upon our operations to determine quickly what the original instructions were and what actually happened so that the transaction could be set right.

Parameters of the application were:

- o 11,000 source documents input per day
- o 49 customer service managers each requiring a terminal
- o locations requiring terminals at two locations, several miles apart in New York City
- o sufficiently rapid access to serve customer telephone inquiries
- o at least six months activity (approximately one million images) maintained on line.

Digitized image filing storage and retrieval was intensively studied for this application. The analysis went roughly as follows. Recall that 1,500,000 bits were required adequate representation of an image. With compression of 4 to 1, 50,000 bytes per document or 50,000 megabytes for a million document file would have been required. Ordinary data discs could be used for image storage (they were used as buffers in several actual systems), but the cost of that quantity of disc storage, to say nothing of control, rendered discs an unrealistic alternative. Several of the exotic memories did offer such capacity at lower unit cost, but were not field proven and had to be considered high risk. Next, very high band-width lines would have been required for reasonable image transmission times, especially if images were transmitted uncompressed to avoid the cost decompressor at each terminal. Speeds of over 250,000 bits per second were considered. Finally, because low unit cost for storage was achieved only with very large mass storage devices input-output problems were anticipated.

By contrast, photographic systems delivered the same functional parameters with few drawbacks. Photographic film with readily available lenses routinely provide twice the resolution specified for the proposed digitized system. Owing to the relatively low cost of photographic film, the image file could be affordably duplicated to be available at many locations. This in turn solved the problems of wide band-width transmission and contention for a centralized memory. Communication to a centralized computer index was feasible by by ordinary telephone lines.

A decision was made to install a photographic system based upon Microform Data Systems hardware. The system, requiring over $1,000,000 capital investment, has been operational since late 1976. Forecasts predict an annual net saving in excess of $100,000 achieved through greater labor efficiency. More significantly, the system enables real time response to many customer inquiries.

A second opportunity to weigh photographic technology against digitizing presented itself with a similar application in Citibank's Investment Management Group. Again, after rigorous analysis, digitizing offered insufficient functional advantage to outweigh cost, complexity and technological risk. The application is currently being installed with a photographic based system.

Two major systems now being implemented further confirm the trend towards computer indexed document image files. The first of these systems, employing the Ragen equipment described earlier will file, store, and retrieve masses of stock certificates and related work papers. The second, using Kodak equipment will provide computer indexed retrieval for a set of the check transactions described in TRACE. This set represents our international customers.

5.3 CITICOM

The previous section re-counted Citibank's analysis of digitized image systems versus photographic systems. The conclusion was that with readily available technology and even peering somewhat into the future computer-assisted photographic file management systems offered all the functionality of digitized systems at significantly less cost and complexity was further confirmed by considering computer output microfilm (COM).

COM systems translate computer digital code from magnetic tape into printed microfilm formats. Several technologies are used in printing on photographic film. One system, the Minnesota Mining and Manufacturing Laser Beam Recorder employs a galvo controlled laser to "write" characters at up to 48X reduction of standard computer character size. A more usual technique is simple photography of a CRT screen face. This technique as employed by Bell and Howell is illustrated in Figure 8. The mini-computer controls various system functions such as report formatting, indexing of frames, and control of the systems electromechanical activities. Because the photographic process captures the CRT page image in a fraction of a second, COM systems typically "print" at up to 10,000 pages an hour, or seven to 10 times the rate of a paper line printer.

The economics of COM confirm the efficiency of photographic film as a data storage medium. The capital cost of a computer controlled COM system ($100,000 to $150,000) is somewhat less than the equivalent productive capacity for line printers and respective controllers. Therefore, only the marginal cost of media and apparatus required to present human readable images will be considered.

	$ per 1000 pages	Capital $ to present human readable image
COM		
One copy	$2.50	
Add'l.copies	.80	Negligible
PAPER		
One copy	$ 5.00	
Two copies	$ 10.00	
Three copies	$ 18.00	
Four copies	$ 30.00	Negligible
MAGNETIC TAPE		
Off line	$ 7.00	$10-25,000
FLEXIBLE DISC		
Off line	$ 60.00	$10-25,000

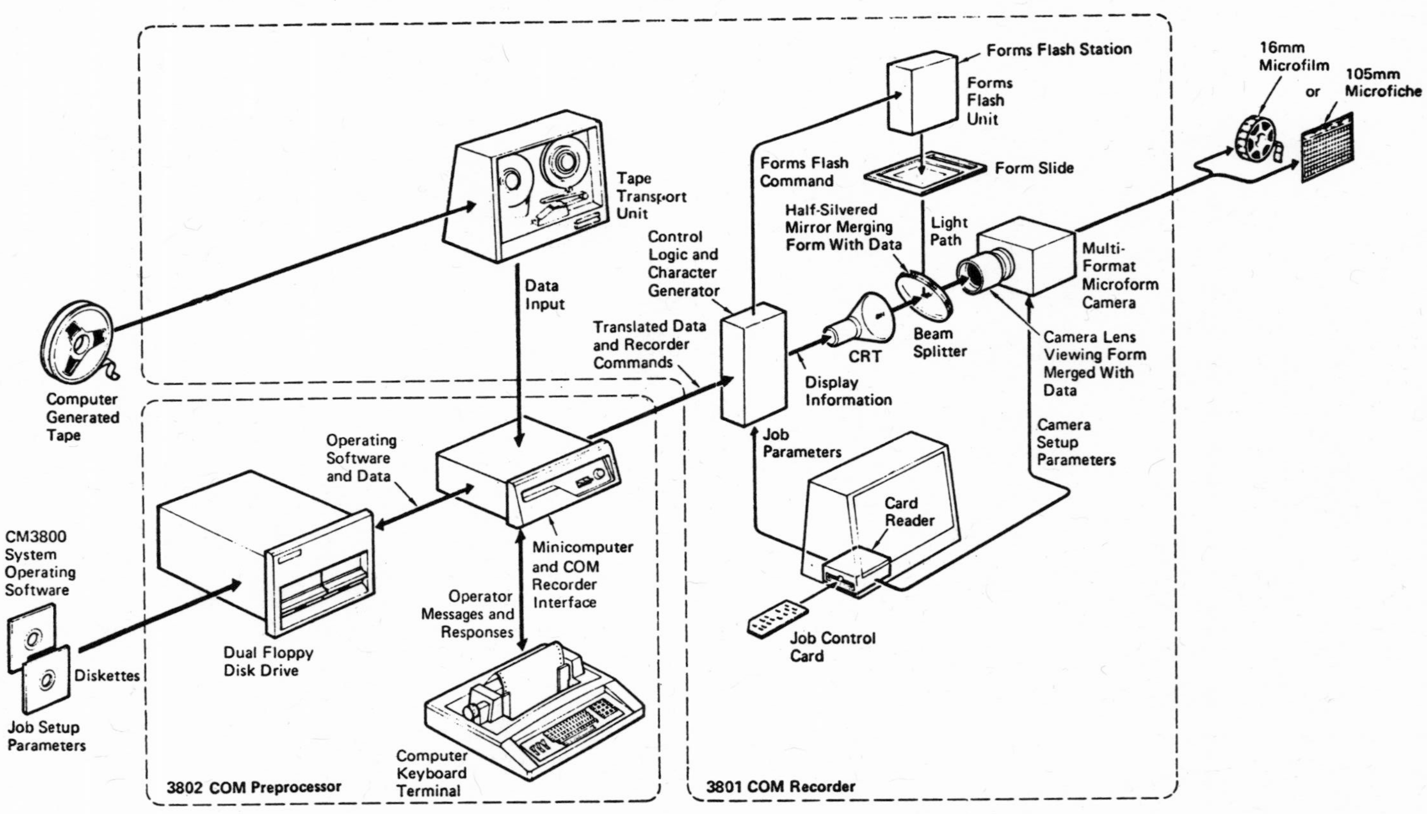

Figure 8. Bell & Howell 3800 COM system data flow.

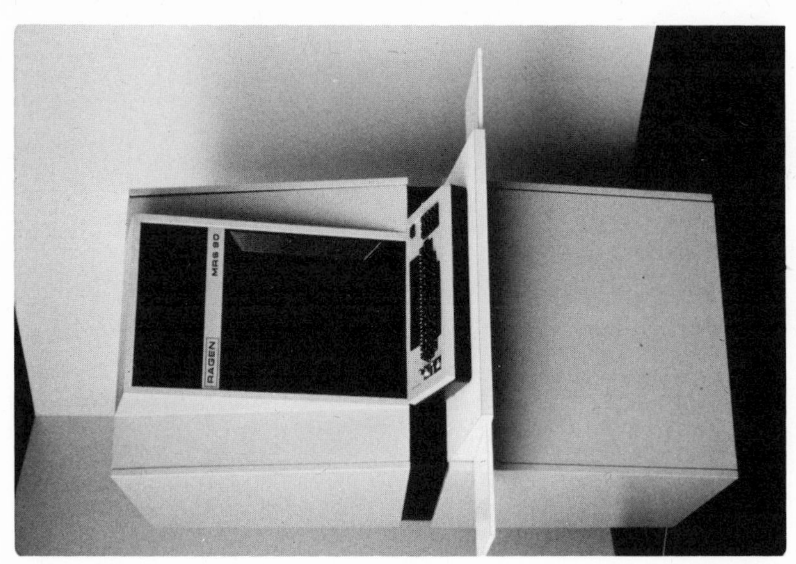

Figure 9a. Ragen microimage retrieval station.

Figure 9b. Carousel containing 1,000,000 images.

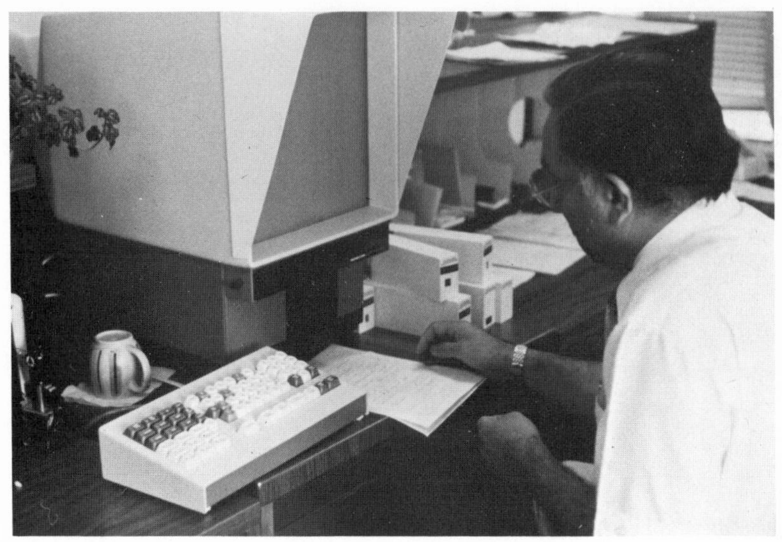

Figure 10a. Microform data system image retrieval station.

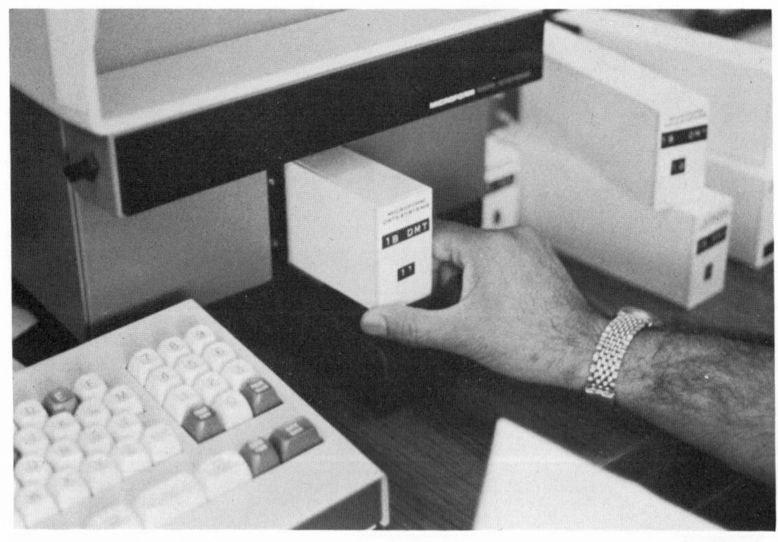

Figure 10b. Cartridges containing up to 100,000 images each.

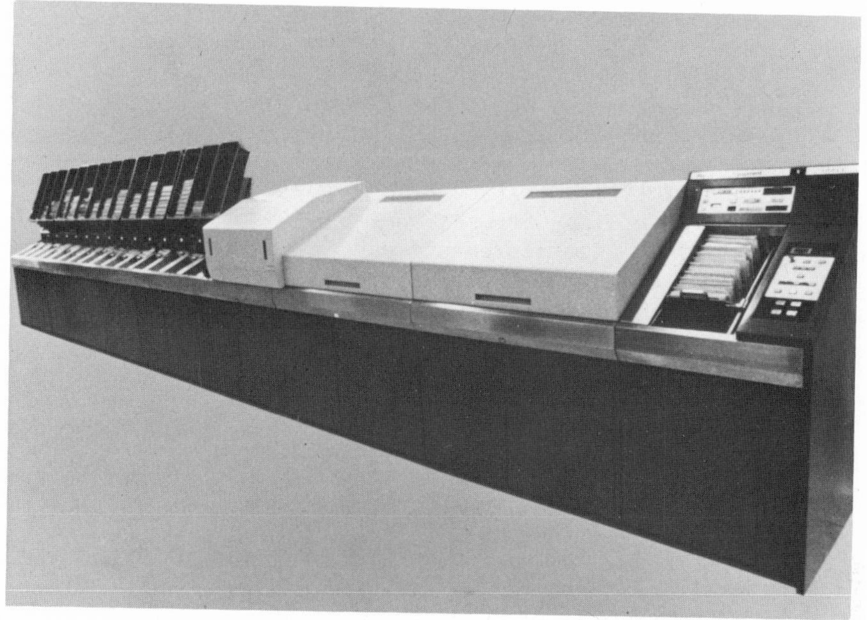

Figure 11. Recognition Equipment, Inc. TRACE.

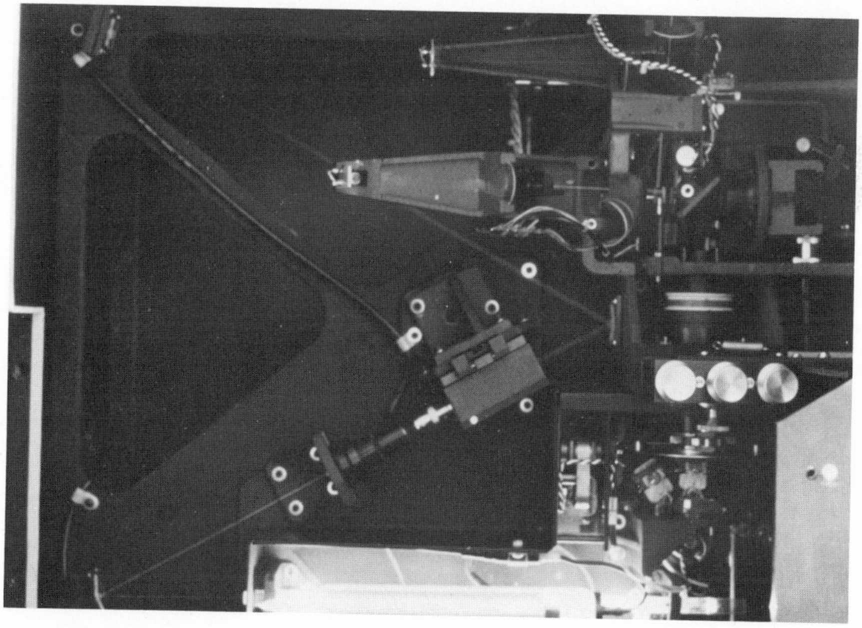

Figure 12. Laser in 3M laser computer output microfilm recorder.

Of course, one might correctly argue that photographic film is not machine readable, and thus not suited to data processing or dynamic file updating. This is true. However, COM, as a human readable memory, can provide real time access to static files through the techniques described in the previous section.

By the end of 1976, Citibank produced COM at an annual running rate of 80 million frames, an increase from 50 million at the end of 1975. This represents annual transference of 320,000 megabytes of magnetic based data in coded format which is at least 15 times more efficient than digitized image sampling, including compression, to photographic "memory". Photographic storage is obviously more cost effective than electronic storage of even coded data.

HOLOGRAPHIC BULK MEMORIES USING LITHIUM NIOBATE CRYSTALS FOR DATA RECORDING

A.L.Mikaeliane

President of QE Commission A.S.Popov Society

Moscow , U S S R

ABSTRACT

Physical processes ocurring when holograms are recorded in lithium niobate crystals are considered. The data are recorded with the use of a He-Ne laser. The erasing of data by means of a He-Cd laser and the rewriting of information are studied. The proposed holographic memory differs from the known arrangements in that it is provided with a control transparency. The results of investigating an orthoferrite transparency are presented.

Introduction

The advances in lasers and holography have made it possible to design a new type of data memories having high capacities (about 10^9 bits) and reduced cycles of data search and readout (about 10^{-6} s). Due to the fact that the theory of holographic memories was quite thoroughly developed [1,2], it has become possible to determine maximum performance of such memories and to chart the most feasible ways of their practical designs. Quite a number of papers have recently been published dealing with experiments involving holographic memories of various types. The most attractive of them seen to be the memories where an image-intensifier [2] is substituted for a discrete photodetector array. The data search and readout cycle in such memories does not exceed 10 s even when low power He-Ne lasers are used. Experimental set-ups of hign-speed 10^7 bit memories of this type and methods of increasing this value up to 10^9 bit are described in [1-3]. An Increase of the memory capacity upto 10 bits is achieved by means of employing an electromechanical device to change the data carrier [3] which is rather slow.

All the holographic memories mentioned above do not allow any part of the recorded data to be rewritten since the function of the data carrier is performed by a photoemulsion.

The possibility of rewriting data [4,5] can broaden the applications of holographic memories since errors that occur in the process of recording bulks of data can be quickly rectified. Such memories seem to be useful for storing archive data which can be regularly updated.

Our first attempt to use a carrier that would permit the rewriting of data in a holographic memory was made in 1968 [6]. It was however unsuccesful since the material used for the purpose suffered from a number of drawbacks.

1. Hologram Read-Write Processes in Crystals

The peculiarities of a holographic memory allowing the rewriting of data are determined mainly by the data carrier properties. Investigations of the characteristics of Fe-doped $LiNbO_3$-crystals have demonstrated that they can be used to design holographic memories allowing the rewriting of data.

We have made the investigations involving the determination of the sensitivity and the diffraction efficiency of crystals, the duration of hologram storage and readout cycles in relation to the conditions of crystal growth, to the concentration of dopes and the ratios between them in crystals, to the conditions of hologram recording and recovering [7,8,9]. All these characteristics affect the performance of a holographic memory (information capacity, speed of operation, storage cycle).

It is known 8 that the principle difficulties while using $LiNbO_3$:Fe crystals as well as the majority of other reversible materials consist in that they have a relatively low sensitivity.

The sensitivity of samples having the concentration of Fe from 0.02 to 0.03 mass percent that was found to be almost optimal is equal to 50 J/cm^2 and 1 J/cm^2 at the wavelengths of 0.63 μm and 0.44 μm respectively. these values correspond to a 25% diffraction efficiency of a hologram of two flat waveforms (the effect of surface reflectivity is neglected). The maximum efficiency of such holograms recorded in certain samples exceeded 70% (90% when the reflection is taken into account). In this case the exposure is much higher (which corresponds to a sensitivity which is several times lower). The sensitivity in the course of data recording as well as the maximum diffraction efficiency may be different for the case of one and the same energy level. It depends on the recording radiation power (power density). Fig.1 presents curves that characterize the relationship between the diffraction efficience and the duration of the write cycle when a He-Ne laser (λ = 0.63 μm) is used at different power density levels. It is evident that the sensitivity grows with the power. The curves show also that power growth brings about an increase of the maximum diffraction efficiency. The hologram recording process is affected also by an external electrical field applied along the C-axis of the crystal. Fig.2 shows how diffraction efficiency variations depend on the exposure in case of different electrical field intensities. Two peculiarities should be pointed out here. The first is the effect produced by the polarity of the electrical field applied along the C-axis of the crystal. The second in the dependence of the effect produced by the electrical field on the level of the recording radiation energy. It has been noted that at high energy levels (above 0.2

W/cm^2) the external electrical field will not affect the hologram recording process in a marked way. Obviously it may be attributed to high (and non-uniform) internal electrical fields that are produced in the crystal by a high entensity light beam used for hologram recording. Therefore, to increase the sensitivity and the diffraction efficiency it is necessary either to use an external electrical field or to raise the radiation power of the beam used for hologram recording. The information recorded in $LiNbO_3$ crystal can be erased

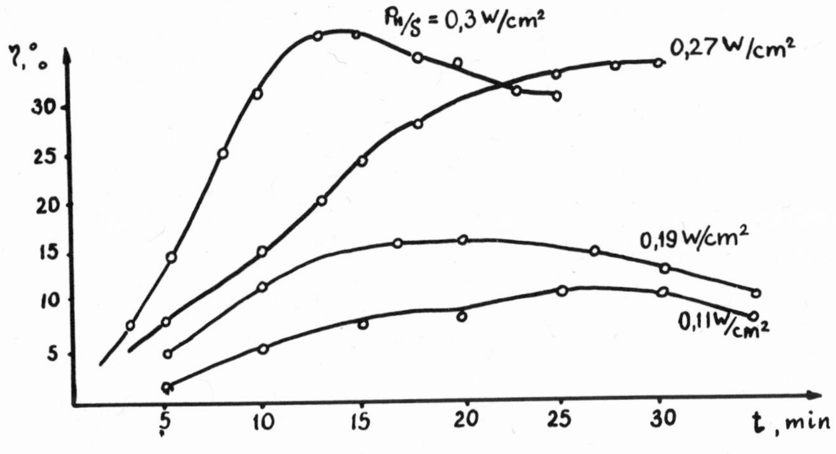

Figure 1

either thermally (by heating it to a temperature of 150°C) or optically. No doubt, the latter is more feasible since it provides for a high speed (cpmparable with the speed of the write process) and local erasure. The sensitivity to optical erasure is about one order of magnitude lower than to optical recording. It is feasible that data should be erased with the use of shortwave and sufficiently powerful (but not necessarily) light sources such as He-Cd lasers (λ = 0.44 μm). The amount of energy required for the data erasure process will be about 20 J/cm^2.

The recorded data storage time depends, in the first place, on the composition of the crystal (types and amounts of dopes) [10] as well as on the storage conditions (temperature and illumination level) [7]. The storage time will be small when the Fe^{3+} ion concentration is low, the ions serving as charge traps. An increase of the Fe^{2+} ion concentration results in the growth of absorption and the sensitivity of the crystal to data recording and erasing processes. Our samples in which the Fe ion concentration was about 0.02 mass percent exhibited data storage time of from two to three months. In this case the diffraction efficiency level dropped from 25% to 20% at room temperature. In contrast to the temperature, the ordinary room illumination level does not affect the hologram storage time. If the crystal temperature is lowered to 0°C the hologram can be conserved for about a year. The data storage time may be increased considerably by means of fixing to be carried out in the following manner [8]. The crystal containing a hologram recorded in it is heated upto about 75% within 15-20 min and then cooled to room temperature within

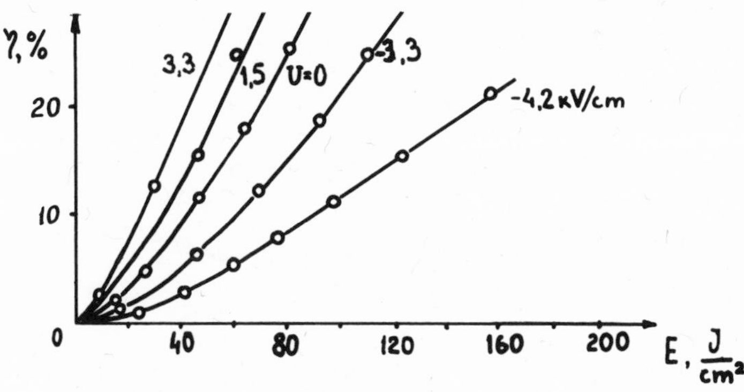

Figure 2

20-30 min. The maximum temperature should be selected so as to avoid the spreading of the hologram. Our hologram fixed in this way was preserved from about a year with no decrease of the diffraction efficiency.

The multiple **readout** process results in the optical erasure of recorded data. The shorter the wavelength the speedier the erasure. Therefore, for the readout purposes a laser operating at $\lambda = 0.63\ \mu m$ seems to be most suitable. A longer wavelength will result in a decrease of the diffraction efficiency in the course of data readout due to a weak electrooptical effect in the crystal.

2. Holographic Memories Using $LiNbO_3$ Crystals and Allowing Data Rewrite

The fundamental principles of holographic memory design were developed earlier [1-3]. A diagram of such a memory using $LiNbO_3$:Fe crystals is presented in Fig.3.

The data from a transparency having a capacity of 10^4 bits is recorded in a crystal in the form of holograms with a diameter of about 1 mm. A certain amount of such holograms recorded in the crystal as a matrix represents a data carrier of the memory which can store from 10^7 to 10^8 bits of information.

The write-read processes are carried out with the use of one and the same He-Ne laser ($\lambda = 0.63\ \mu m$) while a He-Cd laser ($\lambda = 0.44\ \mu m$) is used for data erasing purposes. The sensitivity of the material makes it impossible to reduce the duration of write-erase cycle to below several seconds when the average power of the lasers is about 100 mW. Hence, an electromechanical device may be used to move the crystals while rewriting individual holograms.

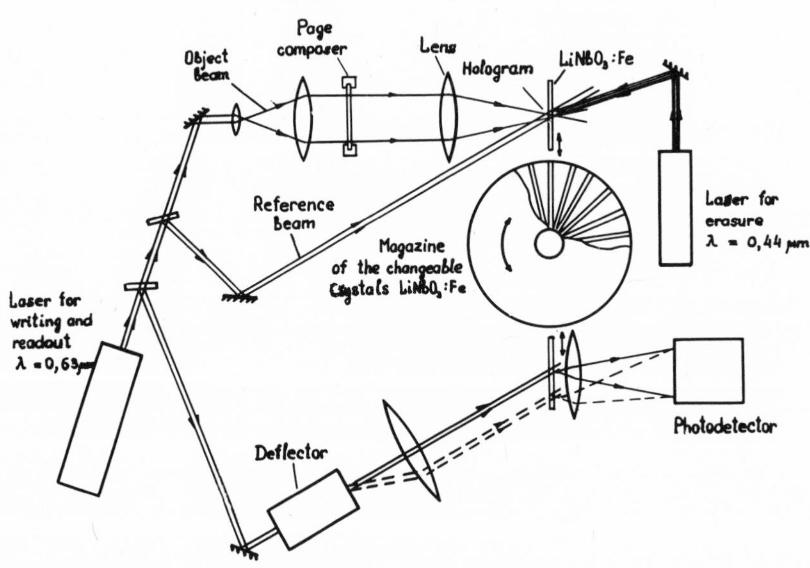

Figure 3

The search and readout of the required holograms in the matrix are carried out with the use of an electro-optical deflector, i.e. at a sufficiently high speed of about 10^{-6} s [3] . As it was mentioned above, holograms are recorded by means of a laser operating at λ = 0.63 μm although the crystal sensitivity in this case is much lower than at λ = 0.44 μm . This is attributed to the fact that in order to obtain a high diffraction efficiency it is necessary to use quite thick crystal plates (about 1-5 mm)*. Holograms in this case will have a high degree of selectivity and the transparency image will be recovered only partially at different write and read wavelengths. For instance, if a hologram is recorded at the wavelength of 0.44 μm (a He-Cd Laser) while the transparency with an area of 20x20 mm^2 is recovered at λ = 0.63 μm , the hologram thickness at which the recorded image could be recovered completely will be above 40 μm.

The device used crystal whose size was 25x25x1 mm^3. Hologram groups consisting of 5 by 5 individual holograms 1 mm in diameter (Fig.4) were recorded at various points of each crystal. The hologram pitch in the horizontal direction was 1 mm, i.e. the holograms were arranged close to one another, while the pitch in the vertical direction was 1.5 mm. This difference in pitch is a natural consequence of the fact that the reference and signal beams are located in the horizontal plane and it is in this plane that the hologram exhibits its maximum selectivity. Fig. 5 shows photos of recovered transparency images (20x20 mm^2; 2.5x10^3 bits) reconstructed from such holograms. In the present device the ratio of the intensity of the signal beam to that of the reference beam was about 1:30, the reference beam intensity being about 2 mW and the output power of the He-Ne laser used in the device being 12 mW. The exposure time of the holograms was from 2 to 3 min to obtain the required diffraction efficiency of 1.5 - 2%.

The process of hologram recording at such a low level of the diffraction efficiency has a number of

---

* Further increase of the crystal thickness is not advisable since the light while penetrating the crystal loses its energy and its intensity drops. Hence, the diffraction efficiency of holograms recorded on the respective sections of the crystal becomes low, which finally sets a limit to the effective hologram thickness in the crystal.

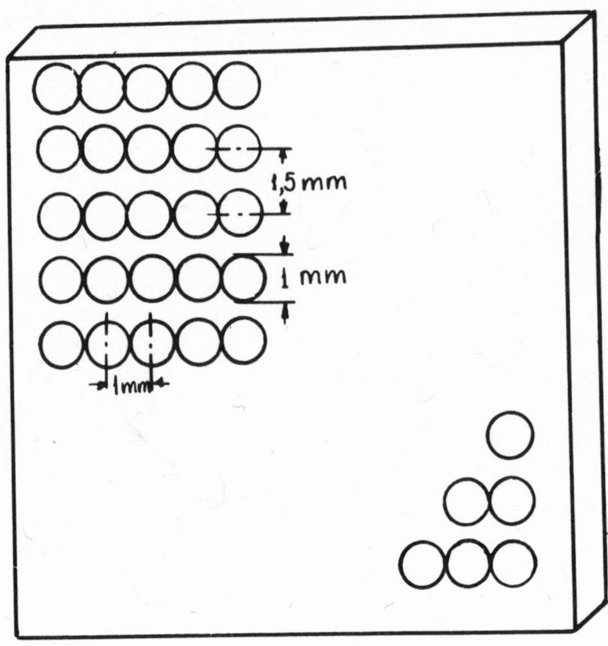

Figure 4

advantages. First, the write process is carried at a comparativly high speed (when a 50 mW He-Ne laser is used the recording time is less than 20 s), which is quite important for writing bulks of holograms. Second, in case no fixing is provided in the course of hologram storage or reading the diffraction efficiency of the hologram decreases exponentially [7]. Therefore, holograms can be preserved for longer periods of time at low levels of diffraction efficiency. Fig. 6 ("a" and "b") presents the dependance of the hologram (transparency) diffraction efficiency on the storage time and the duration of a continuous readout procedure respectivly.

A hologram recorded in $LiNbO_3$ crystal can be erased both thermally and optically. The thermal erasing of holograms produces no changes in the properties of the material. For instance, after more than 200 rewrite cycles (with thermal erasure) the quality of thr restored image was the same as after the first cycle.

The optical erasing was performed with the use of a He–Cd laser ($\lambda = 0.44\ \mu m$). With a power density of about 10 mW/mm^2 the duration of a complete erasure cycle was found to be approximately 1 min, i.e. 60 J/cm^2.

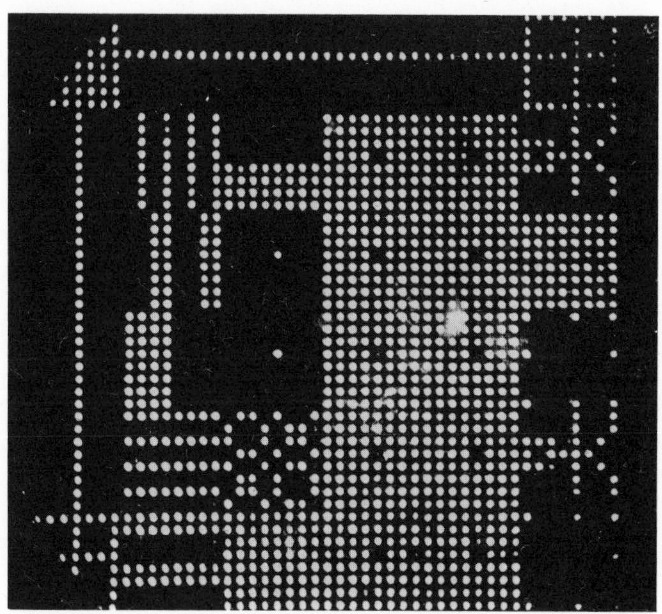

Figure 5

The high quality of the reconstructed image has been retained after ten write–erase cycles. When a number of cycles become more than that figure, the image has been deteriorated because of the nonuniform energy distribution in the erasing beam and corresponding noise in reconstructed image. The magnetic page composer can be used to convert the incoming data into pattern to be recorded on the hologram. The page composer contains the orthoferrite platlet of 60μm thick which is pasted on the glass substrate. Both sides of the substrate are coated by the system of conducting loops which are ortogonal to each other (Fig.7). When the current pulses of a proper value are applied to the conducting loops the state of magnetization of the cell which is disposed on the intersection of these loops is changed. If the properly polarized light is passed through page composer one can obtain two-dimensional intensity pattern when adjusting the analyser of polarization is used.

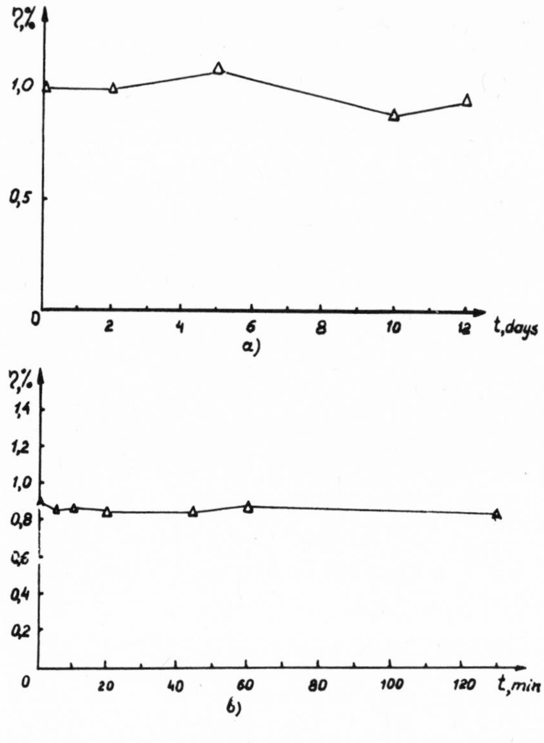

Figure 6

Design of such page-composers because possible owing to developing of the above mentioned magnetic materials which have sufficiently high megneto-optical figure-of-merit. The ratio of Faraday rotation to the losses is about 15 degrees per dB. Because of sufficiently high coercitivity of the domain walls the medium posseses the hysteresis peculiarity and switching threshold, which is defined by the coercitivity field H_c. Therefore it is possible to fill up and remember the matrix pattern. The time of storage of the recorded information is practically unlimited. The active medium can be switched only when the applied field exceeds the coercitivity of the domain wall. Therefore only that cell is switched which is on the intersection of the excited loops if the total magnetic field exceeds the coercitivity field. This field can be from 10 to 50 oe and can be adjusted by the proper pressure during polishing of the orthoferrite. Therefore as compared

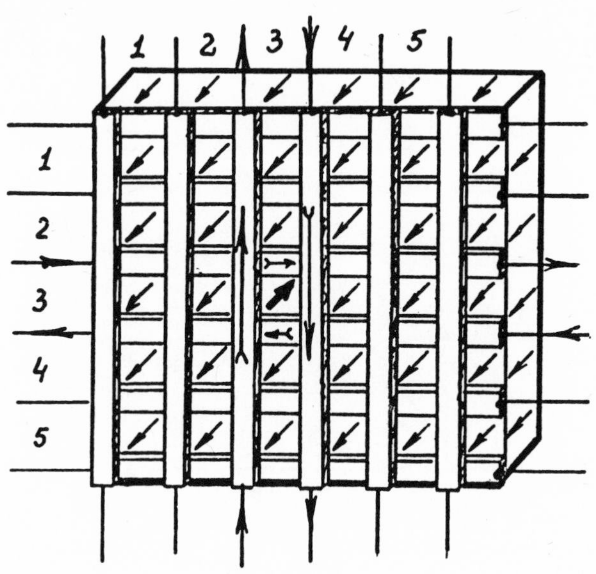

Figure 7

to the bubble films these media have more high coercitivity and more high speed of the domain wall under influence of the external field. It was confirmed by the experiments as is shown in Fig.8. Thanks to this peculiarity speed of operation of the considered page-composer is of order 10^8 bits per sec. Therefore transparency of 10^4 bits can be filled up for 100 sec. After recording of this information on the hologram the pattern on the transparency can be erased by the magnetic field and it can be used for recording of a new information. The chess-board pattern is shown on Fig.9 This pattern was obtained on the page-composer of 40x40 emelents. The distance between centers of adjecent elements is 200 μm the element dimensions are of 130x130 μm. The measured contrast between "0" and "1" was of 300 for the ittrium orthoferrite (λ = 0.63μm)*. Up to 10% of the incoming light passed through transparent cells. Because of high index of refraction of the medium (n = 2.2) this figure can be increased upto 25% by using the untireflection coating.

*) The same figure for the garnets of $Y_2BiFe_{3,8}Ga_{1,2}O_{12}$ was about 100 (λ = 0.53 μm)

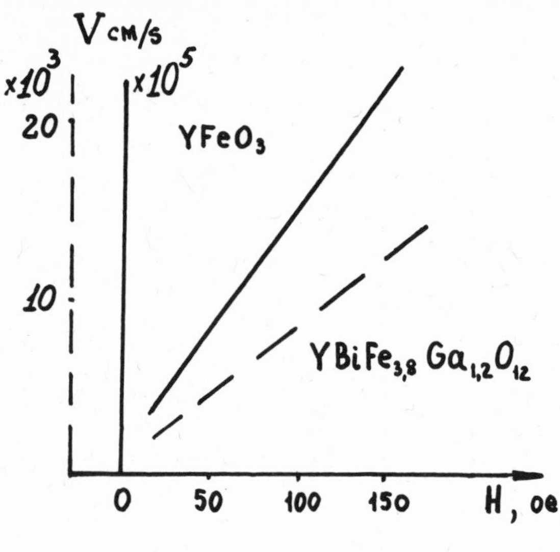

Figure 8

The experiments have shown that holographic memory system can be designed on the base of Fe-doped $LiNbO_3$ crystals with high density of information storage. More than $1.5 \cdot 10^3$ holograms each of them containing up to 10^4 bits can be recorded on the crystal with dimensions $50x50x1\ mm^3$. The total write-erase time can be of 30 - 40 sec if recording is performed by the He-Ne laser of 50 mW and erasing by the He-Cd laser of the same power. Speed of operation for readout can be the same as in readout-only memory [3], i.e. about 10^{-6} sec.

The storage time for the holograms without fixing can be some months; the hologram can be illuminated by the readout beam for some hours without appreciable erasure of the recorded information. The diffraction efficiency is diminished during this process not more than on 10% from initial value.

There are possibilities to increase the total memory capacity. If the relatively slow chanee of the information blocks is allowed (each block contains up 10^7 bits on one crystal) than the electromechanical

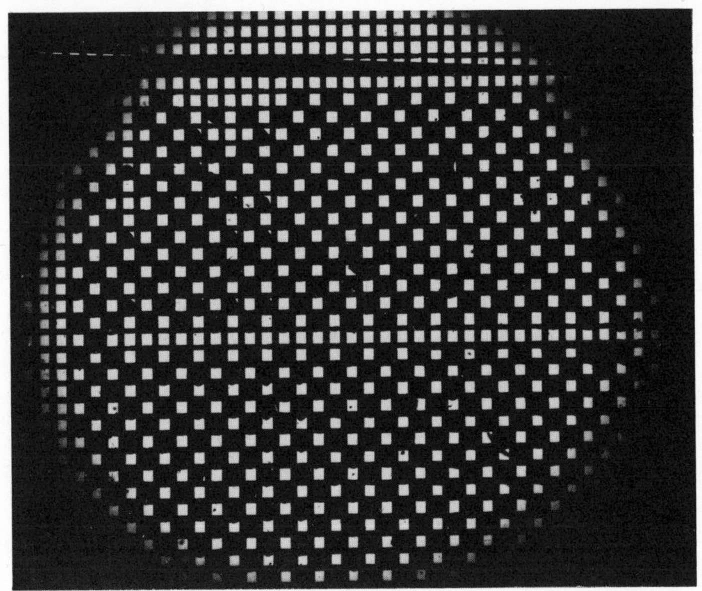

Figure 9

system which changes the crystals can be used with the reserve of the crystals in the special magazine (for, example in form of drum as is shown in the Fig.3). Than the memory capacity can be enlarged up to 10^9-10^{10} bits.

The other possibility is connected to the high selectivity of the volume holograms. This peculiarity can be used to record many holograms on the area on the crystal using the reference beams with the different angles of incidence. The corresponding readout beams can reconstruct the recorded images independent by using this method up to 5-6 holograms were recoeded with the diffraction efficiency of 5%. When more holograms are recorded the diffraction efficiency of the first holograms diminished. By the hologram superposition the memory capacity can be enlarged tenfolds but the optical setup and the system construction becomes a complicated one (see, for example [3]).

3. The Reflection Holograms in the $LiNbO_3$ Crystals

We have performed the experiments on the recording

and readout of the reflection holograms using the method proposed by Yu.N.Denisjiuk [11]. In this case the hologram has been recorded along crystal C-axis, as is shown in the Fig.10a. The holograms were recorded in the 5 mm-thick crystal on the wavelengths 0.63 μm and 0.44 μm. The angle between the reference and signal beams was about 170°. The maximum diffraction efficiency on the red light was of 65%, and that on the blue light was of 1% (because of strong lighr absorption in the crystal). Therefore, resolving power of the crystal is more than 10^4 l/mm. The curve of Fig.10b illustrates the sensitivity of such holograms to the angle deviation of the readout beam. This dependence can be transformed into wavelength sensitivity this last can be estimated to be about 3Å for this hologram. Reflection holograms of the transparency have been recorded too; their diffraction efficiency was 9%. The experiments on the hologram superposition have been performed. The obtained results were analogous to that of the oridinary hologram recording by the converging beams.

The obtained results show that the reflection holograms can be used in the holographic memory with the

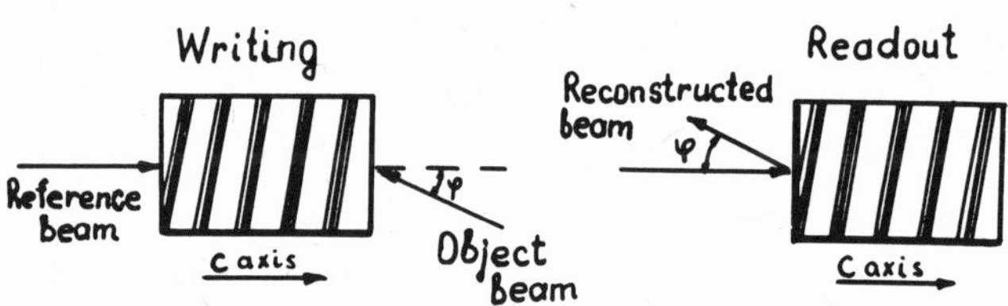

Figure 10a

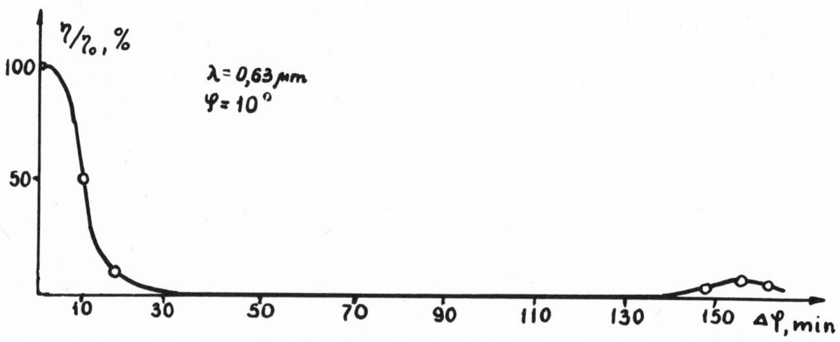

Figure 10b

frequency-tuned laser. It was shown that for the tuning range of 100 Å up to 10 holograms can be recoeded and reconstructed independently on the different wavelengths. Such method is more convenient as compared to the ordinary hologram superposition because it doesn't require complicated setup to readout the different holograms.

It is interesting to note that some peculiarities of the reflection holograms are very evidently because of very fine structure of the interference pattern (very small period of the recorded grating). In Fig's 11a,b the dependence of the diffraction efficiency of the holograms during recording and readout are shown for two opposite orientation of the C-axis of the crystal relative to the readout beam. One can see that with proper orientation of the C-axis the diffraction efficiency doesn't diminish during readout for two hours owing to the autoamplification effects.

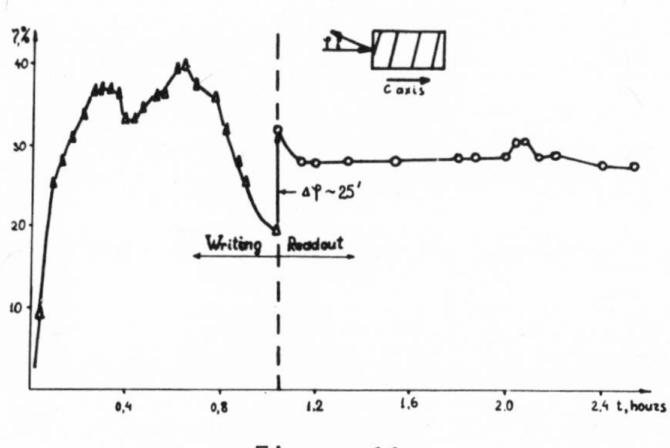

Figure 11a

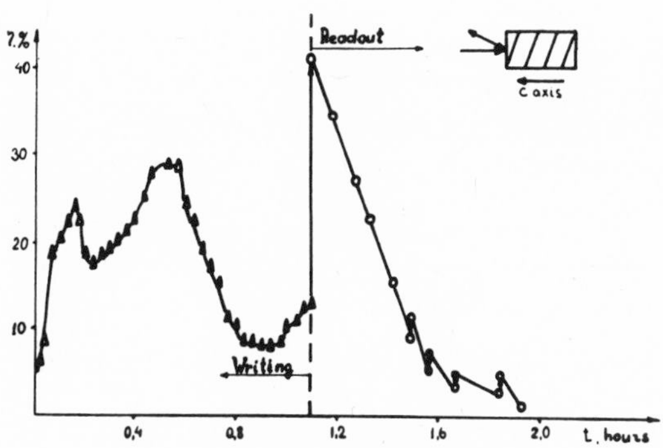

Figure 11b

It is very interesting that magnification of the diffraction efficiency occurs when the readout angle changes from 10' to 30' (relative to the reference beam it is shown by arrow in Fig.10a,b). Therefore during recording and readout there is a change of the period of the diffraction grating. This change can be caused by the piezoeffect or by the change of the index of refraction because of the charge accumulation in the crystal and respective internal electric fields.

It is necessary to say in conclusion that in spite of noted drawbacks the crystal of $LiNbO_3$ can be used in principle for design of their base the read-write-erase holographic memory.

But the further investigations are necessary for the studying of the peculiarities and, in particular for the more defined understanding of the physics of the optical recording in the $LiNbO_3$ crystals. These investigations will allow to realize the full potential possibilities of this material.

References

1. A.L. Mikaeliane, V.I.Bobrinev, S.M. Naumov, L.Z. Sokolova, Radiotehnika i electronika, 1969, v.14,N1,p.115-123

2. A.L.Mikaeliane, V.I.Bobrinev, Opto-Electronics, 1970, N2, p.193-198

3. A.L.Mikaeliane, V.I.Bobrinev, Radio-tehnika i electronika, 1974, v.14,N5,p.898-926 Radiotehnika, 1974, v.29,N5,o.7-18

4. J.C Urbach, R.W. Meier, Appl.Optics, 1966 v.5,N4,p.666

5. D. Chen, J.D.Zook, Proc. IEEE, 1975, v.63, N8, p.1207-1230

6. A.L. Mikaeliane, A.P. Axenchikov, V.I. Bobrinev, E.H. Gulaniane, V.V. Shatun, IEEE J. Quant. El., 1968, QE-4,N11, p.757-762

7. J.J.Amodei, W. Phillips, D.L. Staebler, Appl.Optics, 1972, v.11, N2, p.390-396

8. V.I. B o b r i n e v , E.H. G u l a n i a n e , Z.G. V a s i l i e v a , A.L. M i k a e l i a n e, Pisma J E T P, 1973, v.18, N4, p.267-269

9. D. V o n d e r L i n d e , A.M. G l a s s , K.F. R o d g e r s , Appl. Phys. Lett., 1975, v.26,N1, p.22-24

10. W.P h i l l i p s , J.J. A m o d e i , D.L.S t a e b - l e r, RCA Rev., 1972, N33, p 94

11. Yu.N. D e n i s j i u k , Optica i spectroscopija , 1963, v.15, N4, p.522

HOLOGRAPHIC MEMORY OF HIGH CAPACITY WITH SYNTHESIZED APERTURE

V.N.Morozov

P.N.Lebedev Physical Institute, Ac.Sci.

Lenin prospect 53, Moscow, USSR

INTRODUCTION

The role of holographic memory devices for the development of computer technique is growing in importance /1/. But the capacity of holographic memory with random data access and page organization on thin holograms is restricted by dimensions of an optical system /2/. In practice, the diameter of a high-quality lens may hardly exceed 30-40 cm, and hence a limit for the holographic memory capacity is of 10^8-10^9 bit. On the other hand, the nonoptic memory on new principles, such as charge coupled devices, magnetic bubbles, and the conventional memory elements on magnetic discs can provide the same and more greater memory volume with sufficiently improved engineering and economical properties.

Therefore, the main problem is to increase by 10^3-10^4 times the holographic memory capacity. The principle trends today in increasing the capacity of holographic memory (HM) are as follows:

- unification of a great number of memory modules of 10^8 capacity in a common system /3/;
- the use of a great number of holograms arranged in a series;
- increase in the HM capacity is achieved by using multi-channel schemes of recording and reading /2,5/.

The most promising way of increasing HM capacity is to achieve higher information density on holograms.

Here we discuss the principles of constructing HM devices with a super-high capacity. The density of information recorded in the holograms exceeds a classic ratio $\sim A^2/\lambda^2$, where λ is the wavelength, and A is the relative hole of an objective. The gain in the information density is associated with increased resolution of holograms and is achieved by the methods of aperture synthesis.

IDEA OF THE METHOD

Resolution of any optical system is determined by spatial frequency bandwidth transmitted by the optical system in the image space. But a fundamental invariant of the stationary wave field is a total number of freedom degrees:

$$C = \frac{L_x \cdot L_y}{\lambda/\alpha \cdot \lambda/\beta} = \frac{S \cdot \Omega}{\lambda^2} \tag{1}$$

where S is the area of an object, and Ω is the angular aperture of the optical system. But with the same total number of freedom degrees of the field one can redistribute the information degrees of freedom. Particularly, it is possible to improve the space resolution of an object by a proportional decrease in its effective field /6,7/.

In holographic memory devices the information is formed initially in an intermediate carrier-data matrix, where the binary signs are represented by "units" and "zeros" of the binary code (transmitting and non-transmitting cells). A diffraction pattern in the focal plane of the lens is written in a hologram in the form of a space spectrum. If an elementary data matrix cell is a reactangle of Δ width, i.e. the transmittance $(t(x) = \text{rect}\,(x-x_o)/\Delta$, where x_o is the position of the cell center, then a spectrum of such an elementary cell is

$$\tilde{t}(\xi) = e^{-2\pi i \frac{\xi x_o}{\lambda F}} \cdot \text{sinc}\,\frac{\xi \Delta}{\lambda F} \tag{2}$$

where ξ is the coordinate in the hologram plane, and F is the lens focus. The number of binary bits in the square transparant of L x L dimensions is

$$C' = \left(\frac{L}{\Delta\beta}\right)^2$$

where β is the ratio of distance between the centers of data cells to their linear dimension. If the hologram size is a, and the n-th order of spectrum can be registered, i.e. $a = n\lambda F/\Delta$, then the number of bits recorded in the hologram is

$$C'' = \kappa \frac{S\Omega}{\lambda^2} \tag{3}$$

where $\kappa = (\beta n)^{-2}$ is the contrast coefficient of a reconstructed picture; $S = L^2$ is the transparent area; $\Omega = \Omega_x \cdot \Omega_y$ is the solid angle, at which the hologram from the transparent plane is seen, and $\Omega_x = \Omega_y = \frac{a}{F}$. Equation (3) coincides with a total number of freedom degrees of the wave field determined by Eq. (1). Within the limits of this field one can redistribute the information degree of freedom. From a theorem about conservation of the number of wave field degrees of freedom it follows, that if any number of the information cells may be distributed over the area, which is by N times less than the transmitted field of the area S, then one can increase, by N times, the aperture of the optical system Ω . Since the dimensions of a reconstructed "point" are determined by angular aperture, i.e. $\Delta x \cdot \Delta y \sim \lambda^2/\Omega N$ then the resolution of a hologram increases by N times. As will be shown below, since the distribution of a signal point over the subject field is known a priori, it is possible to increase by N times a total number of points resolved by the hologram, in comparison with Eq. (3).

SCHEME OF HOLOGRAPHIC WRITING

To improve the resolution of optical systems by reducing the object field, a number of schemes is proposed, which makes it possible to exceed a classical limit of resolution /7,8/. A common feature of these schemes is the location of diffraction gratings in the space of an object and images. Due to the light diffraction in the grating positioned in the object space, the optical scheme transmits in the space of images a number of waves. These waves undergoing diffraction in the grating of the image plane, obtain a previous direction by producing the image with a greater resolution. Simultaneously, there appear the waves with undesirable directions, which produce additional images. These additional images reduce the efficient field of the image.

Let us discuss the recording of Fourier-holograms represented schematically in Fig. 1a. A transparent illuminated by a parallel light beam is located in the front focal plane of the lens. Space spectrum of the

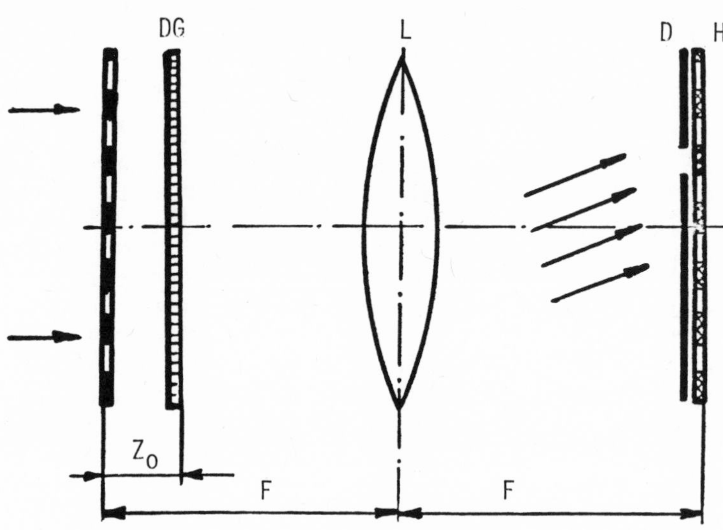

Fig. 1 a. Scheme of Fourier-hologram recording.
M - data matrix, L - Fourier lens, DG - diffraction grating, D - movable diaphragm, H - photoplate.

transparent is registered in the back focal plane. The size of a hologram is restricted by a mask. In contrast to conventional schemes, between the transparent and the lens there is located a diffraction grating in order to increase the information density on the hologram. The diffraction grating is described by transmission coefficient M(x,y).

In x,y,z=0 plane a transparant is located with the transmission U(x,y). The diffraction grating is in the $z=z_0$ plane. Light propagation from z=0 plane to the parallel $z=z_0$ plane is characterized by a transfer function

$$H = e^{j\frac{2\pi}{\lambda} z_0 \sqrt{1-(\lambda K_x)^2-(\lambda K_y)^2}}$$

where k_x and k_y are the space frequencies. Fresnel approximation corresponds to the transfer function

$$H = e^{jKz_0} \cdot e^{-j\pi\lambda z_0 (K_x^2+K_y^2)} \quad (4)$$

Therefore, if $g(K_x, K_y, z=0) = \int \mathcal{U}(x,y,0) e^{-2\pi j (K_x x + K_y y)} dxdy$

is the space spectrum of the data matrix, then the field at $z=z_o$ is

$$\mathcal{U}(x,y,z_o)=\int g(K_x,K_y,z=0)e^{jKz_o}e^{-j\pi\lambda z_o(K_x^2+K_y^2)}e^{j2\pi(K_x x+K_y y)}dK_x dK_y \quad (5)$$

Let us expand the grating function M(x,y) in Fourier series:

$$M(x,y)=\sum m_{\ell,n}e^{2\pi j(\ell\beta_x x+n\beta_y y)}$$

then the field after the grating is

$$\mathcal{U}'(x,y,z_o)=\mathcal{U}(x,y,z_o)M(x,y)$$

and its space spectrum takes the form:

$$\mathcal{F}[\mathcal{U}'(x,y,z_o)]=\mathcal{F}[\mathcal{U}(x,y,z_o)M(x,y)]=\sum m_{\ell,n}\,\delta(K_x-\ell\beta_x)\delta(K_y-n\beta_y)\oplus$$
$$\oplus\, g(K_x,K_y,z=0)\,e^{-jKz_o}e^{-j\pi\lambda z_o(K_x^2+K_y^2)}$$

where F denotes Fourier transform, and $\oplus$ is the convolution. By multiplying the spectrum $\mathcal{F}[\mathcal{U}'(x,y,z_o)]$

over the transfer function $H^*=e^{-jKz_o}e^{-j\pi\lambda z_o(K_x^2+K_y^2)}$ one can derive an expression for the space spectrum for the signal in the initial plane z=0, producing the same field distribution at $z>z_o$, as the transparent with the diffraction grating:

$$\tilde{g}(K_x,K_y,z=0)=\sum_{\ell,n} m_{\ell,n}\,g(K_x-\ell\beta_x,K_y-n\beta_y)\,e^{-j\pi\lambda z_o[(K_x-\ell\beta_x)^2-K_x^2+(K_y-n\beta_y)^2-K_y^2]} \quad (6)$$

From Eq. (6) it follows that a spectral component $m_{\ell,n}$ of the grating shifts the spectrum in the frequency plane by $\ell\beta_x$, $n\beta_y$, and multiplies it by a phase term depending on the value of this shift. The effect of the grating is as follows. Instead of the initial transparant in the object plane, there appears a superposition of transparants shifted by $\lambda z_o\beta_x\ell$, $\lambda z_o\beta_y n$. The number of transparants is determined by the number of spectral components of the diffraction grating.

If the spectrum of Eq.(6) is registered in the back focal plane of the lens with the point coordinates x_f, y_f then $K_x=x_f/\lambda F$, $K_y=y_f/\lambda F$. A joint action of the field (6) and reference wave $E=\exp jk\sin\theta\, x_f$ incident at θ angle to x_f-axis, produces the hologram transmittance:

$$t(K_x,K_y)=P(\lambda FK_x,\lambda FK_y)[\tilde{g}(K_x,K_y,z=0)e^{-jK\sin\theta x_f}+c.c.] \quad (7)$$

P is the transfer function of the optical system pupil determined by dimensions of a hologram, or in the general case, by the function of mask transfer.

IMAGE REPRODUCTION

There are different schemes of holographic memory. They are distinguished by the images, imaginary or real, which are projected on the photodiode matrix. We shall discuss a definite case of a real image.

The scheme of the hologram reconstruction is shown in Fig. 1,b. Between the lens, which realizes Fourier transform (Eq. (7)) and its back focal plane, a diffraction grating M'(x,y) is installed, which produces the image of increased resolution over the object field.

If in its back focal plane z=0 the lens produces the field, whose spectrum is

$$g'(K_x, K_y, z=0) = \int v(x,y,z=0)\, e^{-2\pi j(K_x x + K_y y)}\, dx\, dy$$

then the field at $-Z_0'$ distance from the focal plane is

$$v(x,y,-z_0') = \int g'(K_x,K_y,z=0)\, e^{-jkz_0'}\, e^{j\pi\lambda z_0'(K_x^2+K_y^2)}\, e^{j2\pi(K_x x + K_y y)}\, dK_x\, dK_y$$

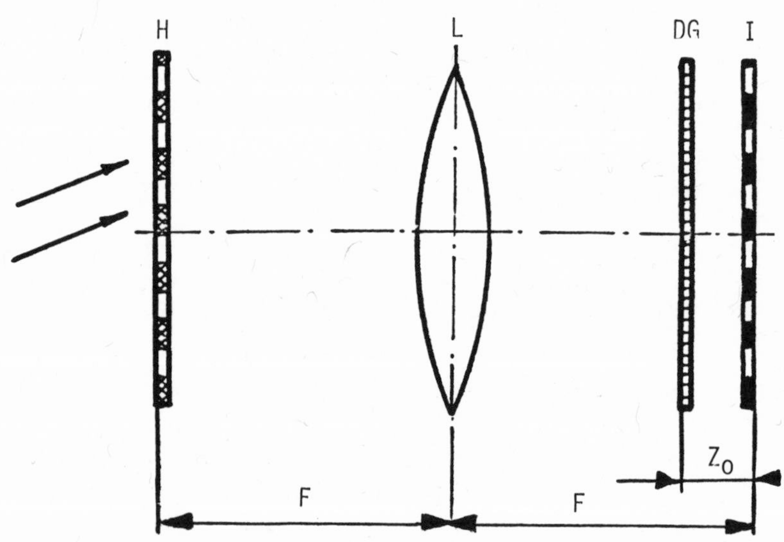

Fig. 1,b. Scheme of Fourier-hologram reconstruction. H - hologram, I - reconstructed image, DG- diffraction grating.

The field behind the grating $v'(x,y,-z_0') = v(x,y,-z_0')M'(x,y)$ has the space spectrum

$$\mathcal{F}[v'(x,y,-z_0')] = \sum_{\ell',n'} m'_{\ell',n'}\,\delta(K_x-\ell'\beta'_x)\,\delta(K_y-n'\beta'_y) \oplus g'(K_x,K_y,z=0)\,e^{-jKz_0'+j\pi\lambda z_0'(K_x^2+K_y^2)}$$

where $m'_{\ell',n'}$ are the coefficients of Fourier expansion M'(x,y). From this it follows that by analogy with Eq. 6, initial field spectrum $g'(K_x,K_y,z=0)$ is transformed by diffraction grating

$$\tilde{g}'(K_x,K_y,z=0) = \sum_{\ell',n'} m'_{\ell',n'}\,g'(K_x-\ell'\beta'_x, K_y-n'\beta'_y)\,e^{j\pi\lambda z_0[(K_x-\ell'\beta'_x)^2-K_x^2+(K_y-n'\beta'_y)^2-K_y^2]} \quad (8)$$

By reconstructing the real image in the scheme of Fig. 1,b without diffraction grating, the field in the back focal plane will be

$$v = \mathcal{F}\Big[P(\lambda FK_x,\lambda FK_y)\cdot\sum_{\ell,n} m^*_{\ell,n}\,g^*(K_x-\ell\beta_x, K_y-n\beta_y)\,e^{j\pi\lambda z_0[(K_x-\ell\beta_x)^2-K_x^2+(K_y-n\beta_y)^2-K_y^2]}\Big]$$

and its spectrum, respectively, is

$$g' = \mathcal{F}[v] = P(-\lambda FK_x,-\lambda FK_y)\sum_{\ell,n} m^*_{\ell,n}\,g^*(-K_x-\ell\beta_x,-K_y-n\beta_y)\,e^{j\pi\lambda z_0[(K_x+\ell\beta_x)^2+K_x^2+(K_y+n\beta_y)^2+K_y^2]}$$

Assuming that the gratings are equal, and located in the optically conjugate planes, we finally obtain the spectrum of the reconstructed data matrix image

$$G = \sum_{p,q} P_{p,q}\,g^*(K_x-p\beta_x, K_y-q\beta_y)\,e^{j\pi\lambda z_0[(K_x-p\beta_x)^2-K_x^2+(K_y-q\beta_y)^2-K_y^2]} \quad (9)$$

where

$$P_{p,q} = \sum_{\ell,n} m_{\ell,n}\,m^*_{\ell+p,n+q}\,P[(K_x+\ell\beta_x)\lambda F, \lambda F(K_y+n\beta_y)] \quad (10)$$

For convenience, in Eqs. (9) and (10) the coordinates $K_x \to -K_x, K_y \to -K_y$ are substituted. From Eqs. (9,10) it follows that the reconstructed image consists of a number of symmetrically positioned "images" of the transparant. The term with p=0, g=0 corresponds to the central image, whose optical system is spatially invariant (Fig. 1).

Below we shall use pulse response of the optical system. From Eq. 9 it follows that field distribution in the reconstructed image of the point source with x^0, y^0 coordinates has a form

$$u^0(x,y) = \iint\sum_{p,q} P_{p,q}\,e^{2\pi i[(K_x-p\beta_x)(x-x_0-\lambda z_0 p\beta_x)+(K_y-q\beta_y)(y-y_0-\lambda z_0 q\beta_y)]}\cdot e^{2\pi i[p\beta_x(x-\frac{p\beta_x}{2}\lambda z_0)+q\beta_y(y-\frac{q\beta_y}{2}\lambda z_0)]}\,dK_x\,dK_y \quad (11)$$

By introducing

$$\xi_p = x-x_0-\lambda z_0 p\beta_x$$
$$\eta_q = y-y_0-\lambda z_0 q\beta_y$$

we get $\overset{\circ}{U} = \sum_{p,q} \overset{\circ}{U}_{p,q}$

$$\overset{\circ}{U}_{p,q} = \tilde{P}\left(\frac{\xi_p}{\lambda F}, \frac{\eta_q}{\lambda F}\right) e^{2\pi i\left[p\beta_x\left(x - \frac{\lambda Z_0 p\beta_x}{2}\right) + q\beta_y\left(y - \frac{\lambda Z_0 q\beta_y}{2}\right)\right]} \times \sum_{\ell,n} m_{\ell,n} m^{*}_{\ell+p,n+q} e^{-2\pi i[(\ell+p)\beta_x\xi_p + (n+q)\beta_y\eta_q]} \quad (12)$$

where $\tilde{P}\left(\frac{\xi}{\lambda F}, \frac{\eta}{\lambda F}\right)$ is the Fourier image of the pupil function. The sum

$$L_{p,q} = \sum_{\ell,n} m_{\ell,n} m^{*}_{\ell+p,n+q} e^{-2\pi i[(\ell+p)\beta_x\xi_p + (n+q)\beta_y\eta_q]}$$

is periodical over ξ_p, η_q, with D_x and D_y periods, respectively. If $|\xi| \le \frac{D_x}{2}, |\eta| \le \frac{D_y}{2}$, then

$$L_{p,q}(\xi,\eta) = \int_{-\frac{D_x}{2}}^{\frac{D_x}{2}}\int_{-\frac{D_y}{2}}^{\frac{D_y}{2}} T(\xi',\eta')\, T^{*}(\xi+\xi',\eta+\eta')\, e^{2\pi i[p\beta_x\xi' + q\beta_y\eta']}\, d\xi' d\eta' \quad (13)$$

and $L_{p,q}(\xi+mD_x, \eta+nD_y) = L_{p,q}(\xi,\eta)$ if m, n=0, ± 1, ... Let us analyze central image of the point source where

$$\overset{\circ}{U}_{00} = \tilde{P}\left(\frac{x-x_0}{\lambda F}, \frac{y-y_0}{\lambda F}\right) L_{00}(x-x_0, y-y_0) \quad (14)$$

For the case of p=g=0 the function L_{00} coincides with that for the autocorrelation of grating transmission. From Eq. (14) it is seen that without gratings $L_{00}=1$, $L_{pg}=0$. p,g≠0, and resolution of the optical system is determined in the regular way by the system pupil. By using a modified optical system the resolution is determined by the product of two terms, namely, Fourier image of the pupil transmission function, and autocorrelation of the diffraction grating. If extension of the autocorrelation function is less than diffraction image width of the point source, the resolution of the modified optical system exceeds a classical diffraction limit.

CAPACITY OF HOLOGRAPHIC MEMORY WITH SYNTHESIZED APERTURE

The scheme of holographic memory on the basis of super-resolved optical system requires a reduction, by N times, of the data matrix field, and a similar decrease in the image field. This requirement does not cause any principle difficulty. The transparants, whose signal field occupies a small part of the total data matrix area, are already used in holographic memory devices. Figure 2 shows, as an example, a scheme on hologram recording with the lens raster /9/.

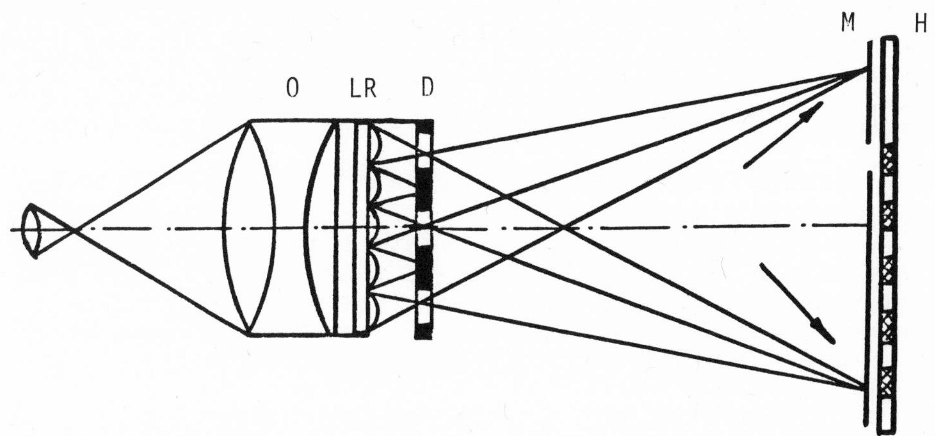

Fig. 2. The scheme of HM with the lens raster as a "point" signal source. O - telescope, M - data matrix, LR - lens raster, D - movable diaphragm, H - hologram.

In the signal beam a lens raster is positioned, whose number of elements is equal to the total number of binary signs of a page. The light from each raster illuminates the entire surface of recording medium. To restrict dimensions of the hologram in the plane of registration a movable diaphragm is located, whose coordinates determine the hologram address. The raster elements unlocked by the transparant, produce during exposure the point sources of spherical waves in the focal raster plane. These waves are registered by the hologram.

If each binary sign in the transparant has a raster lens with diameter d_o and focus length f_o, then the signal spot area is $d_o' \sim \frac{\lambda f_o}{d_o}$ in diameter. The ratio of areas is

$$N = \left(\frac{d_o}{d_o'}\right)^2 \sim \frac{d_o^2 B_o^2}{\lambda^2} \tag{15}$$

where $B_o = \frac{d_o}{f_o}$ is the relative hole of a single raster lens. From Gabor theorem about conservation of the total number of wave field degrees of freedom it follows, that the resolution of each signal "spot" of the transparant may be increased by N times in comparison with the case where resolution is uniform over the total area of the transparant.

Here is a numerical example. Let d_o=0.5 mm, B_o=0.3 and λ =0.6 mc, then N is about 10^4. Hence, at the expense of reduced image field, the optical super-resolution system produces a principle possibility for increasing the information density on hologram, and thus a total memory volume, by 10^4 times. A super-resolution optical system differs from that depicted in Fig. 2 by the presence of diffraction grating located between the transparant and recording medium.

Let us estimate the parameters of diffraction grating and conditions for the observation of reconstructed image of the signal transparant.

The case of a rectangular amplitude grating with the period $D_x=D_y=D=\beta^{-1}$ and width of the slit Δ , can be estimated from Eq. (13):

$$L_{p,q}=\beta^2 e^{-\pi i\beta(p\xi+q\eta)} \frac{\sin\pi p\beta(\Delta-|\xi|)}{\pi p\beta}\cdot\frac{\sin\pi q\beta(\Delta-|\eta|)}{\pi q\beta} \tag{16}$$

For p=0, g=0 we get $\qquad |\xi|,|\eta|\leq\Delta$

$$L_{00}=\beta^2\Delta^2\Lambda\left(\frac{\xi}{\Delta}\right)\Lambda\left(\frac{\eta}{\Delta}\right) \tag{17}$$

Data cells in the transparant should be placed so that the signal from each cell did not go to the structure maxima $L_{p,g}$ of the other cell.

Let us assume that the signal transparant is a combination of regularly located, at distance d, point sources, i.e.

$$U(x,y)=\sum b_{m,n}\delta(x-nd)\delta(y-md)$$

where b_{mn} is equal to 0,1. The conditions where a real extended source may be a point source, will be defined below. Hence, $(L/d)^2$ of data "point" lies in the square transparant with the side L.

Field distribution in the reconstructed image of a single data point is determined by Eq. (11), and consists of a sum of relatively shifted, at distance $\lambda z_0/D$, images, each of them being a diffraction image of the point modulated by autocorrelation function $L_{p,g}$ with the period D.

Let us choose the distance from the object plane to diffraction grating so that to fulfill the equality

$$\frac{\lambda z_0}{D}=D \tag{18}$$

By fulfilling Eq. (18) the maxima of periodic functions $L_{p,q}(x-x_0-\frac{\lambda z_0}{D}p,\ y-y_0-\frac{\lambda z_0}{D}q)$ at any p and g, lie in rectangular regions with the sides 2Δ x 2Δ distributed around the points

$$x_p = x_0 + Dp \qquad y_q = y_0 + Dq$$

The number of signal points which can be positioned in the data matrix depends on the width of each structure maxima, and the rate of their intensity decreases. Intensity distribution over the structure maxima in the point source image may be evaluated if we neglect the oscillating multipliers at the summation of $U^o_{p,g}$. Let us consider a definite signal point with $x_o=y_o=0$. From a deliberately increased value the intensity distribution over the structure maxima in the point image is:

$$I^{\bullet}_{p,q}=\left\{\sum_{p'q'}\tilde{P}\left(\frac{D[p-p']}{\lambda F},\frac{D[q-q']}{\lambda F}\right)L^{max}_{p'q'}\right\}^2 \approx \left\{\iint_{-\infty}^{+\infty}\tilde{P}L^{max}_{p'q'}\,dp'dq'\right\} \qquad (19)$$

For a rectangular grating

$$L^{max}_{p,q} = \operatorname{sinc} p\beta\Delta \cdot \operatorname{sinc} q\beta\Delta$$

In the case of a rectangular hologram from Eq. (19) it follows that

$$I^{\bullet}_{p,q} = \operatorname{sinc}^2 p\Gamma \operatorname{sinc}^2 g\Gamma \qquad \Gamma = \max\left\{\begin{array}{l}\frac{aD}{\lambda F}\\ \frac{\Delta}{D}\end{array}\right. \qquad (20)$$

The intensity of structure maxima $I^o_{p,g}$, as shown by Eq.(20), is determined by the greatest value of $\tilde{P}(\frac{pD}{\lambda F},\frac{qD}{\lambda F}), \operatorname{sinc}^2\frac{p\Delta}{\lambda F}\operatorname{sinc}^2\frac{q\Delta}{\lambda F}$. This conclusion is valid in the common case.

The secondd information point with n=1 can be located in the data matrix at the distance of d=2(D-Δ) from the initial point x_o=0. By the same distance all the structure maxima of the point image will be shifted. The third information point with n=2 will be located at 2d= 4(D -Δ) distance from the initial one etc. Since the distance between all the structure image maxima is equal to D, and the width does not exceed 2Δ , it is evident that at a definite distance between data points x_o and x_n up to $n=n^*=\frac{D-\Delta}{\Delta}$, additional maxima of each image are not superimposed on each other and on the "central" information signal. At $\tilde{n}=n^*+1=\frac{D}{\Delta}$ the central maximum of the point image $x_{\tilde{n}}$ is superimposed by a structure maximum $I^{\bullet}_p$ of x_o=0 point, which causes the noise effect. Noise intensity in the point of a signal is determined by Eq. (20). Central maxima of $x_{\tilde{n}}$ point is at $\tilde{n}2(D-\Delta)$ distance from the initial point x_o=0, and at the same point the initial structure maximum with $p = 2(D-\Delta)/\Delta$ lies. Therefore, the intensity of crossover noises for the signals at point $x_0=0$ and x_n is

$$\sim \mathrm{sinc}^2 \frac{2(D-\Delta)}{\Delta} r$$

From Eq. (20) it follows that for the maximum density of holographic information the size of a hologram and the grating period should be chosen from

$$\frac{aD}{\lambda F} = \frac{\Delta}{D} \tag{21}$$

therefore, maximum level of crossover noises does not exceed 5.10^{-2}.

Physical sense of Eq. (21) is in the fact that to resolve two points lying at the distance D, in the super-resolution system, the size of the hologram may be by $K \sim \frac{D}{\Delta}$ times less than its size in the classical case.

After defining a relationship between diffraction grating parameters and the hologram size, and showing that the level of crossover noises is small in the chosen distribution of signal points, the density of holographic information density can be determined as

$$C = \left(\frac{L}{2Da}\right)^2$$

where L is the side of a square transparant, which includes $m^2 = \left(\frac{L}{2D}\right)^2$ signal points, and a is the side of a hologram. If the hologram size is greater than it is necessary for the maximum data density, then Eq. (21) yields:

$$\gamma \frac{\lambda F}{a} = DK \tag{22}$$

where $K = \frac{D}{\Delta}$, and $\gamma > 1$, which shows by how many times the information density is less than the maximal, wherefrom

$$C = \left(\frac{L_0 - 2DK}{2\gamma\lambda F} K\right)^2$$

From each signal point in the transparant the diffraction grating produces k-images of the point, with spacing D. Therefore, total size of the image recorded by the hologram is $L_0 = L + 2Dk$, and

$$C = \left(\frac{L_0 - 2DK}{2\gamma\lambda F} K\right)^2 \tag{23}$$

has maximum in the function from degree of information compression. From Eq. (23) it follows that information density is maximal at $L_0=2L$ and $L = 2Dk$. If $A=L_0/F$ is the relative objective hole, then

$$C = \left(\frac{A}{4\gamma\lambda}K\right)^2$$

Maximum gain of capacity is $K^2_{max} = \left(\frac{DB}{\lambda}\right)^2$ where B is the relative hole of a single raster lens, which produces a signal "point", and therefore

$$C_{max} = \left[\frac{A \cdot B}{4\gamma\lambda^2}D\right]^2$$

In the optimal case the value of block data is $m^2_{max} = K^2_{max} = BL/2\lambda$. Total capacity of holographic memory (HM) with superresolution

$$C_{o\,max} = C_{max} L_o^2 = \frac{A^2 B}{8\gamma^2\gamma_1^2}\left(\frac{L}{\lambda}\right)^3 \quad (24)$$

is proportional to the volume of a cube with the side equal to a linear size of the transparant.

If the block data $m \leq m_{max}$ and the compression factor $k \leq k_{max}$ are defined, then the size of the hologram is chosen from the relation

$$\frac{\lambda F}{\gamma a} = \frac{LK}{2m}$$

where L is the transparant size. The information density is

$$C = \left(\frac{m}{a}\right)^2 = \left[\frac{A}{2\gamma\lambda}\,\frac{mK}{m+K}\right]^2$$

and a total HM volume is

$$C_o = \frac{A^2 L_o^2}{4\gamma^2\gamma_1^2\lambda^2}\left(\frac{mK}{m+K}\right)^2 \quad (25)$$

Equation (25) develops into Eq. (24) at $m=k=k_{max}$. Let us determine the maximum limit of HM capacity with the superresolution. At A=B=0.25, λ = 0.63 mc, $\gamma = \gamma_1 = 1$, L_o=10 cm, the maximal memory volume C_{max} = 8.10^{12} bit, and the number of bits in the data matrix is $m^2=2.10^4$ bit, the size of a single hologram a=16λ, compression factor k = 1.4 x 10^2. In the classical case at k =1 the density of reconstruction is

$$C' = A^2/4\lambda^2$$

and the memory volume

$$C_o' = A^2 L_o^2/4\lambda^2\gamma_1^2$$

At A = 0.25 the writing density is ~ 4.10^4 bit/mm^2, and at L_o=10 cm, γ_1 = 1 the maximum limit of memory volume C_o = 4.10^8 bit.

From Eq. (25) it is seen that at $m^2=10^4$ and k=50 the HM capacity with super-resolution increases by $\left(\frac{mK}{m+K}\right)^2 \sim 10^3$ times and amounts to 4.10^{11} bit.

ENERGY RATIOS AND REQUIREMENTS TOWARD OPTICAL ELEMENTS

The amplitude diffraction gratings have a significant drawback. With increasing number of diffraction order of a two-dimensional grating the intensity of light transmitted by the grating quadratically decreases. It is therefore more expedient to use phase gratings. The most gain in resolution is observed in the case of uniform energy distribution over the spectral components. The most appropriate means in this respect is the diffraction grating, which comprises scattering or focusing lens as one of the elements, since the Fourier image module of lens transmission is a constant value. Calculation of an indefinite function $L_{p,g}$ in this case shows that maximal data packing is reached in the case of fulfilling the ratio $\lambda F/a \approx kD$, which is equivalent to Eq.-(21) and produces a satisfactory ratio of a signal to crossover noises. Here $K=\frac{D^2}{\lambda f}$, where D is the diameter, and f is the focus of the raster lens. By using the second lens raster as diffraction grating it is possible to reach maximum gain in resolution since the signal dimensions are minimal for the raster.

The volume of information, which may be reproduced from a single hologram, and the gain in resolution depend, apart from the optical scheme, on the energy ratios connecting the basic elements of holographic memory. Before the photorecorder matrix a mask is placed, which interlocks "excessive" structural maxima.

If P_o is the laser power, then each element of the photorecorder is supplied with the power

$$P = P_o \cdot \eta_1 \cdot \eta_2 \cdot \eta_3 \cdot \eta_4 \frac{1}{m^2 N}$$

where η_1 is the deflector transmission; η_2 , the optical transmission; η_3 , the diffraction efficiency of holograms; η_4 , the diffraction grating transmission; m^2, the number of binary signs in the data matrix,; $N=k^2$, and k is the gain resolution due to reduction of the image field. If $P=P_{min}$, where P_{min} is the detector photosensitivity, then

$$N = \frac{P_o \cdot \eta_1 \cdot \eta_2 \cdot \eta_3 \cdot \eta_4}{m^2 P_{min}}$$

To obtain a satisfactory signal/noise ratio in the image convertors, the minimal energy per image element is $W_{min}=10^{-15}$J /2/. Minimal power per photodetector is $P_{min}= W_{min}\nu$ and

$$N = K^2 = \eta_1 \cdot \eta_2 \cdot \eta_3 \cdot \eta_4 \frac{P_o}{m^2 W_{min} \nu}$$

where ν is the frequency of the page access.

By using ion or solid-state lasers it is possible to obtain 1 Watt of radiation power in the regime of CW single-mode operation. The most efficiency can be obtained from acousto-optical deflectors. Assuming that

$$\eta_1 = \eta_2 = \eta_4 = \eta = 0.8, \eta_3 = 0.2, m^2 = 10^4, W_{min} = 10^{-15} J, P_o = 1 W$$

and $\nu = 10^6$ s^{-1}, we obtain that energy ratios in the above example allow increasing of information density by 10^4 times. With increase in the memory capcity the requirements toward random access time decrease. With memory volume of 10^{12} bit the time of page selection may be of 10^{-4}–10^{-5} s, and the energy ratios, therefore, allow obtaining of the maximal packing.

One of the factors affecting HM capacity is the granularity of photosensitive material. The light scattered in discrete photolayer structure produces a background noise signal in the image plane. If the laser power is P, and the block of m^2 signal points is written in the hologram with diffraction efficiency η, and information density is by N times greater than the classical limit, then the signal intensity over the (2Δ) area is

$$I_s = \frac{\eta P}{m^2 N 4\Delta^2}$$

If we assume that the light is scattered by the photoemulsion uniformly in all directions, then the noise intensity is

$$I_n = \frac{\sigma P}{2\pi F^2}$$

where σ is the coefficient of photoplate light-scattering. Hence,

$$I_s/I_n \sim \frac{\pi\eta}{2\sigma A^2} \cdot \frac{L^2}{m^2 N \Delta^2}$$

and since $N = \frac{L^2}{m^2\Delta^2}$, then $\frac{I_s}{I_n} = \frac{\pi\eta}{2\sigma A^2}$

Assuming that A = 0.5, $\eta/\sigma = 5$ /10/, we get $I_s/I_n = 30$, which is a sufficient value for reliable reading of information. By applying photoplates with a small level of light scattering or using non-haloido-silver materials, such as photochromic glass or semiconductor-metal structures, one can avoid the noise problem associated with the granularity of photomaterials.

The degree of data packing in the hologram depends on the number of orders produced by diffraction grating. But at large angles the deformations may appear due to distortion and chromatic abberations of the grating. The distortion can be neglected if the diffraction angles are not greater than 10-15° /11/. Hence, $K_{max} = \frac{\varphi_{max} D}{\lambda}$

and since L=m(2D), then

$$k_{max} = \frac{\varphi_{max} L}{2m\lambda}$$

At $\varphi_{max}=\frac{\lambda}{12}$, L= 10 cm, m = 100 the value of k_{max} is about $5\cdot10^2$. The maximal diffraction angle restricts the distance from the transparant to the grating. In the case of HM with the lens raster, this distance should satisfy the condition $z_0 > L/B$, where is the relative hole of a single lens or $B \leq A$, since $z_0 \leq F$.

Since the transparant image is formed by k-orders of the grating, the relative spectral width should not exceed the resolution of diffraction grating in $k=k_{max}$ order of diffraction, i.e. $\frac{\Delta\lambda}{\lambda} < \frac{1}{m k_{max}}$

At $k_{max}=10^2$, m=100, the relative spectral width $\frac{\Delta\lambda}{\lambda} \leq 10^{-4}$.

For a profiled grating one can neglect refraction on the surface of the grating, if the profile depth is less than the grating period.

Requirements to the resolution of HM photomaterial with the information density exceeding the classical limit are the same as to the conventional HM devices, since the maximal space frequency of the object, as seen from Eq. (23) does not exceed $A/2\lambda$. Relative shift of two diffraction gratings should not be greater than the signal size $\sim \frac{D}{\lambda f}$.

CONCLUSION

HM capacity with synthesized aperture is proportional to $\sim L^3/\lambda^3$, where L is the linear size of a transparant, whose limiting value is determined by the optical system diameter. The gain in capacity by L/λ times is due to the fact that the size of a reconstructed signal point is determined by diffraction not on the hologram, but on the synthesized aperture, whose size is by $k = \frac{DB}{\lambda} \sim \sqrt{\frac{L}{\lambda}}$ times greater than the hologram dimensions.

A theoretical limit of data packing by 10^4-10^5 times is determined, first of all, by energy relations (Eq. 29), and by a feasibility of recording the microholograms dimensioned several tens or hundreds wavelengths under conditions of precise addressing of the reading beam at the stage of reconstruction.

The proposed method is distinguished by a possibility of constructing optical memory devices of high volume by means of accessible optical elements-objectives and lenses with decreased requirements to their optical hole. The preformed analysis is also valid for the case of Fresnel holograms.

It should be noted that the possibility of obtaining holographic memory with random access and capacity of memory above 10^{12} bit makes it possible to approach a problem of increasing efficiency of computers /12,13/.

References

1. N.G.Basov, W.H.Culver, B.Shan. "Laser Handbook", p. 1051 (1972).
2. A.L.Mikaelian, V.I.Bobrinev. Radiotekhnika i Elektronika, 19, 5, 898 (1973) (in Russian).
3. H.Kiemle. Appl.Opt. 13, 803 (1974).
4. D.Pole. Appl.Opt. 13, 341 (1974).
5. M.A.Maiorchuk, V.V.Nikitin, V.D.Samoilov. Kvantovaya Elektronika, 1, 2, 302 (1974) (in Russian).
6. A.N.Korolev. Usp.Fiz.Nauk, 96, 2 (1968). (in Russian).
7. W.Lukosz. JOSA, 56, 1463 (1966).
8. A.Bachl, W.Lucosz. JOSA, 57, 2, 163 (1974).
9. A.A.Verbovetsky, V.B.Fedorov. J.Techn.Fiz. 49, 10, 2203 (1972) (in Russian).
10. H.Kiemle. Optics Technology, May, 196 (1969).
11. C.P.Bovin. Appl.Opt. 11, 8, 1782 (1972).
12. N.G.Basov, Yu.M.Popov et al. Possibility of constructing new fast-operating multi-channel optoelectron computer systems. Preprint FIAN(1973) (in Russian).
13. L.A.Orlov, Yu.M.Popov. Avtometry, 6, Nov.-Dec. 14 (1972) (1972)(In Russian).

INVESTIGATION OF A VERSION OF A HOLOGRAPHIC CHARACTER MEMORY DEVICE

I.S. Gibin, M.A. Gofman, S.F. Kibirev, and P.E. Tverdokhleb

Institute of Automation and Electrometry
Novosibirsk, USSR

ABSTRACT

A holographic character memory is a part of data addressed optical storage systems. A version of holographic character memory with bit-sequential search is considered. The input of quest digits is carried out by a deflector, and data processing to volve problems of simple and complex search is performed by parallel electronic processor with an optical input. Experimental system for organic compound search is described. Evaluations show, that the time of solving typical search problems in holographic character memory, being under consideration, is less than that of a system of the "magnetic disk-computer" type by 2-3 orders.

INTRODUCTION

Structures of holographic character memories which are a part of data addressed optical storage systems [2] are proposed in the work [1].

The specification of such a storage system (SS) and the results of its investigation are given below. In the working version of a holographic character memory the input of quest digits is carried out by a high speed deflector, and the page data processing is performed by a parallel electronic processor with an optical input, which can be in the form of an integrated circuit. In the storage system under consideration the time of solving typical retrieval problems is 2 or 3 orders less than that in a system of the type "magnetic disk-computer". Experimental investigation of the work of a holographic character memory with a computer was conducted by the example of retrieval of organic compounds according to a single character, i.e. their molecular weight being preset in the range from 0 to 999.9 atomic units. The initial data on organic compounds were given by the Institute for organic chemistry, Novosibirsk (cor. mem. USSR AS V.A. Koptyug laboratory).

Let each of data arrays $D_1, \ldots, D_h, \ldots, D_H$, upon which information retrieval is performed, be presented in character memory by its description, that is by a set of R characters $\{K_{h1}, \ldots, K_{hr}, \ldots, K_{hR}\}$, where $h = \overline{1, H}$, and K_{hr} are N_r bit key-words $\{K_{hr1}, \ldots, K_{hrn_r}, \ldots, K_{hrN_r}\}$. Data arrays themselves are supposed to be stored in cells of basic memory of data addressed storage system, which addresses are univocally connected with data array numbers $1, \ldots, h, \ldots, H$. Besides, let a quest to a character memory be of the structure analgous to data array discriptions $\{Z_1, \ldots, Z_r, \ldots, Z_R\}$, where Z_r is a N_r bit key-word $\{Z_{r1} \ldots Z_{rn_r} \ldots Z_{rN_r}\}$. Then the logic function system $(y_{1r}, \ldots, y_{hr}, \ldots, y_{Hr})$ is calculated for each of characters entering an array discription and a quest in the process of retrieval in character memory; function argument are digits of r corresponding binary words. A form of these functions is defined by a retrieval problem, and their unit values indicate data array addresses fitting the rth character. For example, the problems of search for equality or inequality, nearest above or nearest below, ordered access, and etc. are solved most frequently. The result of a search through the whole complex of characters in the form of logic array $(A_1, \ldots, A_h, \ldots, A_H)$ is found by digit logic multiplication of systems of functions $(y_{1r}, \ldots, y_{hr}, \ldots, y_{Hr})$, computed for all r index values.

The structure of holographic character memory solving the problem of search according to a character system is given in Fig. 1. Herein 1 is a laser; 2 is a deflector for $\sum_{r=1}^{R} N_r$ positions; 3 is a memory module

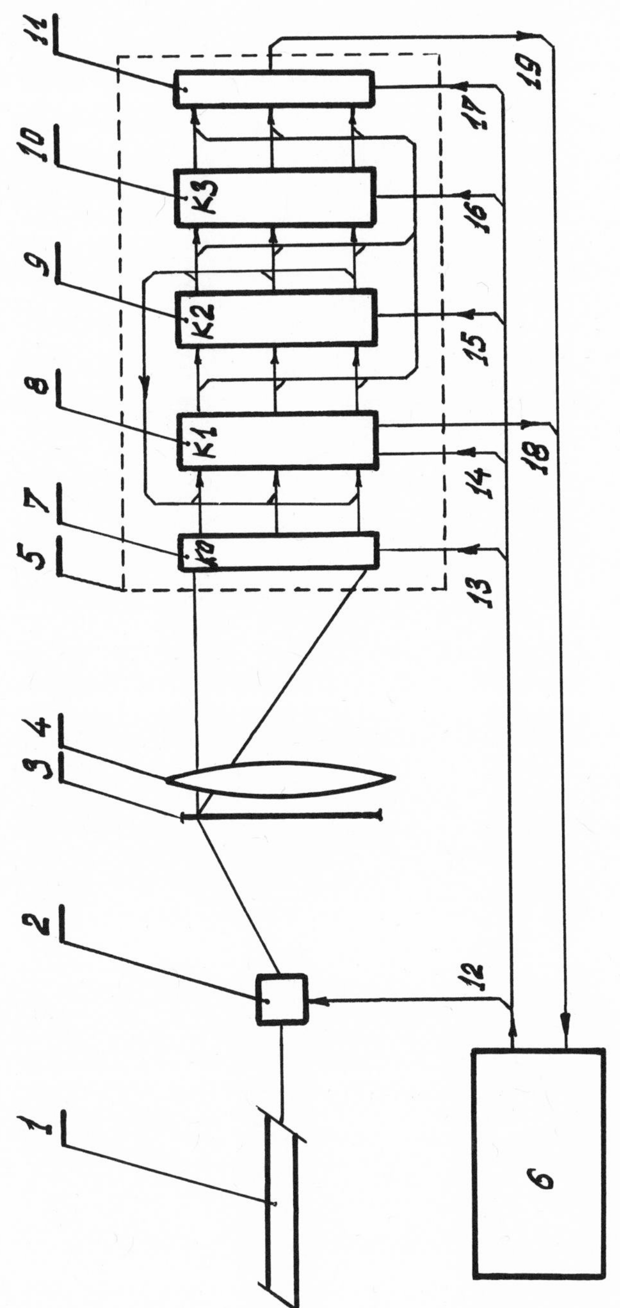

Figure 1. Block-diagram of holographic character memory with sequential input of quest digits.

Table 1

Character memory elements	Commands	Codes	Operations	
			Indications	Meaning
Deflector	$P(n_r)$	n_r	$(K_{1rn_r}, .. K_{hrn_r}, .. K_{Hrn_r})$	Reconstraction of the n_r th memory module hologram
Photo-matrix	α_0	1	$K0 = (K_{1rn_r}, .., K_{hrn_r}, .. K_{Hrn_r})$	Information reception in PPP
First memory level	$\alpha_{11} \alpha_{12} \alpha_{13}$	000	$K1 \rightarrow K1$	$K1$ page storage on the 1st memory level
		001	$K0 \cdot K1 \rightarrow K1$	Logical multiplication of pages $K0$ and $K1$
		010	$\overline{K0} \cdot K1 \rightarrow K1$	Logical multiplication of pages $K0$ and $K1$ with $K0$ inversion
		011	Set"1" $K1$	Setting of all elements of page $K1$ into "1"
		100	$K0 \cdot K2 \rightarrow K1$	Logical multiplication of pages $K0$ and $K2$
		101	$\overline{K0} \cdot K2 \rightarrow K1$	Logical multiplication of pages $K0$ and $K2$ with $K0$ inversion
		110	$K0 \rightarrow K1$	$K0$ page recording on the 1st memory level
		111	in computer code U	00, when $K1$ of 0 01, when $K1$ of 0 and 1 11, when $K1$ of 1

Character memory elements	Commands	Codes	Operations	
			Indications	Meaning
Second memory level	$\alpha_{21}\alpha_{22}$	00	$K2 \rightarrow K2$	$K2$ page storage on the 2-nd memory level
		01	$K1+K2 \rightarrow K2$	Logical addition of pages $K1$ and $K2$
		10	$K1 \rightarrow K2$	$K1$ page recording on the 2-nd memory level
Third memory level	$\alpha_{31}\ \alpha_{32}$	00	$K3 \rightarrow K3$	$K3$ page storage on the 3-d memory level
		01	$K2 \cdot K3 \rightarrow K3$	Logical multiplication of pages $K2$ and $K3$
		10	Set"1" $K3$	Setting of all elements of page $K3$ into "1"
Scheme of data readout		01	in computer $K1$	Page $K1$ is entered the computer
		10	in computer addresses $K2$	Numbers of unit elements $K2$ are entered the computer
		11	in computer addresses $K3$	Numbers of unit elements $K3$ are entered the computer

Note: An arrow in operation indications shows the memory level, on which the result of operation fulfilment is recorded.

from $\sum_{r=1}^{R} N_r$ hologramms, corresponding similar digits of binary words, that is, characters of all data array descriptions $(K_{1rn_r}, \ldots, K_{hrn_r}, \ldots, K_{Hrn_r})$ written in the form of binary images in each of them; 4 is an objective; 5 is a parallel page processor with an optical input (PPP); 6 is a control computer. PPP contains: photomatrix 7, three levels of page memory 8,9, 10, and readout circuitry 11. The computer controls a character memory through lines 12 to 17 by commands to the deflector $P(n_r)$, to the photomatrix α_0 , to memory levels $\alpha_{11}\,\alpha_{12}\,\alpha_{13}$, $\alpha_{21}\,\alpha_{22}$, $\alpha_{31}\,\alpha_{32}$ and to readout circuitry $\alpha_{41}\,\alpha_{42}$. Operations being carried out according to these commands in character memory are given in Table 1. The binary page KO correspond to the photomatrix in the Table, and binary pages $K1$, $K2$ and $K3$ - to three memory levels. The code U characterizing a zero, unit or mixed page state enters the computer from the character memory through line 18 from the first PPP memory level, and the page $K1$ itself, or the addresses of the unit elements of pages $K2$, $K3$ - through line 19 from the PPP readout circuitry. Similar digits of the rth character of all data array descriptions are intered the PPP electronic unit with the deflector and photomatrix by $P(n_r)$ and α_0 commands. The addresses of the arrays which fit separate characters are calculated in the form of logical functions $(y_{1r}, \ldots, y_{hr}, \ldots, y_{Hr})$ with the help of operations $K0 \cdot K1 \rightarrow K1$, $\overline{K0} \cdot K1 \rightarrow K1$, Set. "1" $K1$, $K1 + K2 \rightarrow K2$ and $K1 \rightarrow K2$ at the first and second memory levels. The required address array fitting the whole complex of characters is formed with the operations $K2 \cdot K3 \rightarrow K3$ and Set. "1" $K3$ from the mentioned systems by their digit by digit logical multiplication at the third memory level. The commands $K0 \cdot K2 \rightarrow K1$, $\overline{K0} \cdot K2 \rightarrow K1$ and "in computer code U " were entered the PPP to speed up complex search of the type greater than, less than, ordered access. The commands $K0 \rightarrow K1$ and "in computer $K1$ " provide the PPP operation within the usual page buffer photometric regime.

The optical system of the character memory being under discussion doesn't differ from that of a common type holographic memory. It is used for storage and readout of data array descriptions. All parallel logical operations and intermediate result storage operations necessary for solving the problems of address array $(A_1, \ldots, A_h, \ldots, A_H)$ retrieval are carried out by a PPP digital method. The possibilities of quick data access from holographic storage system, and those of page organization of their memory are used in full measure in this case. It allows to eliminate the fulfilment by the computer of parallel logical operations over multidigital codes, which are unusual for it; and besides it reduces information volume being transfered through the holographic storage system-computer (HSS - computer) channels; the channel capacity of which is limited, as a rule.

A holographic character memory allows to solve various kinds of retrieval problems according to the independent character system. According to each character one can find data array addresses which binary key-words coinside with a quest-word, are within prescribed limits, ordered in magnitude, are extreme, and etc. The set of PPP operations is functionally complete, and it allows to carry out arbitrary logical functions, quest character digits and array discriptions being their arguments.

Experimental investigation of the proposed character memory version has been implemented. The problems of complex and simple search for addresses (numbers) of 100 organic compounds being defined for ease by single character, that is, their molecular weights, presented by 16-digit binary-decade codes $(K_{h1} \ldots K_{hn} \ldots K_{H16})$ where $h = \overline{1,H}$. Typical algorithms of search for the compound addresses which molecular weights exactly coinside with the quest $(Z_1 \ldots Z_n \ldots Z_{16})$, are within prescribed limits $(z_1 \ldots z_n \ldots z_{16})$ below $\div (Z_1 \ldots Z_n \ldots Z_{16})$ - above and are ordered in the magnitude, were realized.

The block-diagram of the experimental "HSS-computer" system is given in Fig. 2. It comprises a ЛГ-38 laser 1; HI35-3 [3] two-dimentional galvanometer deflector 2; a memory module containing 16 character holograms 3, a И-37 objective 4, a ЛИ 604 dissector with a controle unit 5; a Э-100 computer 6; a VC-340 alphameric display 7. The external appearance of the system (without a computer) is shown in Fig. 3.

Information pages containing similar digits of all organic compounds molecular weights $(K_{1n}, \ldots, K_{hn}, \ldots, K_{Hn})$

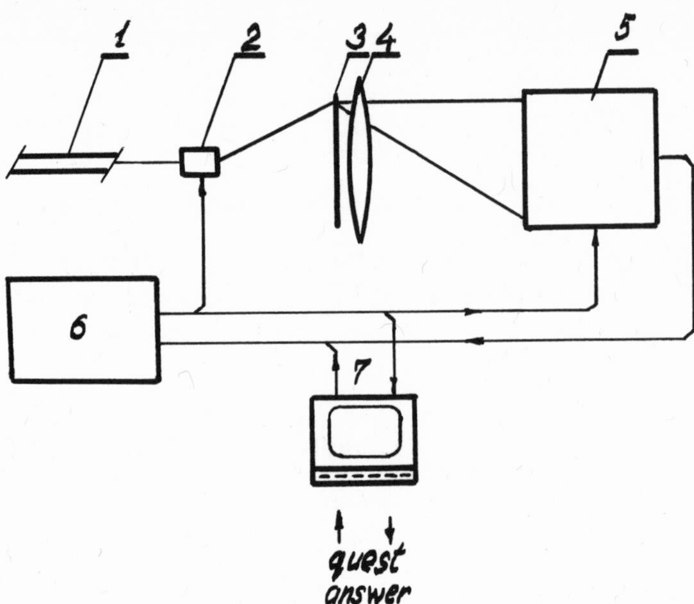

Figure 2. Block-diagram of the experimental "holographic memory-computer" system.

Figure 3. External appearance of the experimental "holographic memory-computer" system (without computer).

were recorded in the nth matrix hologram 3. Page images of 10x10 bits corresponding to masked off quest word digits are reconstructed in the dissector plane, and entered the computer, when a quest is entered with the deflector.

The computer controls the deflector according to a subprogram "DEFLECTOR" realizing a command of the $P(n)$ type on reconstruction of a hologram with a number " n " (see Table 1). The work of a parallel page processor with an optical input which is a part of a hologram character memory was simulated in the system by an apparatus-program method with the help of the dissector and a special subprogram "PPP". The dissector is used for formation of a logical array $K0$, which is an input array for PPP. The subprogram "PPP" being realized in the computer according to commands of the $\alpha_0, \alpha_{11}\ \alpha_{12}\ \alpha_{13}, \alpha_{21}\alpha_{22}, \alpha_{31}\alpha_{32}$ and $\alpha_{41}\ \alpha_{42}$ type, makes operations over logical arrays $K0, K1, K2, K3$ (see Table 1).

Reliable readout of character holograms with a dissector was achieved by scanning of a bit of the reconstructed page image by 3 x 3 raster, and selecting of choosing threshold of their unit values. The image reconstructed from a character hologram (a), and the result of representation of the array on the display screen (b) are shown in Fig. 4.

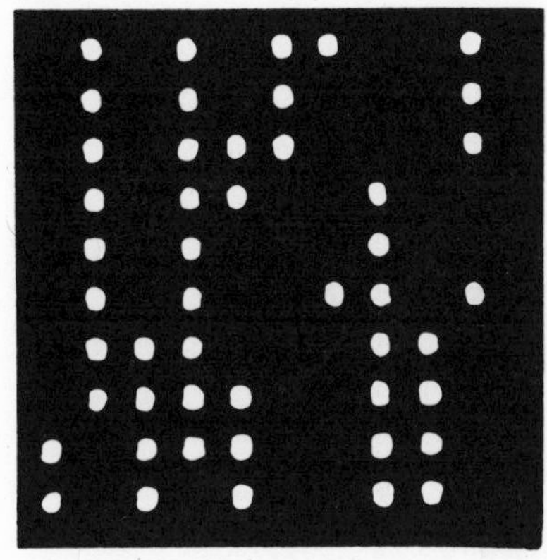

a

ГОЛОГРАММА N: 16

	01	02	03	04	05	06	07	08	09	10
01	0	1	0	1	0	1	1	0	0	1
02	0	1	0	1	0	1	0	0	0	1
03	0	1	0	1	1	1	0	0	0	1
04	0	1	0	1	1	0	0	1	0	0
05	0	1	0	1	0	0	0	1	0	0
06	0	1	0	1	0	0	1	1	0	1
07	0	1	1	1	0	0	0	1	1	0
08	0	1	1	1	1	0	0	1	1	0
09	1	0	1	1	1	0	0	1	1	0
10	1	0	1	0	1	0	0	1	1	0

b

Figure 4. a) The image reconstructed from character hologram; b) the result of representation of the $K0$ array on the display screen.

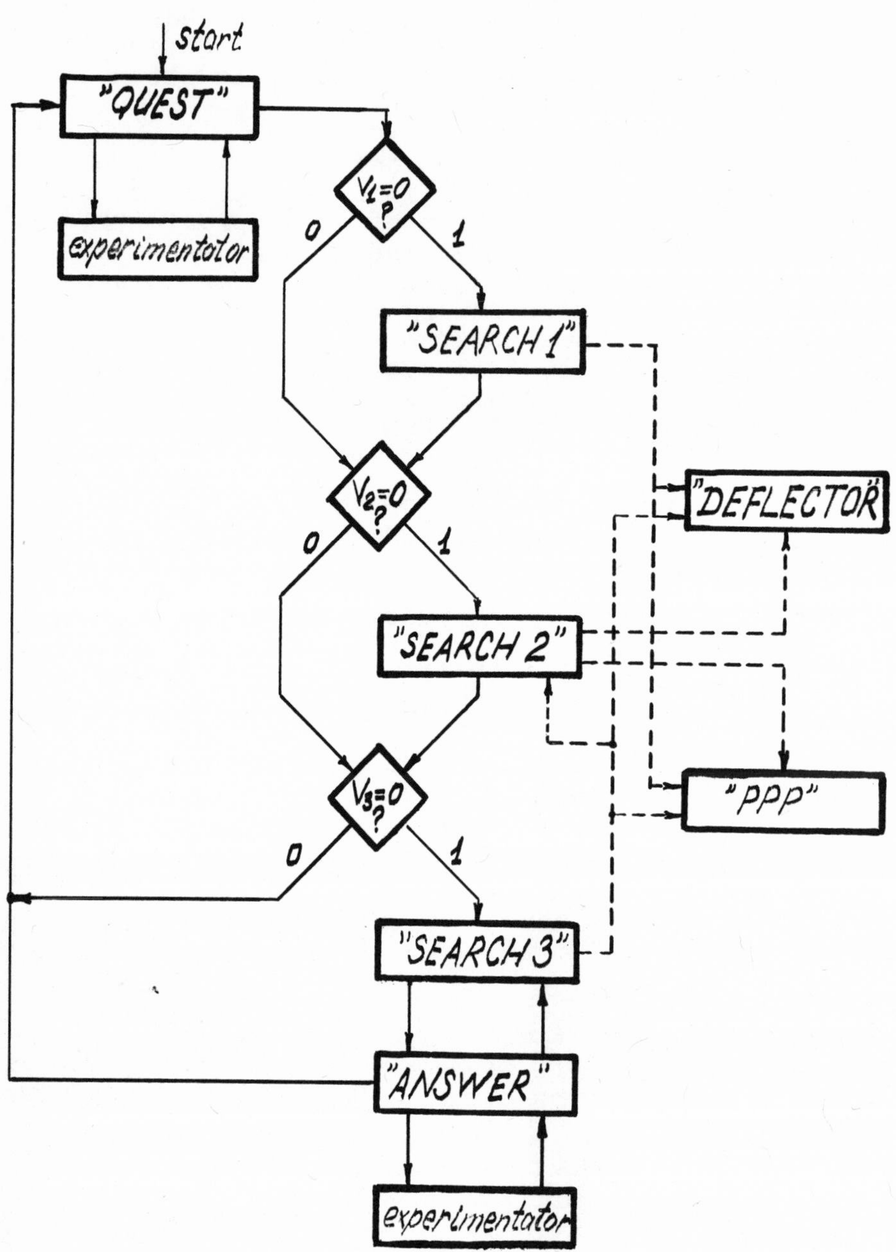

Figure 5. Block-diagram of the program "DIRECTOR".

The quest content and a required type of search (simple search, between limits search, ordered access) are entered the system with the alphameric display. The addresses of required organic compounds being the result of retrieval are displayed. In case of ordered access, compound addresses are displayed in succession according to experimentor commands.

Experimental system element control when solving retrieval problems is carried out with a software, represented by the symbolic coding langquage PAL-III. The software comprises service subprograms ("QUEST", "ANSWER"), retrieval subprograms ("SEARCH"-1,2,3"), control subprograms ("DEFLECTOR", "PPP"), and a program "DIRECTOR", uniting the listed above subprograms as a single shole. The block-diagram of the program "DIRECTOR" is given in Fig. 5. Broken lines correspond to mutual subprogram accesses. The subprogram "QUEST" services the experimentator while making a quest. As soo soon as the quest is over the subprogram "QUEST" allows to determine a "DIRECTOR" program configuration (by giving a unit value to one of the markers U_1, U_2 and U_3), and it prepares the initial data for the subprograms "SEARCH-1", "SEARCH-2". As soon as the retrieval is over the subprogram "ANSWER" inputs the found organic compounds addresses on the display sceen. The subprograms "SEARCH-1", "SEARCH-2", and "SEARCH-3" which model the work of the character memory when solving problems of simple retrieval, between limits retrieval, and ordered access, respectively, are the basic subprograms of the system software.

Now let us consider the algorithms which are in the basis of these subprograms.

1. Simple search for addresses of compounds is carried out according to coincidence of molecular weights, both in their descriptions and the quest. This is made by calculation of equivalence logical functions in the memory:

$$y_h = \bigcap_{n=1}^{N} (z_n K_{hn} + \bar{z}_n \bar{K}_{hn}) \qquad (1)$$

where $\bigcap$ is a sign of logical product; $h=\overline{1,H}$. Quest digits $(z_1 \ldots z_n \ldots z_N)$ can take values 0,1,M (mask). The functions $(y_1, \ldots, y_h, \ldots, y_H)$ depend

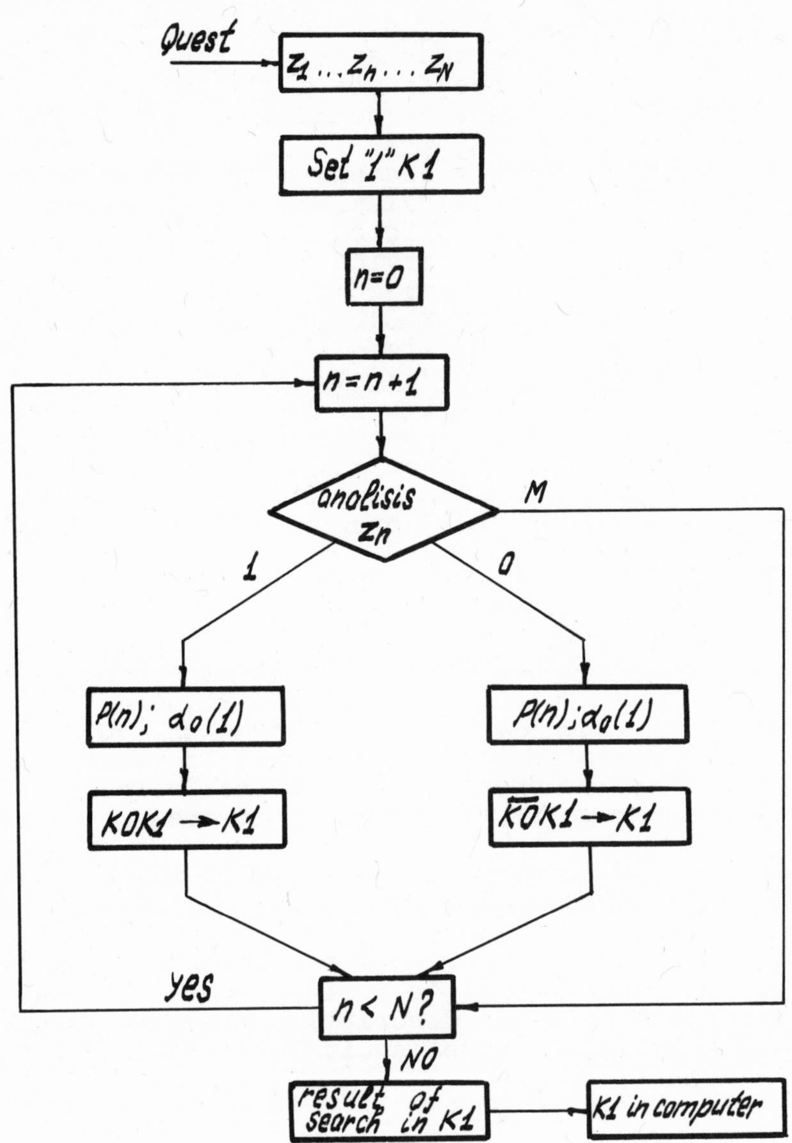

Figure 6. Block-diagram of the subprogram "SEARCH-1".

equally on the corresponding digits of molecular weights $(K_{h1}, \ldots, K_{hn}, \ldots, K_{hN})$ at the given values of quest digits. They are expressed through the logical product of their masked off digits, inversed or not. In this connection a simple retrieval problem can be solved not more than for "N" access commands to the deflector and PPP operations of the $K0 \cdot K1 \rightarrow K1$ and $\overline{K0} \cdot K1 \rightarrow K1$ type. The block-diagram of the simple retrieval algorithm is given in Fig. 6.

2. The problem of retrieval of the compounds which molecular weights are within the limits $(Z_1 \ldots Z_n \ldots Z_N)^{\text{down}} \div (Z_1 \ldots Z_n \ldots Z_N)^{\text{up}}$ is solved by calculation of:

a) functions $(y_1, \ldots, y_h, \ldots, y_H)^{\text{up}}$ corresponding to addresses of compounds, which molecular weights comply with the inequality $(K_{h1} \ldots K_{hn} \ldots K_{hN}) < (Z_1 \ldots Z_n \ldots Z_N)^{\text{up}}$;

b) functions $(y_1, \ldots, y_h, \ldots, y_H)^{\text{down}}$, corresponding to addresses of compounds, which molecular weights comply with the inequality $(K_{h1} \ldots K_{hn} \ldots K_{hN}) > (Z_1 \ldots Z_n \ldots Z_N)^{\text{down}}$;

c) logical product of these functions, whereby we have the array indicating the addresses of the regired compounds.

Algorithms of greater than or less than search are similar and based on the comparison of their numbers according to the relationship of their senior digits. The block-diagram of the algorithm of function calculation $(y_1, \ldots y_h, \ldots y_H)^{\text{up}}$ is given in Fig. 7. The method of retrieval algorithm realization follows from the iterative function representation. A starting form of functions is determined according to the rule: a quest character is larger than that of an array description, if its senior digit is larger than the corresponding description character digit $(Z_1^{up}\overline{K}_{h1})$, or they are equal $(Z_1^{up}K_{h1} + \overline{Z}_1^{up}\overline{K}_{h1})$ but the next according to the weight digit is larger, and etc. As the result of successive application of this rule to all the digits of character pairs being compared we shall have a system of functions of the kind:

$$y_h^{up} = Z_1^{up}\overline{K}_{h1} + (Z_1^{up}K_{h1} + \overline{Z}_1^{up}\overline{K}_{h1}) + \ldots + \quad (2)$$

$$+ (Z_1^{up}K_{h1} + \overline{Z}_1^{up}\overline{K}_{h1}) \ldots (\ldots) \cdot Z_n^{up}\overline{K}_{hn}$$

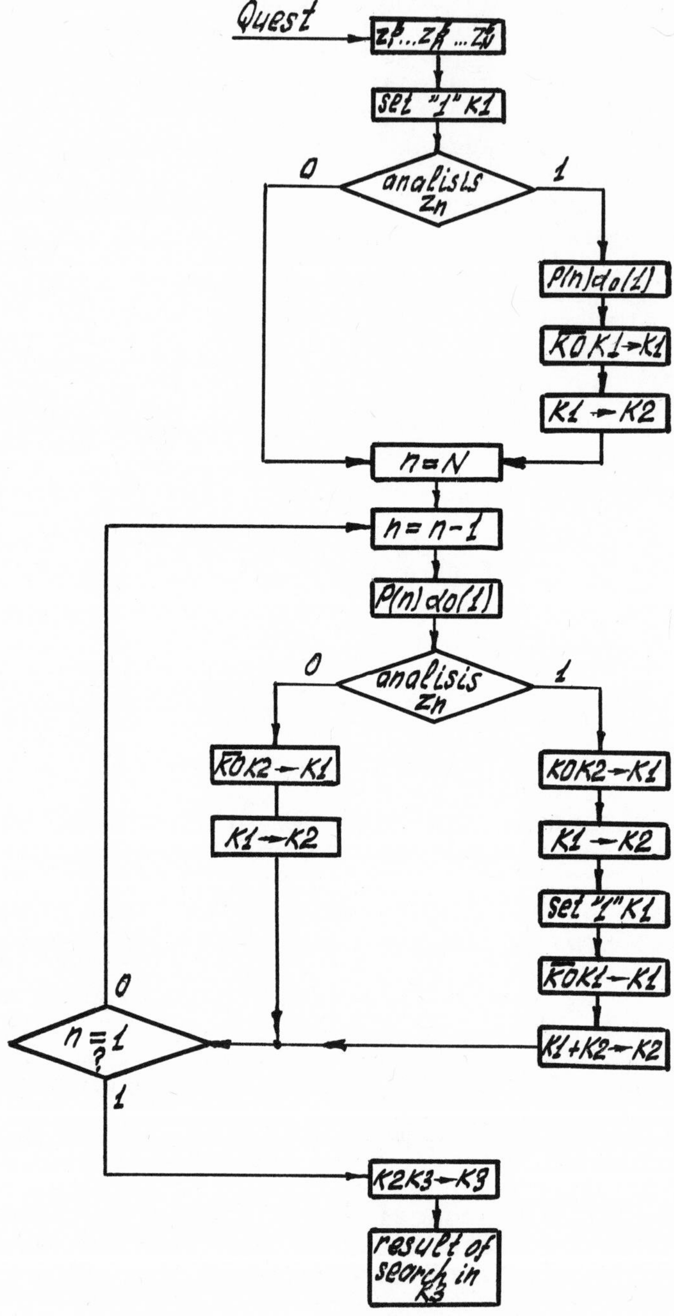

Figure 7. Block diagram of the subprogram "SEARCH-2".

In the iterative form each function y_h^{up} is calculated successfully for $(N-1)$ steps:

$$y_h^{up}(N-1) = Z_{N-1}^{up} K_{hN-1} + (Z_{N-1}^{up} K_{hN-1} + \bar{Z}_{N-1}^{up} \bar{K}_{hN-1})\, y_h^{up}(N)$$

$$y_h^{up}(n) = Z_n^{up} \bar{K}_{hn} + (Z_n^{up} \cdot K_{hn} + \bar{Z}_n^{up} \cdot \bar{K}_{hn})\, y_h^{up}(n+1) \qquad (3)$$

$$\vdots$$

$$y_h^{up}(1) = Z_1^{up} \bar{K}_{h1} + (Z_1^{up} K_{h1} + \bar{Z}_1^{up} \bar{K}_{h1})\, y_h^{up}(2)$$

where $y_h^N(N) = Z_N^{up} \bar{K}_{hN}$; $y_h(1) = y_h^{up}$ At the "nth" step at $Z_n^{up} = 1$ the functions $y_h^{up}(n) = \bar{K}_{hn} + K_{hn} \cdot y_{hn}^{up}(n-1)$ are calculated in the form of logical arrey $K2$, and at $Z_h^{up} = 0$ - the functions $y_h^{up}(n) = \bar{K}_{hn} \cdot y_h^{up}(n-1)$ where $h = \overline{1, H}$ in PPP.

The iterative form of representation of required functions (3) is more convenient in comparison with the initial form (2), because it is possible to have access to each character hologram only once during less than search $(K_{h1} \ldots K_{hn} \ldots K_{hN}) < (Z_1 \ldots Z_n \ldots Z_N)^{up}$ with the result that the retrieval time is reduced.

In a like manner the system of functions $(y_1, \ldots, y_h, \ldots, y_H)^{down}$ is calculated. The result of the retrieval is formed at the third PPP memory level in the form of the logical array $K3$.

The described retrieval of compounds with molecular weights being within the given ranges requires 2 x N access operations to the deflector, and, on average, (proceeding from the equal number of 0 and 1 in the quest), 7 x N PPP logical operations.

3. The ordered access algorithm (Fig. 8) is based on a search for compound addresses with minimal molecular weights among the molecular weights which are greater than the quest word in their absolute value. In this connection the search for addresses and values of molecular weights of compounds is devided into two parts during the ordered access, that is the search for addresses of compounds which molecular weights are greater than a given quest word, and among them the search for addresses of compounds with the minimal molecular weight. Access of the address of the first organic compound and finding out of its molecular weight value begins from the second part of this procedure Later on the found molecular weight value becomes a quest word, and a search for the address of the next organic compound and its molecular weight is carried out through the completement of both parts of the ordered access procedure.

The algorithm of greater than search is given above. The search for a minimal value among the given array of molecular weights is performed by a successive analysis of their similar digits (code U into computer), and elimination of the compounds, which molecular weights have unit senior digits. Equal digits are not taken into consideration in binary representations of molecular weights. To find one organic compound address and its molecular weight it is required not more than 2 x N deflector access operations, 5 x N PPP logical operations, and N operations of the analisis of molecular weight similar digits in a developed algorithm of the ordered access.

A quest for retrieval of an organic compound with molecular weight of 580.5 and answer of the system in the form of the required compound address on the display screen are given, as an example, in Fig. 9. In the experimental system the retrieval time was primarily defined by the deflector high-speed response (switching time is 100 sec), and by that of the dissector (image input time into the computer is 20 msec). It took about 0.3 sec to solve problems of a simple retrieval, and about - 0.5 + 1 sec to solve problems of a complex retrieval.

An important problem rises when using the proposed version of a character memory, that is the problem of its efficiency in comparison with other possible ana-

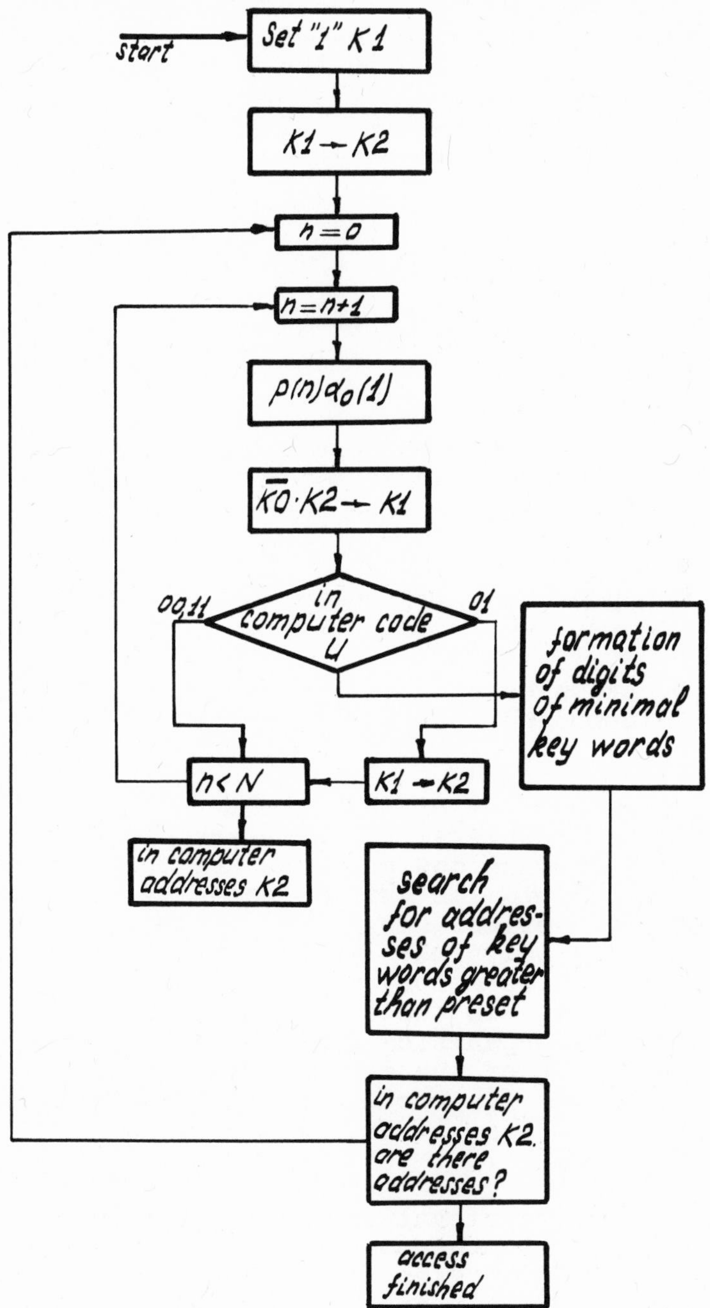

Figure 8. Block-diagram of the suprogram "SEARCH-3".

logues of such memory. The memory module capacity of a holographic random access memory can account for 10^8 bits, and it corresponds to the modern magnetic disk capacity, being used as external memory. That's it is true to compare the information retrieval time in a holographic memory to that in a system of the "magnetic disk-computer" type.

ЭКСПЕРИМЕНТАЛЬНАЯ ИНФОРМАЦИОННО-ПОИСКОВАЯ СИСТЕМА
ОРГАНИЧЕСКИХ СОЕДИНЕНИЙ

ЗАПРОС № IO
МОЛЕКУЛЯРНЫЙ ВЕС: 580,5.

ОТВЕТ НА ЗАПРОС № IO
АДРЕСА СОЕДИНЕНИЙ: 39.

Figure 9. Examples of quest for retrieval of organic compound and answer of the system.

The evaluations of the time necessary to solve typical retrieval problems are given in Table 2, where the following designations are used: R_{ch} is the channel digit; V_{ch} is channel capacity of a computer with external devices (Kbyte/sec); V_d is the magnetic disk readout time; T_a is the access time to the computer operative memory; T_s is time of summation operations; T_{lc} is the time of fulfilment of logical commands, when organizing retrieval programs; T_d is the switch time of the deflector; T_{ph} is the photomatrix time; σ is the number of the addresses being retrieved; and T_{an} is the time of analysing of the logical array K_l .

The whole retrieval array (N x H) was supposed to be transfered into the computer of the system "magnetic disk-computer" (as an example a computer EC 1030 comprising a magnetic disk EC 5056 was considered).The transfer is carried out by bytes (R_{ch} = 8) and is limited by the data readout speed from the disc

Table 2

Type of system	Simple search	Between limits search	Ordered access
Magnetic disk--computer	$\frac{NH}{R_{ch}} \cdot \frac{1}{v_d} + H(2T_a + T_s + T_{lc})$;	$\frac{NH}{R_{ch}} \frac{1}{v_d} + 2H(2T_a + T_s + T_{lc})$;	$\frac{NH}{R_{ch}} \cdot \frac{1}{v_d} + \sqrt{H}(T_0 + 0{,}5T_s + T_{lc})$
	1.4 s	2.5 s	0.3 s
"Holographic character memory--computer	$N(T_d + T_{ph} + T_{lo}) + 2\delta \frac{1}{v_{ch}}$;	$N(2T_d + T_{ph} + 7T_{lo}) + 2\sigma \frac{1}{v_k}$;	$N(2T_d + 2T_{ph} + 5T_{lo} + T_{an}) + (N+1) \frac{1}{v_{ch}}$
	$0.6 \cdot 10^{-4}$ s	$0.1 \cdot 10^{-4}$ s	$0.1 \cdot 10^{-3}$ s

(v_d = 156 Kbyte/sec). Only the addresses being retrieved (let σ = 0.01H), which are preset by the word, which is not longer than 2 bytes, are transfered into the computer through the channel (v_{ch} = 900 Kbyte/sec) in the system "holographic storage system--computer". Besides readout of the answer the code is transfered N times as a preliminary per each sample, to solve the problem of ordered access. The computing time (T_a = 1.25 sec, T_s = 10 sec, T_{lc} =100 sec) is proportional to the number of data arrays being processed H when searching for addresses in the system "magnetic" disk-computer". The ordered access algorithm is based on the Shell method [4], according to which the time being spent to get a single answer is proportional to $\sqrt{H}$. In holographic character memory the capacity of calculation operations is proportional to the number of character digits N , since in this case the input of quest digits is accomplished in succession with the deflector. Retrieval time depends on the deflector fast operation, photomatrix, and the time of making logical operations in PPP.

A correlation between the "holographic character memory-computer" system and that of "magnetic disk--computer" at the given values of parameters for the retrieval in an array of 10^4 32 digital characters of data arrays shows, that the time spent to solve typical problems of retrieval is 2-3 orders less in the first case than in the second one.

The efficiency of information retrieval organization in holographic memory systems based on page data processing is thereby confirmed.

Further effective retrieval time reduction in holographic storage systems can be obtained at the expense of parallel-successive or parallel input of quest digits, which is connected with the use of lines or matrices of light modulators.

In appreciation for their contributions to this work, the authors thank Yu.V. Vovk and Yu.A. Popov.

REFERENCES

1. I.S. G i b i n, M.A. G o f m a n, E.F. P e n and P.E. T v e r d o k h l e b. Assotiativnaya viborka informatsii v gologrammnikh zapominayutshikh ustroistvakh. - Autometria, N 5, 1973.

2. P.E. T v e r d o k h l e b. Opticheskie sistemi s viborkoi po soderzhaniyu. - Avtometria, N 6, 1976.

3. G.I. G r o m i l i n, G.E. K a s p e r o v i t c h, S.F. K i b i r e v, G.S. P r o k o p e n k o, A.N. C h e r n i s h o v. Dvukhkoordinatnii shleifovii deflector s kvazioptimalnim po vremeni upravleniem.- Avtometria, N 6, 1976.

4. V.M. G l u s h k o v, V.P. G l a d u n, L.S. K o s i n s k i i, S.B. P o g r e b i n s k i i. Obrabotka informatzionnikh massivov v avtomatizirovannikh sistemakh upravlenia. Izd-vo "Naukova dumka", Kiev, 1970.

OPTICAL INTERPOLATIVE MEMORY AND OPTICAL PROCESSOR ARCHITECTURE

I.V. Prangishvili, A.K. Glotov, A.E. Krasnov
and V.K. Bykhovsky
Institute of Control Sciences
Academy of Sciences
Moscow 117279, USSR

There are several known approaches to an optical holographic memory including memory with superposed holograms (see, for example,[1]). The basic aim of research on holographic memory is usually the development of a reliable, large-capacity, high-speed memory directly interfaced with a large computer. We believe that the problem of interfacing a digital computer which operates serially with a parallel unit such as an optical memory is very complex, requiring the development of a buffer memory system with sophisticated control circuitry.

In papers[2,3] a method of combining electronic and optical systems was proposed which provides an information flow balance; this method presents further development of the LATRIX[4]. In papers[2,3] an architecture was suggested based on a silicon array microprocessor chip. As a result, the interface between a parallel optical unit (e.g., optical memory) and an electronic array microprocessor was developed which doesn't require any intermediate devices and provides parallel processing (control and processing microprocessor circuits are also arrays).

In this paper, we describe a parallel optical processor based on the processing capabilities inherent to a set of superposed holograms. A holographic memory with superposed holograms may be used, for system/process control whose system process states are coded as images. When control processor output is also an image, one may use it for parallel control of many active devices.

In section 2, a simple model of read out of a superposed set of holograms is considered. In section 3 a control method using holographic 'table' is discussed and in section 4 experimental results are presented which support the holographic 'table' control concept. In section 5 a discussion of results is presented.

2. PROCESSING PROPERTIES OF A SET OF SUPERPOSED HOLOGRAMS

Two concepts of holographic memory design are known. The so-called 'page' memory provides sequential access based on a deflecting addressing device. A memory using a superposed set of holograms provides parallel access and can be content addressable. This access method may be implemented by means of 'orthogonal' (in features space) codes. Such 'non-overlapping' linear or matrix codes allow selection among superposed pages in the read out process.

Let's consider a simple model of an interaction between an input signal (wavefront) and a superposed hologram set (SHS). This model explains the processing properties of SHS (more details of this read out process are given in ref. 5). Each hologram (diffraction lattice) as a filtering element may be described by means of 'projection' operator (matrix) $P_i \equiv |i\rangle\langle i|$, i=1, 2,..,N, which 'projects' the input signal $|s\rangle$ in subspace P_i (defined with i-th diffraction lattice parameters). A SHS defines a 'generating' vector space $R_N:\{P_i\}$, i=1,2,..,N (N is the number of holograms in SHS) and the corresponding expansion of the unit operator:

$$I = \sum_{i=1}^{N} P_i \equiv P_N, \quad P_i^2 = P_i, \quad P_i \equiv |i\rangle\langle i| \tag{1}$$

$$|s\rangle = P_N|s\rangle = \sum_{i=1}^{N} |i\rangle\langle i|s\rangle = \sum_{i=1}^{N} c_{is}|i\rangle \tag{2}$$

Here $c_{is} = \langle i|s\rangle$ is the overlapping (correlation) integral between signals (wavefronts) $|i\rangle$ and $|s\rangle$ ($|s\rangle$ is the arbitrary signal from a space R_N). Relation (2) describes a SHS read out process in which the signal $|s\rangle$ is the content address of the information searched. Thus, if the input signal $|s\rangle$ correlates only with the lattice i=k but not correlates with all other lattices i≠k (in other words, $\langle i \mid s\rangle \equiv 0$ if i≠k) then all projections $c_{is}|i\rangle$ (i≠k) are nullified: $\sum_i c_{is}|i\rangle = |k\rangle$. Thus, if input signals (codes) are reference ones ($|s\rangle = |i\rangle$, i=1,2,..,N) then SHS 'projects out' reference outputs $|i\rangle$; however, if an input signal $|s\rangle$ is an arbitrary one (from space P_N) then SHS forms a 'new' signal $\Sigma'_i c_{is} |i\rangle$ which is a linear combination of reference signals $|i\rangle$. Prime symbol means coherent summation of wave fronts $c_{is}|i\rangle$.

In such a way, with non-reference inputs a SHS memory operates as a generator of 'new' signals (from P_N space). From the mathematical point of view a SHS carries out an interpolation of the input signal $|s\rangle$ over the set of reference lattices P_i. In other words, the SHS relates the input $|s\rangle$ and the set $\{P_i\}$ with output $\Sigma'_i c_{is} |i\rangle$. This output equals 'reference values' $|i\rangle$ at 'interpolation nodes' and gives 'interpolated values' at intermediate points.

Besides this one-step option, the SHS may operate as multiple-step device, generating a temporal signal chain[2,5]. For an implementation of this option the SHS is built into an optical resonator; the resonator provides a feed-back of output signal $P_N|s\rangle$ at the SHS input:

$$P_N P_N |s\rangle = \Sigma_i \Sigma_{i'} |i\rangle\langle i|i\rangle\langle i|s\rangle \tag{3}$$

If the reference lattices are othogonal ($\langle i' | i\rangle = \delta_{ii'}$) then $P_i^2 = P_i$ and double (and multiple) passes through the SHS don't change the input signal. However, if $\langle i'|i\rangle \neq \delta_{ii'}$, then each pass changes the signal and the SHS generates a temporal signal chain (mulitple 'echo'):

$$P_n |s\rangle, \ldots, P_2|s\rangle, P|s\rangle,$$

$$P_q |s\rangle = \underbrace{\Sigma_a \Sigma_b \ldots \Sigma_z}_{q} |a\rangle\langle a| b\rangle \ldots \langle z| s\rangle, \tag{4}$$

$$a, b, \ldots, z = 1, \ldots, N$$

One may consider this mechanism as an extension of the so-called Reninger effect in the theory of thick superposed holograms.[6]

3. METHODS OF CONTROL SYNTHESIS WITH THE SHS MEMORY

A further development of an optical interpolating memory is dictated by the requirements of the process control implementation. As it is well known[5], any process control operation includes two basic stages[10]:

i) measurement (or identification) of the current process control states;
ii) synthesis of control signals (this operation gives controls which direct the process controlled in reference state s_o).

Control systems have been designed[7,11] using the correlational unit for a measurement of the difference $d(\phi)$ between a current process state $s(r,\phi)$ and reference state $s(r,0)$, where

$$d(\phi) = \int 2\Pi r dr . s(r,\phi) . s(r,0) \tag{5}$$

(we assume that the process state may be coded as an image). In conventional correlational difference measurement devices only max $[d(\phi)]$ is used; $d(\phi)$ containing image difference details is not used because process control is directed by a scalar quality criterion.

A developed 'table' optical memory with an interpolation capability may be used as a base for a more general approach to both measurement and synthesis processes. In this case:

i) in measurement (identification) a set of reference signals s_i^o, $i=1...N$ (and not a single s^o) is used and for synthesis process the entire correlation plane $d(\phi)$ is used;
ii) instead of a control program $P(ds)$, developed in advance, connecting the state difference ds with a corresponding control $P(ds)$, the 'table' control algorithm is used; this algorithm is based on the accumulation in the SHS memory of 'control examples' as associative holograms $ds_k-P(ds_k)$, $k=1,...,N$; at the state of control synthesis the SHS memory will operate in the interpolation mode, producing a 'combined' programs $P(ds)$.

The synthesis of the program $P(ds)$ with the optical table data can be carried out by different methods;

a) one can use the algorithm of finding the 'most near' program:

$$P(ds)=\arg(\min_k L(ds,ds_k)),\ ds\neq ds_k, \tag{6}$$

where $L(ds,ds_k)$ is the 'distance' (in differences space) between ds and ds_k, or one may use the generalized interpolation algorithm[8]:

$$P(ds)=\mathrm{int}(ds,T),\ T\equiv \{ds_k-P_k\},\ ds\neq ds_k, \tag{7}$$

where the 'int' function depends upon the hologram writing scheme and writing medium parameters;

b) one can also use the 'hardware' dynamic correction of the content controlled SHS memory by a current difference ds (different optical circuits for search or optimization are possible, see[9]).

The synthesis of control images of the method of ref. (7) can be carried out with the optical interpolating SHS memory. This can be implemented in hardware using an optical resonator with a built-in SHS memory and boundary conditions controlled by bounded control system[5]. For hologram writing, a rewritable medium must be used; this provides the possibility of dynamic correction of the SHS memory and, therefore, allows simulation of the vector space R(t) with a dimension and metric properties depending on quality criteria. Contrary to the usual mathematical synthesis algorithms, hardware (optical) synthesis compares trial and reference process states in parallel (without sequential search)[5,8]. The synthesis time is defined by the time which is necessary for completion of the transition process in feedback optical circuits; usually this transition time is about $10^{-3} \div 10^{-1}$ s.

4. THE EXPERIMENTAL TABLE HOLOGRAPHIC CONTROLLER SET.

In order to check the holographic table controller idea, an experimental controller was made (fig. 1). The device includes the block A for simulation of system plane rotation, the block B for correlational (holographic) measurement of the difference between the current orientation $s(r,\phi)$ and the reference orientation $s(r,0)$, and the block C of program synthesis. By using an optical circuit (fig.1) a current object state (an image rotated at angle ϕ) is transfered to the holographic measurement unit, the SHS with 12 holograms each of which corresponds to a reference position of the object. Holograms are written with reference waves coded with reference object images (rotated images). As signals point beams with different positions were used. Cw He-Ne laser and plates LO1-2 were used. The diffraction efficiency of the SHS with 12 holograms written with 5° steps in -30 to 30° interval was measured as about 2% (for each reference wave).

At the output of the measurement block the correlation matrix (image) is restored which correlates a current input image (rotated at angle ϕ) with a reference image. This is the result of the measurement $ds(\phi)$. The image in the correlation plane is the set of 12 light spots with different intensities. This image is registered with a light-sensitive array of the control synthesis block. In this block an optical interpolation with linear weighting for the different positions is carried out. The system control ouput is the control signal for a rotation system.

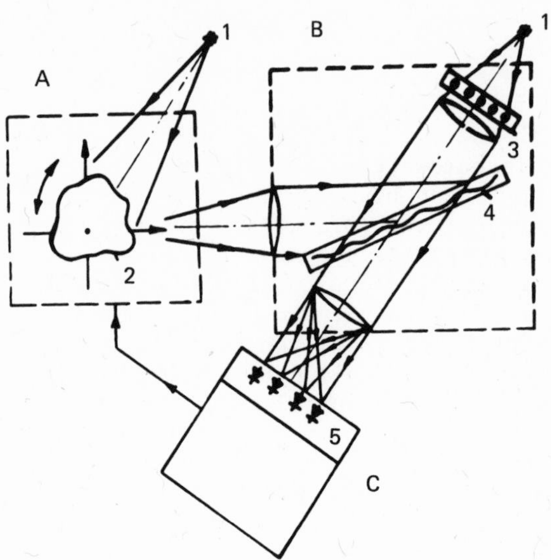

Fig. 1 Block diagram of holographic process controller (see text).

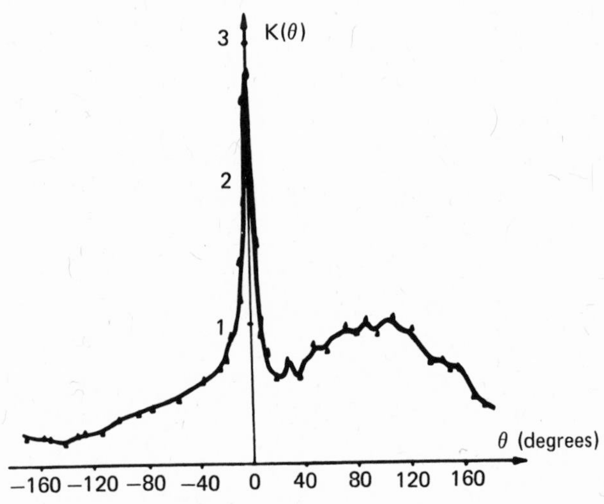

Fig. 2. Angular dependence of auto-correlation function of an image used.

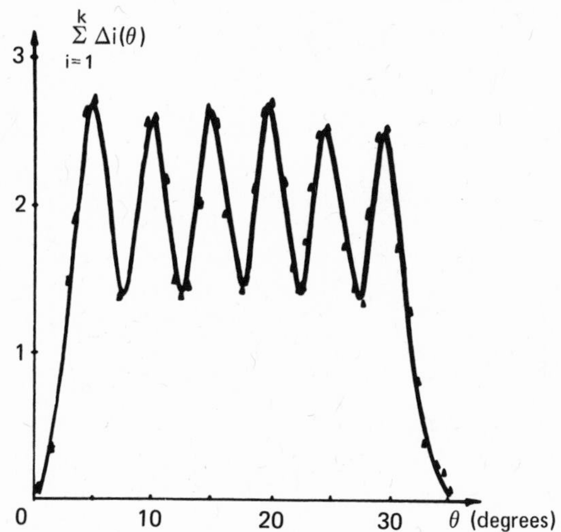

Fig. 3. Plane inegrated output correlation plane signal presented as a function of an image rotation angle (in interval 0÷30°).

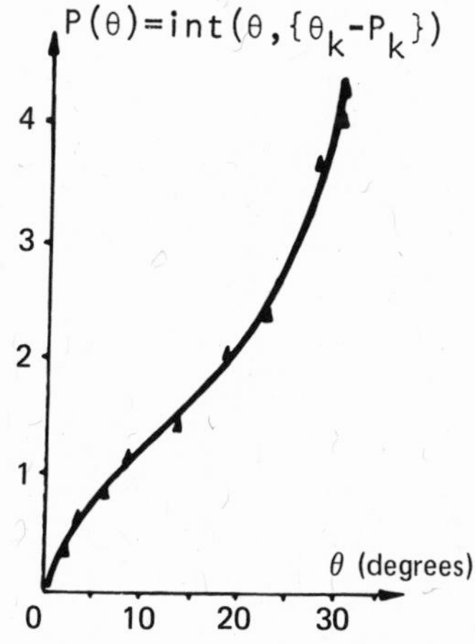

Fig. 4. Angular dependence of output control signal obtained as a result of interpolation.

The correlation function of the image used is presented in fig. 2 and angular dependencies of the output control signal before and after interpolation are shown in fig. 3 and fig. 4 respectively. As is shown, although the number of reference images is not great, the control signal is a continuous and monotonic function of rotation angle ϕ.

5. DISCUSSION OF RESULTS

The table look up approach to control and information processing is very simple and, therefore, quite attractive. However, this classical approach has an essential disadvantage: for storage of reference pairs x_k-y_k (input-output pairs) for real systems a very large memory is necessary especially, as is often the case, when the signals x_k and y_k have an image format.

In papers[9,10] the table method was exended to continuous control of dynamic systems and to computing processes for parallel solution rather than step-by-step method. Such an extension implies the implementation of a table operation algebra with tables containing complex data structures such as trajectories, fields, images, etc. A common operation with a table is, for example, interpolation. An interpolation relates a table $T=\{x_k-y_k\}$ with a continuous set of inputs x and a continous set of outputs y: y=int(x,T). As a result, the above-mentioned disadvantage is overcome: one can store only a limited set of reference pairs and generate an unlimited set of outputs using an interpolation.

Such a new method requires new hardware. Implementation of these operations can be accomlished with wavefronts (not only electromagnetic ones) and media in which the different operations over wavefronts may be carried out. Thus, a hologram can perform the counterpart of the logical operation 'AND' over two two-dimensional operands, and the result is written in the memory with parallel access. Other important operations such as a table interpolation, may also be implemented with an optical SHS memory.

A difference-controlled dynamic correction of 'holographic tables' (SHS) may be implemented using the feedback (resonator) optical circuit with built-in rewritable memory and a controllable transparency (for setting of reference output signal).

For extensions of the processing capabilities of holographically-implemented tables, microprocessor array chips with parallel optical input-output[2,3,8] can also be used. In this communication are presented only the simplest possibilities of an implementation of the table control and processing algorithms. In future papers we will present further experimental results.

REFERENCES

1. Colier R., Berhardt C., Lin L. Optical Holography, Academic Press, N.Y.

2. Bykhovsky V. K., Mirsoyan G. A., Sonin M. S., Prangishvili I. V., Uskatch M. S. "Methods of optical information input in integrated microoptoelectornic structures." In: Problems of Holography, issue N2, MIREA Publishing, 1973, p. 165 (in Russ.).

3. Bykhovsky V. K., Prangishvili I.V., Sonin M.S. Uskatch M. A. "Optoelectronic parallel processors for problem analysis and system control." In: Autometry (Russ.) 1976, N6, p.00 (in Russ.).

4. Stewart A. L., Cosentino L. C. "Holographic readwrite optical memory." Appl. Opt., 1970, v.9, p. 2271.

5. Bykhovsky V. K. "Holography and Future Control Methods." In: Proc. of Vii All-Union Holography School (Rostov-Great, Jan. 1974). Leningrad Nucl. Phys. Institute Publ., 1975, p. 134 (in Russ.).

6. Aristov V. V., Schenchtman V. S., "Theory of volume holograms." Uspeckhi fiz. nauk. 1971, v. 104, N1, p. 51 (in Russ.).

7. Krasovsky A. A., Pospelov G. S., "Foundations of automatics and technical cybernetics." Moscow, Gosenergoizdat, 1962 (in Russ.).

8. Bykhovsky V. K. "Control and Information Processing in Asynchronous Processor Networks." In: Proc. of Soviet-Finnish Symposium on Minicomputers and Distributed Data Processing (Helsinki, Nov. 1974), vol. I, 1974.

9. Bykhovsky V. K., "Methods of problem analysis and system control with a talbe interpolating processor." In. Array Computing System and Structures. Part I, Naukova Dumka, Kiev 1975 p. 234 (in Russ.).

10. Bykhovsky V. K., "Design Principles for Language and Processor for Process Control." In: Dynamic Modelling and Computer Process Control. v.2, USSR Acad. of Sci. Computing Center Publ., 1974, p. 441 (in Russ. and Engl.).

11. Rabinovich V. J. "Application of Holography for Process Automation." Proc. of ALL-UNION Conf. on Control Sciences, Institute of Control Sci. Publ., 1975, v. 2.

OPTICAL MEMORY SYSTEM WITH CONTENT ACCESS

P.E. Tverdokhleb

Institute of Automation and Electrometry
Novosibirsk, USSR

ABSTRACT

Functional model of the optical memory system is decribed in which the information output step by request is preceeded by that of its adresses retrieval. The model was developed to illustrate a general approach to the content access organization and to clarify a type of the transformation made in process of such an access. Possibilities have been discussed for realization of these transformation by opto-electronic methods. When developing this model concepts of the theory of decisions have been used.

I. INTRODUCTION

In the past decade investigations have been carried out in the field of optical (including holographic) storage devices SD with small up to 10^8 - 10^9 bit and bulk up to 10^{12} - 10^{13} bit storage capacities, considered primarily as internal and external computer memories[1]. A possibility has been supported for the development of such SD's, and their application has been restricted primarily to the external archival fixed computer memory. The specification of the optimistic predictions of the end of the 60's (USA:Bell Laboratories [2-4], IBM[5]and

RCA [6,7]; USSR [8-10]) has occurred in connection with significant achievements in magnetic and semi-conductor memory and difficulties encountered when developing the reversive light-sensitive materials.

We believe that further development of optical methods of storing information and widening the scope of their application is closely connected with the development of optical (or, most probably, opto-electronic) multichannel memory systems where access is performed not according to the address likewise in most SD known, but according to the "content" request. The content of such a request can be presented by parts of mass data stored in the structural memory units, by a complex of characters characterizing these mass data and their functional relations. Such memory systems were referred to as SD with content access or as associative SD [11 - 14].

The process of content access is known to include a step of information search complying with an input request, and a step of this information display to the user. The former is realized by the comparison of the request content with that of the memory; the latter - by the addressed memory access. As a result the associative optical SD (AOSD) prove to be an adequate instrument to solve common problems of retrieval, recognition, diagnostics, classification, etc. Such problems most commonly arise in processing mass scientific-technical, economical and other forms of information.

Most of the known associative SD's [11,12,14] involve two memory levels: character and basic. The former involves storing of the information characters and search of this information addresses according to a request ;the latter involves storing and address access. Such division of the memory is also reasonable in the case of AOSD, since significantly different requirements are placed upon the character and basic memory levels in respect of fast access and capacity. The character SD is an optical SD of capacity up to 10^8-10^9 bit with fast access (1-10 mksec) and data processing; the basic SD is an optical archival storage of capacity up to 10^{12} - 10^{13} bit with the access time 10-100 sec. Thus, when developing AOSD, the same scientific-technical experience can be used available at present in the field of optical memory, including the possibilities of multichannel retrieval and access, page-oriented processing, storage and display of mixed (numeric, alphanumeric, grapical, etc.) information.

It might seem that the application of character SD in optical memory systems results in an increase of the data access time and in an inadmissible rise of their cost. However, this is far from how matters stand. In SD utilizing page-oriented processing the time required for retrieval can be negligibly small as compared with the average access time to the optical arhival storage functioning as the basic memory. Such processing, as is shown below, on the analog and digital levels may be performed by opto-electronic techniques. As for the AOSD cost, it might be expected that additional expenditures required for their production will be obtained (due to economy) in the field of mass application of such devices (information-inquiry services, libraries, industrial and scientific archives, foreign-language translation services, etc.).

If the beginning of the development of addressed optical SD with paged data recording is considered to coincide with the time of the first publications [2,3], then the first concepts on the advent of a solid-body associative optical SD were suggested somewhat earlier [15,16]. More recent studies [17-20] extended the assumptions on possible ways for such SD structure, but the concepts given could hardly be applicable to the structure of addressed holographic SD being developed at the time. This disadvantage has been taken into consideration in [21-24], where methods of the associative access organization in the available holographic SD structures have been suggested. Attempts are known to use the optical associative methods for data access in solving problems of the IR-spectroscopy [25-28], alpha-numeric information retrieval [29], multiparameter state diagnostics [30], pattern recognition [31], etc.

From this it follows that on the problem under discussion a rather impressive methodical and applied scientific-technical experience exists gained when carrying out several specific investigations. An idea suggests itself on the necessity to generalize the available results.

Below a functional model of the optical memory system is described in which the information output step by the request is preceeded by that of its address retrieval. The model was developed to illustrate a general approach to the content access organization and to clarify a type of the transformations made in the process of such an access. Possibilities have been discussed for the realization of these transformations by optoelectronic methods. When developing this model concepts of the theory of decisions have been used [32].

II. A MODEL OF THE MEMORY SYSTEM

The scheme of the system is shown in Fig. 1. It comprises three functionally-different parts: optical memory, a data retrieval device and a processor. A request is introduced into the system in the input processor language (a high-level lanquage). In specific cases the request can be presented by an image pattern and be introduced directly to the data retrieval device. The system responces are: information on the presence or absence of these data in the storage; the addresses of memory cells where they are stored; the content of these cells. A structural unit of information stored in a memory cell we consider a multibit page (or file) which constituent is a description - "abstract" of this page (file).

The optical memory is used to store data array B, where $B = \|b_{ij}\|$ is a matrix containing I vectors B_i, $i = 1, 2, \ldots, I$, with components b_{ij}, $j = 1, 2, \ldots, J$. In other words, the matrix B describes I patterns of pages, each containing J samples. The specific feature of the memory used (as compared with other known types) consists in that the samples b_{ij} take more than two values.

Consider the first of J_1 $(J_1 < J)$ samples of each page to form their descriptions, and the rest $J - J_1$ samples to form information words to be sampled from the memory. Then a matrix of characters $B' = \|b'_{ij}\|_{i = 1, 2, \ldots, I}$ can be composed of the page descriptions, $j = 1, 2, \ldots, J_1$, and a matrix $B'' = \|b''_{ij}\|$ can be formed of information words, $i = 1, 2, \ldots, I$, $j = J_1 + 1, J_1 + 2, \ldots, J$, which will be referred to as an information one.

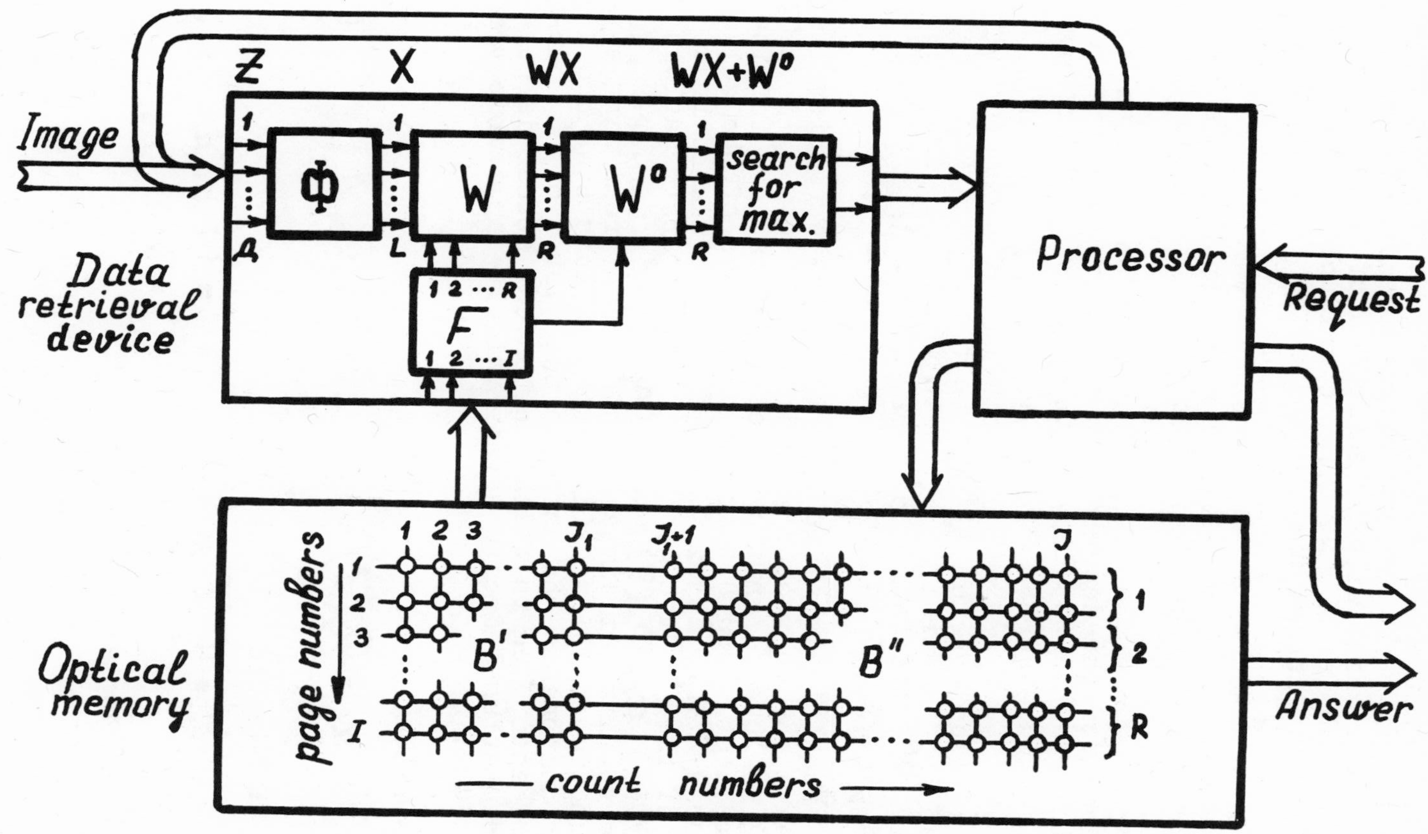

Fig. 1. Functional model of AOSD.

Admit that among J_1 descriptions forming rows of the matrix B' there are coincident or "close" in content ones which within the region of their samples are localized in R $(R \leq J_1)$ overlapping areas, thus allowing division of their associative information words into R classes. It is evident that the number of these classes will determine that variety of answers which can be received on the output of the system memory under consideration. Each of the classes will be characterized by an $L+1$ set of $W_{r\ell}$ and W_r^0 parameters, $\ell = 1, 2, \ldots, L$, which can be obtained by processing of the descriptions of their associated words. In this case the totality of vectors W_r with components $W_{r\ell}$ forms a weight matrix $W = \| W_{r\ell} \|$, $r = 1, 2, \ldots, R$, $\ell = 1, 2, \ldots, L$, and the totality of parameters W_r^0 forms a column-vector W^0. The transformation (non-linear in a general case), with which help from the matrix B' the matrix W and vector W^0 are obtained, in Fig. 1 is shown by the symbol F.

A request signal is sent to the data retrieval device as a column-vector Z with components z_d, $d = 1, 2, \ldots, Д$ characterizing a set of characters forming a word-request. Assume that the components of the column-vector X $(x_\ell,\ \ell = 1, 2, \ldots, L)$ are calculated from those of the Z -vector by realizing transformations of the type $x_\ell = \Phi_\ell(Z)$, where $\Phi_\ell(Z)$, $\ell = 1, 2, \ldots, L$, is the totality of linearly-independent real single-value functions. It is evident that in this case the X vector components characterize the system of L characters according to which the content access is performed. Let us indicate several widely-applicable transformations:

- Linear

$$x_\ell = \Phi_\ell(Z) = z_d\ , \quad \ell = 1, 2, \ldots, Д\ ; \tag{1}$$

- Quadratic

$$x_\ell = \Phi_\ell(Z) = z_\gamma^\alpha z_\varepsilon^\beta\ , \quad \gamma, \varepsilon = 1, 2, \ldots, Д, \quad \alpha, \beta = 0 \text{ or } 1\ ; \tag{2}$$

- Polynomial of the type

$$x_\ell = \Phi_\ell(Z) = z_{\gamma_1}^{\alpha_1} \dots z_{\gamma_\pi}^{\alpha_\pi}, \quad \gamma_1, \dots, \gamma_\pi = 1, 2, \dots, Д, \quad \alpha_1, \dots, \alpha_\pi = 0 \text{ or } 1, \tag{3}$$

where π is the polynomial order;

- Polynomial of the type

$$x_\ell = \Phi_\ell(Z), \tag{4}$$

where $\Phi_\ell(Z)$ is the constituent of a random orthogonal system of the functions $\{\Phi_\ell(Z)\}$, $\ell = 1, 2, \dots, L$.

Transformations of the type (1) - (4) are most frequently performed for the purpose of a change to independent or information characters, data reduction, provision of invariance for uncontrolled perturbations, etc.

The access process consists in calculation of the column-vector components $\Psi_r(Z)$, $r = 1, 2, \dots, R$, for the matrix equation

$$\Psi(Z) = WX + W^0, \tag{5}$$

allowing for weighing of the X character vector by a weight matrix W and algebraic adding of the obtained values with the vector W^0. In other words, the vector components $\Psi(Z)$ are determined from the expressions of the form

$$\Psi_r(Z) = W_{r1}\Phi_1(Z) + W_{r2}\Phi_2(Z) + \dots + W_{rL}\Phi_L(Z) + W_r^0, \tag{6}$$

being linear in respect of the weights.

Then the access algorithm allows for performing a search for the maximum values operation which procedure results from the following geometric considerations. Assume that the weight vectors W_r, $r = 1, 2, \dots, R$, and the character vector X in L-dimensional space determined by their components correspond to certain points. Then the conclusion on the "proximity" of the X and W_r vectors can be drawn, e.g. on the basis of the root-mean-square distance minimum. According to [32] this criterion can be realized by calculating

R - components of the $\Psi(Z)$ vectors from (6) followed by a choice of the maximum value. In the general case the maximum values can be taken by several components, that characterizes ambiquous information sampling from the memory. Functions of Type (6) which arguments is the character vector Z are asually referred to as decision [32] or R-dimensional functionals of "closeness".

The result of the memory access is either the content of a page (class) r, where the decision function takes its maximum value, or its number.

From Fig. 1 it follows that in contrast to the addressed optical SD, the associative memory system comprises a data retrieval device, used to form and store the vectors W_r, $r = 1, 2, \ldots, R$, and W^0, to realize the transformations Φ and F, as well as to calculate functional (6) followed by the maximum values retrieval. For the sake of device simplicity, however, the weight-vectors W_r are formed preliminary and the transform of the type F is not performed.

The retrieval results are fed to a processor which functions are:

- Translation of a request formulated in the user language to an intermediate "system" vector Z search components assembler ;

- Software provision for complex multistep retrieval steps requiring analysis, storing and processing of the intermediate results;

- Change (if necessary) of the Φ and F transformation type in the process of data access;

- Addressed memory access (content output) and forming answers in the user language.

The memory system model admits generalization for a case of parallel access by H various key words. Thus, it gains the properties of a multichannel memory system providing an independent data array access for several users. At a multichannel mode of operation the symbols Z and X take a meaning of the matrices composed of H various request vectors Z_h and X_h, $h = 1, 2, \ldots, H$. The X_h vectors comprise components $x_{h\ell}=\Phi_\ell(Z_h)$, $\ell=1,2,\dots,L$, obtained as a result of carrying out the Φ -type transformation of each Z_h vector. In this case the functional can be represented by the matrix equation

$$\Psi(Z)=\|\Psi_{rh}\| = WX+W^0, \qquad (7)$$
$$r=1,2,\dots,R$$
$$h=1,2,\dots,H$$

where $W^0=\|W^0_{rh}\|$ is the matrix of $R\times H$ size with similar columns. From this it follows that in the process of multichannel access each X_h vector is weighed by a weighing matrix. Then the matrices of the calculated values and limiting cofficients are added, and from the columns of the resulting matrix where the column number corresponds to that of the channel requested, elements with the maximum values are selected.

Note that in certain cases the processes of Z -vector transformation and the obtained characters weighing by the matrix W can be combined. Thus, if necessary to express the Z vector in the orthogonal functions (4) space, decision function (6) can be transformed to the form

$$\Psi(Z)=\|\Psi_{rh}\| = W\Phi Z+W^0, \qquad (8)$$

where $L=\Delta$ and $\Phi=\|\Phi_{pq}\|$, $p,q=1,2,\dots,\Delta$, is the matrix which rows describe Δ -tabulated values of decomposition functions (4).

III. SPECIFIC CASES OF PRACTICAL IMPORTANCE

A theory of decisions points to several parametric methods for the selection of a decision function type and, hence, for the specification of the required type of the Z characters vector transformations. These methods can be used when it is known that the page descriptions forming the B' character matrix are the realizations of random processes, and that these processes can be characterized by a system of parameters being unknown. Then the available I realizations of the descriptions are used to evaluate these parameters and in this way to determine the optimal (in the average risk sense) decision function.

If nothing is known about the system of parameters characterizing random processes then the decision function type is chosen heuristically, and the weight matrices are determined by the methods of non-parametric training.

To illustrate the parametric method of choosing a decision function two specific examples will be given. Let it be known that the page description of each R - classes are distributed according to multidimensional normal laws, which co-variant matrices are the same. Then from [32] it follows that the optimal decision function will take the form of the function

$$\Psi_r(Z) = W_{r1} Z_1 + W_{r2} Z_2 + \dots + W_{r\Delta} Z_\Delta + W_r^0, \quad r = 1, 2, \dots, R \tag{9}$$

which is linear relative to the Z components vector. Here

$$W_{rd} = \langle b_{rd} \rangle \tag{10}$$

is the evaluation of the mathematical expectation of the component with the number d among the descriptions of the class r , and

$$W_r^0 = -\frac{1}{2\Delta} \sum_{d=1}^{\Delta} \langle b_{rd} \rangle^2 \tag{11}$$

is the half-value of the squared mathematical expectation of the class r .

In this case the Φ -transformation consists in the multiplication of a single matrix by the column-vector Z , and the F -transformation is a calculation of the mathematical expectation values by formulas (10) and (11).

Let now the classes of descriptions differ not only in their mathematical expectations but also in the covariant matrices G_r, $r = 1, 2, \ldots, R$, which are symmetrical and positively definite. Then the optimal decision function becomes quadratic in respect of the Z vector components and takes the form

$$\Psi_r(Z) = -\frac{1}{2}\left[(Z - \langle B_r \rangle)^T G_r^{-1} (Z - \langle B_r \rangle)\right] + W_r^0, \quad r = 1, 2, \ldots, R \tag{12}$$

where B_r is the vector evaluation of the class r mathematical expectation descriptions (vector components are determined by formula (10)), G^{-1} is the inverse covariant matrix of the class r descriptions, $W_r^0 = \log P_r - \frac{1}{2}\log|G_r|$ is the class r constant, expressed via the probability P_r of its occurence and the determinant of the covariant matrix, and T is the transportation sign [32].

Function (12) is transformed to the equivalent but more explicit form

$$\Psi_r(Z) = \sum_{d=1}^{\Delta} g_{dd}^{(r)} \tilde{z}_d^2 + \sum_{d=1}^{\Delta-1} \sum_{k=d+1}^{\Delta} g_{dk}^{(r)} \tilde{z}_d \tilde{z}_k + \sum_{d=1}^{\Delta} g_d^{(r)} \tilde{z}_d + W_r^0, \quad r = 1, 2, \ldots, R \tag{13}$$

where $\tilde{z}_d$, $d = 1, 2, \ldots, \Delta$, are the centered Z vector components, and $g_{dd}^{(r)}$, $g_{dk}^{(r)}$ and $g_d^{(r)}$ are the coefficients at the variables of the type $\tilde{z}_d^2$, $\tilde{z}_d \tilde{z}_k$, $\tilde{z}_d$, obtained by matrix multiplication (12). It can be seen that the row r of the weight matrix W in the case under consideration contains Δ weights at quadratic, $\Delta(\Delta-1)/2$ weights at cross and Δ weights at linear terms. The total number of weights in a row $L = 2\Delta + \Delta(\Delta-1)/2$. In addition from (13) if follows that the Φ-transformation is realized to obtain components of the type $\tilde{z}_d^2$, $\tilde{z}_d \tilde{z}_k$ and $\tilde{z}_d$, and the F - transformation is performed to obtain their corresponding weight values $g_{dd}^{(r)}$, $g_{dk}^{(r)}$ and $g_d^{(r)}$ including calculation of the weight W_r^0.

IV. LOGICAL FUNCTIONALS

Functionals of type (5) have logical modifications used in data retrieval by digital methods. Consider several types of such functionals. For this assume that the numerical and independent components of the X and W_r vectors can be presented in the form of N binary bit numbers

$$\begin{aligned} x_\ell &= \{x_{\ell 1}, x_{\ell 2}, \ldots, x_{\ell N}\}, \\ \ell &= 1, 2, \ldots, L \\ W_{r\ell} &= \{W_{r\ell 1}, W_{r\ell 2}, \ldots, W_{r\ell N}\}, \\ r &= 1, 2, \ldots, R, \quad \ell = 1, 2, \ldots, L \end{aligned} \tag{14}$$

where $x_{\ell n}$ and $W_{r\ell n}$ take the values 1 and 0, and $x_\ell = \Phi_\ell(Z)$.

A direct analogy of functional (5) is the logical function

$$\Psi_r(Z) = \overline{\bar{f}_{r1} \vee \bar{f}_{r2} \vee \ldots \vee \bar{f}_{rL}}, \quad r = 1, 2, \ldots, R \tag{15}$$

where " $\vee$ ", " $-$ " are the symbols of logical adding and negation operations, and $f_{r\ell}$ is the logical function dependent on the retrieval type. The arguments of this function are the bits of the binary numbers under comparison (14).

At a simple retrieval of page addresses meeting the condition of coincidence (non-coincidence) of characters of the requests and page descriptions $f_{r\ell}$ has the equivalence function

$$\bigwedge_{n=1}^{N} \left(x_{\ell n} W_{r\ell n} + \bar{x}_{\ell n} \bar{W}_{r\ell n} \right), \tag{16}$$

where " $\wedge$ " is the symbol of the logical multiplication operation.

Function (16) takes the value 1 provided that a request (vector X) coincides with the description of the r-th page (vector W_r) in the binary bits of all similar characters and the value 0 if only one of the bits of any of the characters does not coincide.

More complicated problems of the retrieval are solved by using other special-purpose functions $f_{r\ell}$ specified for these problems. Due to complexity and non-explicity of these functions we will restrict ourselfes to the description of their common features.

If x_ℓ is the search word value ℓ fixed by the request, at the addressed retrieval according to the character inequality the functions

$$f^{б}_{r\ell}(x_\ell; W_{r\ell}) \quad \text{and} \quad f^{м}_{r\ell}(x_\ell; W_{r\ell}), \tag{17}$$

are used which are composed so that they take the value 1 at all the indices r for which the conditions $W_{r\ell} > x_\ell$ ("more") and $W_{r\ell} < x_\ell$ ("less") are met, respectively, and the value 0 unless the above conditions are fulfilled.

Let now Ω_1 is the set of addresses satisfying the condition "more" at $x_\ell = x'_\ell$, and Ω_2 is the set of addresses meeting the condition "less" at $x_\ell = x''_\ell$, in this case $x''_\ell > x'_\ell$. Then the set of addresses satisfying the condition $x'_\ell < x_\ell < x''_\ell$ is found with the help of a logical multiplication of the sets Ω_1 and Ω_2. Thus, the problem of retrieval of page addresses is solved which characters fall within the range given.

The retrieval of pages where the character ℓ has a maximum or a minimum value is realized with the help of the functions

$$\begin{aligned} &f^{max}_{r\ell}(W_{r\ell}; W_{1\ell}, \ldots, W_{(r-1)\ell}, W_{(r+1)\ell}, \ldots, W_{RL}), \\ &f^{min}_{r\ell}(W_{r\ell}; W_{1\ell}, \ldots, W_{(r-1)\ell}, W_{(r+1)\ell}, \ldots, W_{RL}). \end{aligned} \tag{18}$$

The functions $f^{max}_{r\ell}$ and $f^{min}_{r\ell}$ equal to 1 provided that the conditions

$$W_{re} > W_{1e}, \ldots, W_{(r-1)e}, W_{(r+1)e}, \ldots, W_{RL},$$

$$W_{re} < W_{1e}, \ldots, W_{(r-1)e}, W_{(r+1)e}, \ldots, W_{RL}$$

are met, respectively.

Due to the independence of the characters W_{re}, $\ell = 1, 2, \ldots, L$ these functions can be expressed via functions (16) and represented in the form

$$f_{re}^{max} = \bigwedge_{t=1}^{R} f_{re}^{\delta}(W_{re}; W_{te}),$$

$$f_{re}^{min} = \bigwedge_{t=1}^{R} f_{re}^{M}(W_{re}; W_{te}),$$

$$r = 1, 2, \ldots, R, \quad r \neq t.$$

The ordered addressed access in increasing (or decreasing) order of magnitudes of the character ℓ, beginning with a certain given value x_ℓ, is realized by calculating functional (15) with the functions of type (17) which enables us to obtain the set Ω_1 (or Ω_2) of addresses satisfying the condition "more" (or "less") followed by sampling of the adresses with minimum (or maximum) value of the character ℓ from Ω_1 (Ω_2). The address found from the set Ω_1 (or Ω_2) is "crossed out", and the retrieval is repeated for the next minimum (or maximum) value according to (18).

Specific problems of the ordered access are the problems of search for the nearest greatest and least.

By masking the characters in functional (15) (including their bits in functions (16) - (18)), problems can be solved according to both single characters and their random combinations. These digital retrieval methods are advantageously differ from analog ones.

V. HARDWARE TECHNIQUES FOR CALCULATION OF "CLOSENESS" FUNCTIONALS

The possibilities for the development and application of AOSD are greatly determined by the availability of parallel or parallel-series opto-electronic hardware means for the calculation of multi-dimensional "closeness" functionals used for retrieval. It suffies to say that difficulties encountered in the solution of the coupling problem proved to be the main reason for a rather limited application of the associative SD on electronic, semiconductor and cryogen elements.

Recently coherent and non-coheren optical systems have been suggested to perform linear integral general--type transformations, i.e. systems which pulse response is randomly dependent on the coordinates of the input and output planes [33-46]. These systems, in contrast to those known, comprise an optical matrix or linear organization memory. The two-dimensional "cross--sections" of the pulse response are recorded on a photocarrier by optical or holographic means, so that each half-tone (in a general case) element of the input image pattern projected to the optical memory could restore its "cross-section". Then the light intensity distribution in the system output plane is proportional to the result of the weighed summation of "cross-sections".

Systems realizing the operations of matrix multiplication by a row-vector [20,22,23] and of matrix by matrix [22,38,39,40,46] are intended for parallel analog calculation of multidimensional functionals of the scalar product type (9). Possible dimension of the functionals - 10^4, the number of bits in the request pattern 10^3 - 10^4. In this case the weight matrix is recorded in a holographic memory unit. The weight matrix rearrangement is achieved by changing the unit.

Systems realizing the analog multiplication of three matrices [43-45] offer extensive functional possibilities. Besides multidimensional ($\sim 10^4$) functionals of the scalar product type (9) they enable calculations of multidimensional ($\sim 10^2$) functionals of the quadratic form $\mathbf{z}^T G^{-1} \mathbf{z}$ type (two first terms in expression (13)). The weight and the reverse covariant matrices are presented on transparences with the corresponding transmittance functions. Specific situation arises when the transfer from Z - to X -vector occurs by means of decomposing over an arbitrary orthogonal system of the functions. Then the optical systems under consideration are applicable for an analog calculation of multidimensional type (8) functionals.

The possibilities have been studied and experimentally supported for performing a multichannel retrieval in optical systems by matrix multiplication [38, 47].

Functionals requiring more complicated (non-linear) transformations of the Z -vector, can be calculated by optimal redistribution of the processing function between the optical and electronic means and by using iteration methods.

An example is found in [48,49], where, by using an analog-digital method, the calculation of rank and sign functionals was performed. In the developed opto-electronic processor the initial image characteristics are obtained in an optical system, and the operations of these characteristics processing and taking a decision are performed by a computer. The calculation time is approximately an order reduced.

Efficient (in high-speed) opto-electronic digital retricval methods have been developed on the basis of logical multidimensional functional (15) with the account of specialized types (16) - (18) of its constituent functions [24]. The harware realization of these methods allows for the use, besides to the above optical systems, of special-purpose page-organized electron processors with an optical input. The calculation of a logical functional is performed by means of iteration. In this case the hardware means combine relatively large capacity of the optical memory, high-access of opto-electronic and versatility of electronic methods of data processing.

Thus, the modern opto-electronics provides rather various hardware means for the calculation of multidimensional functionals. It became possible due to the recent investigations enabling several new optical and opto-electronic systems to be suggested, combining the functions of storing and parallel processing of data pages on the analog, analog-digital and digital levels.

VI. CONCLUSION

The organization of information recognition in the optical memory systems by a "character" request will enable us to improve the independency level and to extend the field of application of the devices which can efficiently use inique properties of the optical memory: possibilites for mixed storage and information display, multichannel recognition, multichannel memory addressing and page-oriented processing. The experience gained in the development of optical SD (of large and small capacity) and a variety of multidimensional functionals efficiently calculated by opto-electronic techniques makes it possible to believe that the problem of the development of the optical content-access SD is real.

REFERENCES

1. Di Chen, J. David Zook. An Overview of Optical Data Storage. Technology Proceedings of the IEEE, v. 63, N 8, August, 1975.

2. F.M. Smits, L.E. Gallager. Design Considerations for a Semipermanent Optical Memory. Bell Syst. Tech. J., vol. 46, pp. 1267-1278, July 1967.

3. L.K. Anderson. Holographic Optical Memory for Bulk Data Storage. Bell Lab. Rec., vol. 46, pp. 319-325, Nov. 1968.

4. L.K. Anderson. Application of Holographic Optical Techniques to Bulk Memory. IEEE Trans. Magn., vol. MAG-7, pp. 601-605, Sept. 1971.

5. J. Lipp, J.L. Reynolds. A High Capacity Holographic Storage System, In Applications of Holography, E.S. Barrakette et al., Eds. New York: Plenum Press, 1971, pp. 377-388.

6. J.A. Rajchman. An Optical Read-Write Mass Memory. Appl. Opt., vol.9, pp. 2269-2271, Oct. 1970.

7. W.C. Stewart, W.S. Cosentino. Optics for a Read-Write Memory. Appl.Opt., vol.9,pp.2271--2274, Oct. 1970.

8. A.L. Mikaelyan, V.I. Bobrinev, S.M. Naumov, L.S. Sokolova. Vosmozhnosti primenenija metodov holographii dlya sozdaniya novykh tipov zapominayushchikh ustroistv. Radiotechnika i electronika, 1969, N 1, p. 115-123.

9. A.L. Mikaelyan, V.I. Bobrinev, S.M. Naumov, L.Z. Sokolova. IEEE J.Quantum Electron QE-6, 193 (1970).

10. A.L. Mikaelyan, V.I. Bobrinev, A.A. Akselrod et al. Holographicheskie zapominayushchie ustroistva s zapisju informatsii massivami. Quantowaya Electronika, 1971, 1, p.79-84.

11. L.P. Kraizmer et al. Associativnye zapominayushchie ustroistva. Izd. "Energia", 1967.

12. I.V. Prangishvili, G.M. Popova, O.G. Smorodinova, A.A. Chudin. Odnorodnye mikroelektronnye processory. "Sov. Radio", 1973.

13. I.V. Prangishvili. Osnovnye tendentsii razvitiya upravlyayushchei vychislitelnoi techniki. Izmerenia. Control. Avtomatizatsia, N 1, 1975.

14. E.I. Ilyashenko, V.F. Redakov. Associativnye zapominaushchie ustroistva na magnitnykh elementakh. Energia, 1975.

15. Van Heerden P.J. Theory of Optical Information Storage in Solids. Appl.Opt. 1963,April, vol. 2, N 4, pp. 393-400.

16. Van Heerden P.J. Optical Associative Memory System. U.S. Patent, N 3,492.652. Dec. 30, 1966.

17. D. Gabor. Character Recognition by Holography. Nature, 1965, NT-44.

18. D. Gabor. Associative Holographic Memories. IBM J. Res. Devel. vol. 13, pp. 156-159, Mar. 1969.

19. M. Sakaguchi, N. Nishida, T. Namoto. A New Associative Memory System Utilizing Holography. IEEE Trans. Comput., vol. C-19, pp. 1174-1181, Dec. 1970.

20. M. Sakaguchi, N. Nishida. Large Capacity Associative Memory Employing Holography. U.S. Patent, N 3,704.929, Dec. 5, 1972.

21. I.S. Gibin, M.A. Gofman, E.F. Pen, P.E. Tverdokhleb, Associativnaya vyborka informatsii v hologrammnykh zapominayushchikh ustroistvakh. Avtometria, 1973, N 5.

22. G.R. Knight. Page Oriented Associative Holographic Memory, Appl. Opt., vol. 13, N 4, 1974.

23. I.S. Gibin, P.E. Tverdokhleb. Information Processing in Optical System of Holographic Memory Device. The Proceedings of the American-Soviet Seminar an Optical Information Processing. Plenum Publish. Press. N.Y., 1975.

24. I.S. Gibin, M.A. Gofman, S.F. Kibirev, P.E. Tverdokhleb. Issledovanie odnogo varianta holographicheskoi priznakovoi pamyati. Avtometria, 1976, N 6.

25. G.A. Voskoboinik, I.S. Gibin, V.P. Koronkevich, P.E. Tverdokhleb. Ustroistva s hologrammnoi pamyatyu dlya poiska veshchestv po ikh IK-spectram. Optika i spektroscopia, t. XXX, vyp. 6, 1971.

26. G.A. Voskoboinik, I.S. Gibin, E.S. Nezhevenko, P.E. Tverdokhleb. Primenenie coherentnykh opticheskikh vychislitelnykh ustroistv dlya resheniya zadach informationnogo poiska. Avtometria, 1971, N 1.

27. G.A. Voskoboinik, E.S. Nezhevenko, P.E. Tverdokhleb, Yu.V.Chugui.

Opticheskii coherentnyi correlator. Avtorskoe svidetelstvo N 332474. Priority of July 7, 1970.

28. E.S. Nezhevenko, P.E. Tverdokhleb. Sposob polycheniya vzaimno-correlationnykh functsii izobrazheniya. Authorship NO.300883, Priority of February 4, 1970.

29. Koyama Jiro, Sumi Masao. Optical Data Retrieval System. PANDORA. Elec.Commun. Lab. Techn.J. (Japan) 1973, 22, N 8, pp. 2035-2042.

30. V.A. Antonov, Yu.A. Bykovski, A.I. Larkin, A.A. Markilov, S.N. Starikov, A.A. Treshchun. Holographicheskaya diagnostika mnogoparametricheskikh sostoyanii. Summaries of the IId All-Union Conference on Holography p. I, Kiev, 1975.

31. J. Viennot. Analog Methods of Optical Data Processing. Avtometria, 1974, N 1.

32. N. Nilson. Educational computlrs. Izd. "Mir", M., 1967.

33. E.S. Nezhevenko, O.I. Potaturkin, P.E. Tverdokhleb. Lineinye opticheskie sistemy s impulsnoi reaktsiei obshchego vida. Avtometria, 1972, N 6.

34. I.S. Gibin, E.S. Nezhevenko, P.E. Tverdokhleb, O.I. Potaturkin. Sposob obobshchennogo spectralnogo analiza. Authoship No 369587, Priority of August 3, 1971.

35. E.S. Nezhevenko, O.I. Potaturkin, P.E. Tverdokhleb. Sposob obobshchennogo spektralnogo analiza. Authorship No. 391414, Priority of March 17, 1972.

36. I.S. Gibin, E.S. Nezhevenko, O.I. Potaturkin, P.E. Tverdokhleb. Coherentno-opticheskie ustroistva dlya obobshchennogo spektralnogo analiza izobrazhenii. Avtometria, 1972, N 5.

37. I.S. G i b i n, O.I. P o t a t u r k i n, P.E. T v e r d o k h l e b. Sposob obobshchennogo spektralnogo analiza. Authorship No. 428229, Priority of November 25, 1971.

38. E.S. N e z h e v e n k o, P.E.T v e r d o k h l e b. Umnozhenie matrits opticheskim metodom. Avtometria, 1972, N 6.

39. B. B r y n g d a h l. Optical Map Transformations. Optics Communications, 1974, vol. 10, N 2.

40. O. B r y n g d a h l. Geometrical Transformations in Optics. JOSA, August 1974, vol. 64, N 8. pp. 1092-1099.

41. I.S. G i b i n, M.A. G o f m a n,Yu.V.C h u g u i. Obobshchennyi spektralnyi analiz izobrazhenii s ispolzovaniem holographicheskogo metoda formirovania kodiruyushchei plastiny. Avtometria, 1975,N 3.

42. Yu.V. C h u g u i, B.E. K r i v e n k o v, P.E. T v e r d o k h l e b. The Analysis of Images by Hadamard Optical Transform. Appl. Opt., 1975, vol. 14, N 8.

43. B.E.Kr i v e n k o v, S.V. M i k h l y a e v, P.E. T v e r d o k h l e b. Nekogerentnaya opticheskaya sistema dlya vypolneniya matrichnykh preobrazovanii. Avtometria, 1975, N 3.

44. S.V. M i k h l y a e v, Yu.V. C h u g u i, Parallelno-posledovatelnyi analiz izobrazhenii necoherentnym opticheskim metodom. Avtometria,1975.

45. Yu.V. C h u g u i. B.E. K r i v e n k o v, S.V. M i k h l y a e v, P.E. T v e r d o k h l e b. Non-Coherent Optical System for processing of Images and Signals. The Proceedings of the American-Soviet Seminar on Optical Information Processing. Plenum Publish. Corpor., N.Y., 1975.

46. W. S c h n e i d e r, W. F i n k. Incoherent Optical Matrix Multiplication. Optica Achta, 1975, N 11, pp. 879-889.

47. M.A. G o f m a n, S.F. K i b e r e v, B.E. K r i v e n k o v, P.E. T v e r d o k h l e b, Yu.V. C h u g u i. Mnogokanalnyi poisk v necoherentnoi opticheskoi sisteme pamyati. Avtometria, 1976,N 6.

48. O.M. K a r p o v a, E.S. N e z h e v e n k o, G.D. U m a n t s e v. Raspoznovanie izobrazhenii izvestnoi formy na photosnimkakh. Avtometria, 1975, N 3.

49. F.F. V e r y a s k i n, L.V. V y d r i n, V.G.D a v y d o v, T.N. M a n t u s h, E.S.N e z h e v e n k o, B.N. P a n k o v, P.E. T v e r d o k h l e b. Optico-electronnyi processor dlya raspoznavaniya izobrazhenii. Avtometria, 1975, N 3.

MULTICHANNEL INFORMATION RETRIEVAL IN NON-COHERENT OPTICAL STORAGE SYSTEMS

M.A. Gofman, S.F. Kibirev, B.E. Krivenkov, P.E. Tverdokhleb, and Yu.V. Chugui

Institute of Automation and Electrometry
Novosibirsk, USSR

ABSTRACT

An important property of optical memories is the possibility of multichannel access to stored data.

The structure and the principle of operation of a non-coherent optical memory providing a multichannel parallel address retrieval by several quest words, are described in the work. The system can be used as a high-speed catalog of a content addressed optical memory. The evaluation of the information capacity of such a system was carried out, and the results of its experimental investigation are given by the examples of simple and complex searches.

The systems of search for memory cell addresses containing the necessary information are known to be based on the use of storage systems (SS), which make the calculation of the "approximation" functional between the stored data and an inquiry, besides information storage functions. Optical SS (including holographic ones) [1,2] meet the stated demands well.

With the development of information memory access

methods the possibilities of organizing the retrieval upon several parallel inquiries (the simultaneous work of the system with L users) [3,4] attract more and more attention. However, in spite of the fact that optical memory systems are the most adequate to such problems, the questions of creating of a multichannel retrieval system based on them are very little elucidated in literature. In this connection only the works [4,5] can be mentioned, where the possibilities of organizing multichannel access in holographic memory systems are considered. In the present work a non-coherent optical system of the matrix type [6] is suggested to be used for the purposes of multichannel retrieval.

The choice of such a system depends on the fact that the key words storage functions and their parallel comparison with some quest words by the calculation of the corresponding number of "approximation" functionals, as scalar products, are combined quite naturally in it. Besides, when the demands to the optical elements are the same it is possible to calculate the values of these functionals more precisely when using non-coherent systems, than in case of their coherent analogues [7].

The structure and the principle of action of non-coherent optical memory providing the multichannel parallel adress retrieval upon several quest words are given below.

The system can be used as a high-speed catalog of the content addressed optical memory [2]. The estimation of the information capacity for such a system ("address number" x "character number" x "channel number") was carried out, and the results of the experimental investigation are given by the examples of complex and simple searches.

The multichannel information retrieval upon L channels and N memory addresses consists of calculation L approximation functional vectors $\vec{f}_1, \vec{f}_2, \ldots, \vec{f}_L$ between the vectors $\vec{K}_1, \vec{K}_2, \ldots, \vec{K}_N$ which form the character matrix $[K]_{N \times m}$ and the quest vectors $Z_1, Z_2, \ldots, Z_L$ of size m. If we estimate vector approximation in m -dimentional character space by a scalar product, then the matrix expression calculation is carried out during the process of retrieval:

$$\begin{bmatrix} f_{11} & f_{12} & \dots & f_{1L} \\ f_{21} & f_{22} & \dots & f_{2L} \\ \dots & \dots & \dots & \dots \\ f_{N1} & f_{N2} & \dots & f_{NL} \end{bmatrix} = \begin{bmatrix} k_{11} & k_{12} & \dots & k_{1m} \\ k_{21} & k_{22} & \dots & k_{2m} \\ \dots & \dots & \dots & \dots \\ k_{N1} & k_{N2} & \dots & k_{Nm} \end{bmatrix} \begin{bmatrix} z_{11} & z_{12} & \dots & z_{1L} \\ z_{21} & z_{22} & \dots & z_{2L} \\ \dots & \dots & \dots & \dots \\ z_{m1} & z_{m2} & \dots & z_{mL} \end{bmatrix} \quad (1)$$

Further let us suppose that matrix elements $[Z]_{m\times L}$ and $[K]_{N\times m}$ take on values 1 or 0. Thus, looking through the information character matrix, access is accomplished parallel upon L channels according to the algorithm of matrix multiplication, during the process of multichannel retrieval.

Let us consider the structural scheme of the optical system, which performs transformation of the type (1). For this purpose, we call upon Fig. 1, where its anamorphosis integro-projecting units are denoted by symbols 1,2,3. The transparencies T_1 and T_2 used for the formation of L independent light sources and for the parallel quest entering, respectively, are placed in the planes P_1 and P_2 of the system; and the transparency with the matrix image of characters $[K]$ is in the plane P_3.

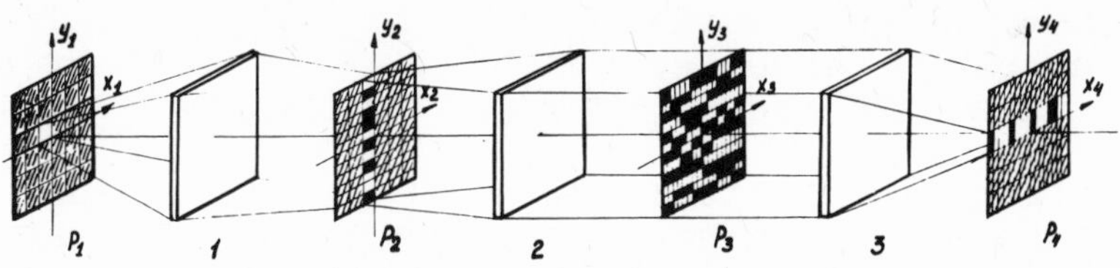

Figure 1. The structural scheme of the multichannel optical retrieval system.

Let the working fields of the system, being in the planes P_1, P_2 and P_3, be of the form of square aperture of $D\times D$ size. Let us divide the field in P_1

into $N \times N$ elements and put one of diagonal elements in conformity with each channel; each element can be 1 ("transparent"), and 0 ("opaque"). Passing through the T_1 transparency a regular light beam is modulated by the transmittance function, which is of the form in general case:

$$T_1(x_1,y_1)=\sum_{i=1}^{L} \omega_i \, rect\left[\frac{x_1-(i-1/2)\delta+D/2}{\delta}\right] rect\left[\frac{y_1-(i-1/2)\delta+D/2}{\delta}\right] \quad (2)$$

where $\delta=\frac{D}{L}$, and $\omega_i = 1$, if the i-th channel is on and $\omega_i = 0$, otherwise.

The transparency T_2 which serves to inquire L quests in the form of binary code combinations contains L independent columns, each of them consists of m resolution elements(or 2m in case of paraphase coding). The quest input is carried out by a preset of a required binary combination along the column. In general case this transparency has the transmittance function:

$$T_2(x_2,y_2)=\sum_{p=1}^{L}\sum_{q=1}^{2m} b_{pq} \, rect\left[\frac{x_2-(p-1/2)\delta+D/2}{\delta}\right] rect\left[\frac{y_2-(q-1/2)\varepsilon+D/2}{\varepsilon}\right] \quad (3)$$

where $\varepsilon=D/2m$; and the factors b_{pq} either coincide with the quest vector components $\vec{Z}_p$ written by a direct paraphase code $\vec{Z}_p=\begin{bmatrix} Z_{ip} \\ \bar{Z}_{ip} \end{bmatrix}$ or they have arbitrary values (zero in particular) for p, corresponding to off channel number. The transmittance of the transparency T_3 corresponding to the information character array is represented in:

$$T_3(x_3,y_3)=\sum_{r=1}^{2m}\sum_{s=1}^{N} k_{rs} \, rect\left[\frac{x_3-(s-1/2)\alpha+D/2}{\alpha}\right] rect\left[\frac{y_3-(r-1/2)\varepsilon+D/2}{\varepsilon}\right] \quad (4)$$

wherein $\alpha=D/N$, and k_{rs} are the factors corresponding to elements of a character matrix $[K]$ of $2m \times N$ size.

The light beams, being formed by L elements (channels) of the T_1 transparency, are projected by unit 1 into P_2 along the X coordinate and defocused along the Y coordinate. As a result of such transformation we have L incident beams illuminating L columns with quest images independently of one another, incidence angles of these beams being defined by the ordinates of on elements in P_1 . The light distribution is further modulated by the T_2 transparency and is projected into P_3 along the Y coordinate and defocused along another one by unit 2. Defocusing along the X coordinate leads to the simultaneous superposition (at different angles) of all quest images on the T_3 transparency. Thus, the summary light intensity distribution in the plane P_3 is proportional to:

$$\sum_{q=1}^{2m} \left(\sum_{p=1}^{L} b_{pq}\, \omega_p\right) rect \left[\frac{y_3-(q-1/2)\varepsilon + D/2}{\varepsilon}\right] \tag{5}$$

The light distribution (5), modulated by the character array image is processed by integro-projecting unit 3. At the same time, the light distribution is projected along the X coordinate into P_4 , and it is summarized along the Y coordinate through all matrix columns $[K]$, with simultaneous space separation of the light beams corresponding to different channels. The latter is achieved by projecting of the plane P_1 along the Y coordinate into P_4 . Taking into account (4), the light intensity in the output plane will be distributed according to the law:

$$I(x_4, y_4) = \sum_{p=1}^{L} \sum_{s=1}^{N} rect\left[\frac{y_4-(p-1/2)\delta+D/2}{\delta}\right] \times$$

$$\times\, rect\left[\frac{x_4-(s-1/2)\alpha+D/2}{\alpha}\right] \sum_{r=1}^{2m} k_{rs}\, b_{pr}\, \omega_p \tag{6}$$

Now, one can readily see that the obtained distribution (6) corresponds to the components of the required vectors $\vec{f}_p$, if one takes into consideration the meaning of the parameters k_{rs} , b_{pr} , ω_p previously introduced.

The principal scheme of the optical system under consideration is given in Fig. 2. An extended light source 1 is an important unit of the system given in two projections. Its characteristics (light regularity, the character of a scatter diagram) define the parameters of the retrieval system to a great extent.

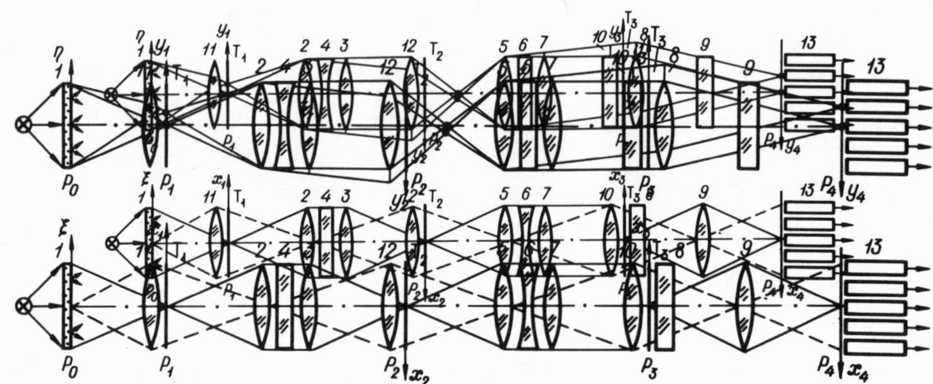

Figure 2. The principal scheme of the multichannel optical retrieval system (in two projections).

The units used for the illuminating light beam formation (elements 2,3,4), and for quest superposition (elements 5,6,7) are projecting systems defocused in orthogonal directions. They have spherical (2,3 and 5,7) objectives and cylindrical (4 and 6) lenses. Cylindrical lenses (8 and 9) are the elements of an output integro-projecting unit. Condenser objectives (10, 11, 12) providing the space invariance of the system, and a multielement photoreceiver 13 are given in the block diagram, as well.

The number of channels, addresses, and characters providing the search of the system is limited by diffraction effects. Thus, the number of addresses is defined by the resolution of the projecting objective 9. If the objective resolution is n line/mm, then the maximum address number is limited by the value $N = knD$, where $k \leq 1$ is the utilization factor. When n = 40 line/mm, k = 0.5, D = 100 mm we have $N = 2 \cdot 10^3$ addresses. More exacting requirements are placed upon the number of character m and the number of channels L. As it follows from [6], the maximum character number is connected with the maximum channel number by the relationship

$$mL = D^2/8\lambda f_0 , \qquad (7)$$

wherein f_0 is the distance between the planes P_3 and P_4 ; and λ is the average wave lenght of the light source being used. Thus, the product mL is a constant for the optical system being chosen. For example, if $D = 10$ mm, $f_0 = 300$ mm, $\lambda = 0.5\cdot10^{-3}$mm, then $mL = 2\cdot10^4$.

The information capacity of such a system can be evaluated according to the formula:

$$I = mNL = \frac{nD^3}{8k\lambda f_0} . \qquad (8)$$

At the values m, N and L given above, the information capacity will be 10^7.

The possibility of the fulfilment of multichannel complex and simple searches by the system was experimentally confirmed. For instance, the problem of the organic matter retrieval according to their molecular weight was considered [8]. Binary-decade digits of their molecular weight representation were the characters of matters. For example, binary-decade word 0011010000100011, each digit of it is considered as a separate character (16 characters all in all), corresponds to the matter of 342,3 molecular weight.

The results of simple search are shown in Fig.3-1. The number of channels was 6, the number of characters was 16 (each character was represented in the paraphase code), and the number of addresses was 32. The picture (a) corresponds to the situation when all the channels work at a time (the same quest are entered into the fifth and sixth channels); the picture (b) corresponds to six quests, each of them has its own channel (except for the sixth one). The quests were entered in the form of the true paraphase code. The picture (c) corresponds to the character matrix the characters being written in the form of the complement paraphase code in each column

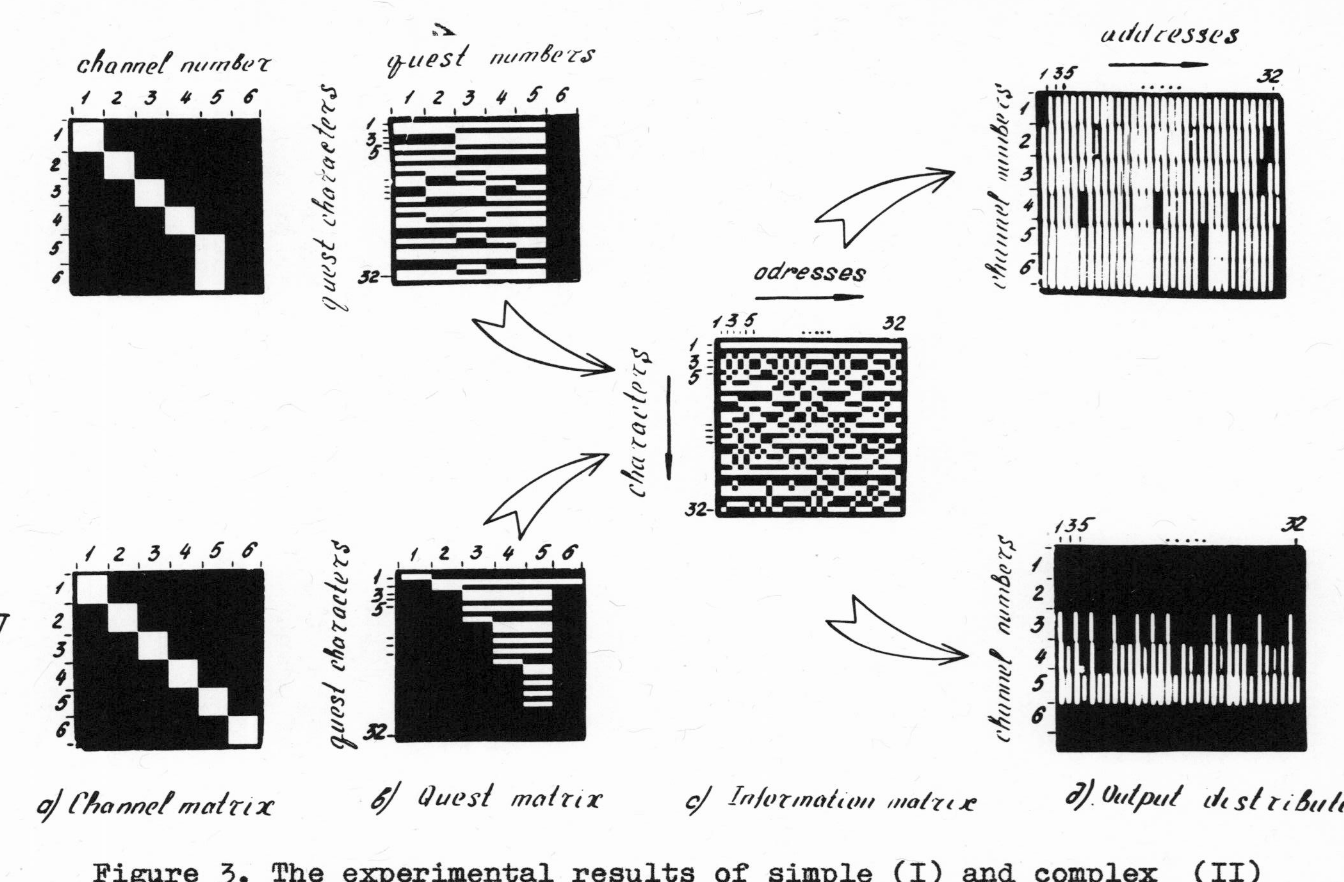

Figure 3. The experimental results of simple (I) and complex (II) searches.

(the column number is the address). The output light distribution is represented in Fig. 3-1 by the picture (d). In this case the number of a channel is defined by the number of a row, and the address - by the number of a column. Addresses of the matters which characters have matched with the quest, are defined by the position of unilluminated elements. Thus, the matter with address 1 fits the 1-st quest, the matter with address 31 fits the 2-nd one, the matters with addresses 6,16, 26 fit the 4-th one, and the matters with addresses 22 and 32 fit the 5-th one. No matter fits the 3-d quest.

The problem of the complex associative address search of matters of μ molecular weights within the $m_1 < \mu < m_2$ range was considered, as well. The algorithm of solving the problem provides for formation of two sets of quests to find the ensemble of addresses A_1 and A_2 with molecular weights $\mu > m_1$ and $\mu < m_2$, respectively. The addresses being found in each set quest are combined with the help of multichannel "OR" operation

$$A_1 = \bigcup_i A_{1i}, \quad A_2 = \bigcup_i A_{2i},$$

where A_{1i}, A_{2i} are address arrays being found upon one quest of a set. The addresses fitting the condition $m_1 < \mu < m_2$ are determined from the expression:

$$A = A_1 \cap A_2,$$

where $\cap$ is the character of the "AND" operation.

It is important to show that the maximum channel number, neccessary for the parallel realization of the retrieval algorithm within any given ranges,is defined by a quest word digit number, and in our example it is equal to 16 per on set of quests [9].

The results of a complex retrieval of matters with molecular weights within the range $377.7 < \mu < 800.0$ are given in Fig. 3-II. In this case a specially made set of quests consists of 5 quests arranged in the first five channels for $\mu > m_1$, and it consists of one quest arranged in the sixth channel for $\mu < m_2$. A retrieval upon the given quest is carried out at a time.The picture (d) in Fig. 3-II corresponds to the light output distribution resultant from a multichannel retrieval. It is clear that no address was found upon the first

Table 1

Channel number	Quest (m_1=377.7; m_2=800.0)	Ranges μ	Addresses of matters being retrieved
1	10 00 00 00 00 00 00 00 00 00 00 00 00 00 00 00	800.0+ +999.9	A_{11} = (—)
2	01 10 00 00 00 00 00 00 00 00 00 00 00 00 00 00	400.0+ +799.9	A_{12} = (1,3,5,8, 11,13,15,21, 23,27,31)
3	01 01 10 10 10 00 00 00 00 00 00 00 00 00 00 00	380.0+ +399.9	A_{13} = (10)
4	01 01 10 10 01 10 10 10 10 00 00 00 00 00 00 00	378.0+ +379.9	A_{14} = (—)
5	01 01 10 10 01 10 10 10 01 10 10 10 10 00 00 00	377.7+ +377.9	A_{15} = (—)
	$\bigcup_{i=1}^{5} A_{1i}$	377.7	A_1 = (1,3,5,8, 10,11,13,15, 21,23,27,31)
6	01 00 00 00 00 00 00 00 00 00 00 00 00 00 00 00	000.0+ +800.0	all the addresses A_2 = (1 + 32)
Re-seult	$A_1 \cap A_2$	377.7+ +800.0	A = (1,3,5,8, 10,11,13,15, 21,23,27.31)

fourth and fifth quests, a number of addresses was found upon the second quest, and only one address was found upon the third quest (10). The combination of a set of results obtained in the first five channels gives values of addresses fitting the condition $m > 377.7$. All the addresses fit the quest upon the sixth channel, that is the molecular weight value is $\mu < 800.0$ for all addresses.

The initial data as well as the intermediate and final experimental results are given in detail in the table.

Thus, the possibility of a multichannel information retrieval in a non-coherent optical memory has been theoretically and experimentally confirmed. The problem of technical realization of such a system requires a separate discussion.

REFERENCES

1. I.S. Gibin, M.A. Gofman, E.F. Pen and P.E. Tverdokhleb. Assotsiativnaya viborka informatsii v gologrammnikh zapominayutshikh ustroistvakh. - Autometria, N 5, 1973.

2. P.E. Tverdokhleb. Opticheskie systemi pamyati s viborkoi po soderzhaniyu. - Autometria, N 6, 1976.

3. E.A. Gulesha, V.Yu. Lidak, V.V. Spiridonov, and M.J. Markov. Voprosi organizatsii obmena informatsiei s zapominayutchimi ustroistvami pryamogo dostupa. - Isv. Vusov, Priborostroenie, N 2, 1976.

4. T. Gaylord. Optical Memories. - Optical Spectra, 1974, vol. 8, N 6.

5. C. Alford, T. Gaylord. The potential of multiport optical memories in digital computing. International Optical Computing Conferenic. 1975, Washington. IEEE Catalog NO 75 CHO941-5e.

6. B.E. K r i v e n k o v, S.V. M i k h l y a e v, P.E. T v e r d o k h l e b, Yu.V. C h u g u i. Non-coherent optical system for processing of images and signals. - In : Optical information processing. New-York - London. Plenum Press, 1975, p. 203-217.

7. E.S. N e z h e v e n k o, P.E.T v e r d o k h l e b. Umnozhenie matrits opticheskim metodom. - Autometria, N 6, 1972.

8. Yu.P. D r o b i s h e v, R.S. N i g m a t u l l i n, V.J. L o b a n o v, I.R. K o r o b e i n i t c h e- v a, V.S. B o c h k a r e v and V.A. K o p- t y n g. Ispolzovanie EUM dlya opoznovania chimicheskikh soedinenii po spectralnim kharakteristikam. - Vestnik USSR AS, N 8, 1970.

9. I.S. G i b i n, M.A. G o f m a n, S.F.K i b i r e v, P.E. T v e r d o k h l e b. Issledovanie odnogo varianta gologramnoi priznakovoi pamjati. - Autometria, N 6, 1976.

PSEUDOINVERSE IMAGE RESTORATION COMPUTATIONAL ALGORITHMS

William K. Pratt

Image Processing Institute

University of Southern California

INTRODUCTION

Digital processing techniques have been widely utilized for the restoration of blurred and noisy images <1>. The major limitations associated with such techniques arise from the large dimensionality of practical images. Large dimensionality not only leads to lengthy processing, but more importantly, creates problems of numerical instability.

Recent research has been directed toward the development of efficient computational algorithms which can reduce the computational requirements and improve the numerical stability of conventional digital restoration methods. Pratt and Davarian <2> have developed a fast computational algorithm for pseudoinverse image restoration which provides significant computational savings compared to brute force computation. Andrews and Patterson <3-5> have introduced the singular value decomposition (SVD) of matrices as a means of performing pseudoinverse image restoration. With this technique the deleterious effects of numerical instability can be avoided by a sequential restoration procedure, but the computational requirements are usually excessive. This paper introduces a fast computational algorithm for SVD image restoration.

PSEUDOINVERSE IMAGE RESTORATION

An imaging process resulting in the observation of a spatially blurred image in the presence of additive noise can be modelled by a set of linear equations. The vector space representation of these equations is given by

$$\underline{g} = \underline{B}\,\underline{f} + \underline{n} \tag{1}$$

where $\underline{g}$ is an M x 1 vector of pixels obtained by sampling and column scanning of the observed image, $\underline{f}$ is an N x 1 vector of points of an ideal image field, $\underline{B}$ is the M x N blur matrix containing elements which are samples of the blur impulse response function, and $\underline{n}$ is an M x 1 vector representing additive observation noise. In most imaging models the sample spacing on $\underline{g}$ and $\underline{f}$ is assumed identical. Consideration in this paper is limited to image blurring which is space invarient. In such cases the blur matrix $\underline{B}$ is of block Toeplitz form. Furthermore, for simplicity in explanation, the presentation is limited to a one dimensional blur. The extension of the fast computational algorithms to two dimensional form is straightforward. Under these conditions the blur matrix assumes the form

$$\underline{B} = \begin{bmatrix} h(L) & \cdots & h(1) & 0 & \cdots & 0 \\ 0 & h(L) & \cdots & h(1) & 0 \cdots & 0 \\ \cdot & & \ddots & & \ddots & \cdot \\ \cdot & & & \ddots & & \cdot \\ \cdot & & & & \ddots & 0 \\ 0 & \cdots & & h(L) & \cdots & h(1) \end{bmatrix} \tag{2}$$

where $\underline{h}$ is an L x 1 vector of points of the impulse response. It should be noted that this system is underdetermined since N = M+L-1.

The pseudoinverse estimate of $\underline{f}$ is given by

$$\hat{\underline{f}} = \underline{B}^{-}\,\underline{g} \tag{3}$$

where $\underline{B}^{-}$ denotes the generalized inverse of $\underline{B}$. The generalized inverse provides a minimum least square error, minimum norm restoration of the ideal image. If

$\underline{B}$ is a matrix of rank M the generalized inverse assumes the form

$$\underline{B}^{-} = \underline{B}^{T}(\underline{B}\,\underline{B}^{T})^{-1} \tag{4}$$

Algorithms requiring on the order of N^3 operations have been developed for computation of the generalized inverse.

The fast pseudoinverse algorithm <2> is based upon the development of an adjoint model

$$\underline{q}_E = \underline{C}\,\underline{f}_E + \underline{n}_E \tag{5a}$$

where the extended vectors $\underline{q}_E$ and $\underline{f}_E$ are defined in correspondence with

$$\begin{array}{r} M\left\{\begin{array}{c} \\ \\ \end{array}\right. \\ J-M\left\{\begin{array}{c} \\ \end{array}\right. \end{array} \begin{bmatrix} \underline{q} \\ \\ ---- \\ \underline{0} \end{bmatrix} = \begin{bmatrix} \\ & \underline{C} & \\ \\ \end{bmatrix} \begin{bmatrix} \underline{f}_T \\ --- \\ \underline{0} \end{bmatrix} \begin{array}{l} \left.\begin{array}{c} \\ \\ \end{array}\right\} K \\ \left.\begin{array}{c} \\ \end{array}\right\} J-K \end{array} + \begin{bmatrix} \underline{n}_T \\ --- \\ \underline{0} \end{bmatrix} \tag{5b}$$

The vector $\underline{f}_T$ is composed of the center K pixels of the ideal image vector $\underline{f}$ where $K = N-2(L-1)$, and $\underline{C}$ is a circulant matrix of the form

$$\underline{C} = \begin{bmatrix} h(1) & 0 & \cdot\;\cdot & 0 & h(L) & \cdot\;\cdot\;\cdot\;\cdot & h(2) \\ h(2) & h(1) & \ddots & & \ddots & & \cdot \\ \cdot & & \ddots & & & \ddots & \cdot \\ \cdot & & & \ddots & & & \cdot \\ \cdot & & & & \ddots & & \cdot \\ h(L) & h(L-1) & & & & \ddots & h(L) \\ 0 & h(L) & & & & & 0 \\ \cdot & \cdot & \ddots & & & & \cdot \\ \cdot & \cdot & & \ddots & & & \cdot \\ \cdot & \cdot & & & \ddots & & 0 \\ 0 & 0 & \cdot\;\cdot & 0 & h(L) & \cdot\;\cdot\;\cdot\;\cdot & h(1) \end{bmatrix} \tag{6}$$

The vector $\underline{q}$ is identical to the image observation $\underline{q}$ over its $\overline{R}$ center elements where $R = M-2(L-1)$. The outer elements of $\underline{q}$ can be approximated by <2>

$$\underline{q} \approx \underline{\tilde{q}} = \underline{W}\,\underline{g} \tag{7}$$

where $\underline{W}$ is a windowing matrix. Combining eqs.(5) and (7) an estimate of $\underline{f}_T$ can be obtained from

$$\underline{\hat{f}}_E = \underline{C}^{-1}\,\underline{\tilde{q}}_E \tag{8a}$$

in correspondence with

$$\begin{matrix} K \\ \\ J-K \end{matrix} \left\{ \begin{bmatrix} \underline{\hat{f}}_T \\ --- \\ \\ \end{bmatrix} \right. = \begin{bmatrix} \\ \underline{C}^{-1} \\ \\ \end{bmatrix} \begin{bmatrix} \underline{\tilde{q}} \\ --- \\ \underline{0} \end{bmatrix} \tag{8b}$$

where $\underline{C}^{-1}$ is the inverse of the circulant blur matrix.

The restoration operation of eq.(8) can be performed indirectly by Fourier domain processing. Letting $\underline{A}$ denote the Fourier transform matrix, eq.(8) can be rewritten as

$$\underline{\hat{f}}_E = \underline{A}^{-1}\underline{\mathcal{C}}^{-1}[\underline{A}\,\underline{\tilde{q}}_E] \tag{9}$$

where both

$$\underline{\mathcal{C}} = \underline{A}\,\underline{C}\,\underline{A}^{-1} \tag{10a}$$

and its inverse

$$\underline{\mathcal{C}}^{-1} = \underline{A}\,\underline{C}^{-1}\,\underline{A}^{-1} \tag{10b}$$

are found to be diagonal matrices since $\underline{C}$ and $\underline{C}^{-1}$ are circulant. The terms of $\underline{\mathcal{C}}$ can be obtained from <2>

$$\underline{\mathcal{C}} = \mathrm{diag}[\mathcal{h}_E(1), \ldots, \mathcal{h}_E(J)] \tag{11}$$

where

$$\begin{bmatrix} \\ \underline{\mathcal{h}}_E \\ \\ \end{bmatrix} = \begin{bmatrix} \\ \underline{A} \\ \\ \end{bmatrix} \left. \begin{bmatrix} \underline{h} \\ ---- \\ \underline{0} \end{bmatrix} \right\} \begin{matrix} L \\ \\ J-L \end{matrix} \tag{12}$$

With the fast pseudoinverse algorithm, the brute force matrix operations of eq.(8) can be replaced by a pair of one-dimensional Fourier transforms and J scalar multiply-add operations. The computational saving is usually at least a factor of 1000:1!

SVD IMAGE RESTORATION

The blur matrix $\underline{B}$ of eq.(1) can be decomposed in the product form

$$\underline{B} = \underline{U}\,\underline{\Lambda}^{\frac{1}{2}}\,\underline{V}^T \qquad (13)$$

by a singular value matrix decomposition where $\underline{U}$ and $\underline{V}$ are unitary matrices composed of the eigenvectors of $\underline{B}\,\underline{B}^T$ and $\underline{B}^T\underline{B}$, respectively, and $\underline{\Lambda}$ is a diagonal matrix whose diagonal terms contain eigenvalues of $\underline{B}\,\underline{B}^T$ or $\underline{B}^T\underline{B}$. Since $\underline{U}$ and $\underline{V}$ are unitary matrices the generalized inverse of $\underline{B}$ can be written directly as

$$\underline{B}^- = \underline{V}\,\underline{\Lambda}^{-\frac{1}{2}}\,\underline{U}^T \qquad (14)$$

and the SVD pseudoinverse estimate assumes the form

$$\hat{\underline{f}} = \underline{B}^-\,\underline{g} = \underline{V}\,\underline{\Lambda}^{-\frac{1}{2}}\,\underline{U}^T\,\underline{g} \qquad (15)$$

The operations of eq.(15) may be viewed as a linear transformation $(\underline{U}^T\underline{g})$ followed by a scalar weighting (multiplication by $\underline{\Lambda}^{-\frac{1}{2}}$), followed by another linear transformation (multiplication by $\underline{V}$). The transformations are into the eigen-space of the blur matrix $\underline{B}$. Estimation by eq.(15) requires N^2+N multiply and add operations.

As a consequence of the orthogonality of $\underline{U}$ and $\underline{V}$ it is possible to express the blur matrix in the series form

$$\underline{B} = \sum_{i=1}^{R_B} [\lambda(i)]^{\frac{1}{2}}\,\underline{u}_i\underline{v}_i^T \qquad (16)$$

where the $\lambda(i)$ are diagonal elements of $\underline{\Lambda}$, $\underline{u}_i$ and $\underline{v}_i$ are the i th columns of $\underline{U}$ and $\underline{V}$, respectively, and R_B is the rank of the matrix $\underline{B}$. The generalized inverse of $\underline{B}$ can also be written in the series form

$$\underline{B}^{-} = \sum_{i=1}^{R_B} [\lambda(i)]^{-\frac{1}{2}} \underline{v}_i \underline{u}_i^T \tag{17}$$

Hence, the SVD pseudoinverse estimate can be expressed as

$$\hat{\underline{f}} = \sum_{i=1}^{R_B} [\lambda(i)]^{-\frac{1}{2}} \underline{v}_i \underline{u}_i^T \underline{g} \tag{18}$$

Equation (18) can be manipulated to obtain a sequential SVD pseudoinverse estimate. The k th estimate is equal to

$$\hat{\underline{f}}_k = \hat{\underline{f}}_{k-1} + [\lambda(k)]^{-\frac{1}{2}} (\underline{u}_k^T \underline{g}) \underline{v}_k \tag{19}$$

for $1 \leq k \leq T$ where T represents the number of terms employed in the series expansion. One of the principal advantages of the sequential formulation is that problems of ill-conditioning generally occur only for the higher order singular values. Thus, it is possible to terminate the expansion before numerical problems occur. The major difficulty associated with the sequential SVD algorithm is the large amount of computation involved in generation of the orthogonal vectors $\underline{u}$ and $\underline{v}$ and the number of operations required to implement eqs.(14) or (17).

FAST SVD ALGORITHM

The SVD pseudoinversion technique can be applied directly to the adjoint model of eq.(5) by the SVD expansion

$$\underline{C} = \underline{X}\, \underline{\Delta}^{\frac{1}{2}}\, \underline{Y}^{*T} \tag{20}$$

where $\underline{X}$ and $\underline{Y}$ are unitary matrices defined by

$$\underline{X}(\underline{C}\,\underline{C}^T)\underline{X}^{*T} = \underline{\Delta} \tag{21a}$$

$$\underline{Y}(\underline{C}^T\underline{C})\underline{Y}^{*T} = \underline{\Delta} \tag{21b}$$

Since $\underline{C}$ is circulant, $\underline{C}\,\underline{C}^T$ is also circulant.

Therefore, $\underline{X}$ and $\underline{Y}$ must be equivalent to the Fourier transform matrix $\underline{A}$ (or $\underline{A}^{-1}$) since the Fourier matrix performs a diagonalization of a circulant matrix. For purposes of standardization let $\underline{X} = \underline{Y} = \underline{A}^{-1}$. As a consequence, the eigenvectors $\underline{x}_i = \underline{y}_i$, which are rows of $\underline{X}$ and $\underline{Y}$, are actually the complex exponential basis functions of a Fourier transform. That is,

$$x_k^*(j) = \exp\{\frac{2\pi i}{J}(k-1)(j-1)\} \qquad (22)$$

It is easy to show that

$$\underline{\Delta} = \underline{\mathcal{C}}\underline{\mathcal{C}}^{*T} \qquad (23)$$

where $\underline{\mathcal{C}} = \underline{A}\ \underline{C}\ \underline{A}^{-1}$ is computed from eqs.(11) and (12).

Then in correspondence with eq.(15) the SVD pseudoinverse estimate of the extended ideal image vector becomes

$$\hat{\underline{f}}_E = \underline{Y}\,\underline{\Delta}^{-\frac{1}{2}}\,\underline{X}^{*T}\,\tilde{\underline{g}}_E \qquad (24a)$$

or

$$\hat{\underline{f}}_E = \underline{A}^{-1}\,\underline{\Delta}^{-\frac{1}{2}}\,\underline{A}\,\tilde{\underline{g}}_E \qquad (24b)$$

Equation (24b) should be recognized as exactly equivalent to the Fourier domain pseudoinverse procedure defined in eq.(9). Hence, it may be concluded that the SVD eigen space for a circulant matrix is the Fourier domain. Computation of the fast SVD estimate by eq.(24b) requires J log J operations each for the forward and inverse Fourier transforms using a fast Fourier transform (FFT) algorithm plus J scalar multipliers for a total of $J(1 + 2\log_2 J)$ operations.

A series form of the fast SVD pseudoinverse estimate can be written directly in correspondence with eq.(18) as

$$\hat{\underline{f}}_E = \sum_{i=1}^{R_C} [\delta(i)]^{-\frac{1}{2}}\,\underline{x}_i^*\,\underline{x}_i^T\,\tilde{\underline{g}}_E \qquad (25)$$

where R_C is the rank of $\underline{C}$. Similarly in correspondence with eq.(18) the sequential pseudoinverse estimate assumes the form

$$\hat{\underline{f}}_{E_k} = \hat{\underline{f}}_{E_{k-1}} + [\,\delta(k)\,]^{-\frac{1}{2}}\,[\underline{x}_k^T\,\tilde{\underline{q}}_E]\,\underline{x}_k^* \tag{26}$$

for $1 \le k \le T$ where T represents the number of terms employed in the series expansion. Using an FFT to pre-compute and store the product $[\underline{x}_k^T\tilde{\underline{q}}_E]$ results in a total of $TJ(1+\log_2 J)$ operations required for the computation. Greater computational efficiency can be realized if the series is successively bisected until the point of ill-conditioning is reached. This strategy requires about $J\log_2 J[1+\log_2 J]$ operations.

FAST MODIFIED SVD ALGORITHM

The series and sequential SVD pseudoinverse estimates of eqs.(24) and (25) can be expressed in equivalent vector space form as

$$\hat{\underline{f}}_{E_T} = \underline{A}^{-1}\underline{\Delta}_T^{-\frac{1}{2}}\;\underline{A}\,\tilde{\underline{q}}_E \tag{27}$$

where

$$\underline{\Delta}_T^{-\frac{1}{2}} = \begin{bmatrix} [\delta(1)]^{-\frac{1}{2}} & & & & & & 0 \\ & [\delta(2)]^{-\frac{1}{2}} & & & & & \\ & & \ddots & & & & \\ & & & [\delta(T)]^{-\frac{1}{2}} & & & \\ & & & & 0 & & \\ & & & & & \ddots & \\ 0 & & & & & & 0 \end{bmatrix} \tag{28}$$

inherently provides a truncation of the high frequency terms of the Fourier domain pseudoinverse operator. Complete truncation to avoid ill-conditioning is often unnecessary. As an alternative to truncation, the diagonal zero elements could be replaced by unity, or the value of $[\delta(k)]^{-\frac{1}{2}}$, or perhaps by some sequence that declines in value as a function of frequency.

EXPERIMENTAL RESULTS

Figure 1 contains an example of the modified fast SVD pseudoinverse image restoration method. A blurred

(a) blurred observation

(b) restoration
T = 58

(c) restoration
T = 60

Figure 1. Examples of sequential SVD pseudoinverse image restoration for horizontal Gaussian blur with $\sigma_b = 3$, L = 23, J = 256.

image has been formed by computer simulation of horizontal blurring with an impulse response of the form

$$h(\ell) = \exp\left\{-\frac{1}{2}\left[\frac{\ell - \left(\frac{L+1}{2}\right)}{\sigma_b}\right]^2\right\} \tag{29}$$

where σ_b is the blur standard deviation and L is the length of the impulse response. The restoration algorithm involves repeated restoration with eq.(26) in which the J-T zero terms along the diagonal are replaced by $[\delta(T)]^{-\frac{1}{2}}$. In figure 1 for blurring with $\sigma_b = 3$ and L = 23, the onset of ill-conditioning effects occurs between T = 58 and T = 60 modified Fourier coefficients. The restoration at T = 58 exhibits a significant improvement in apparent resolution and subjective quality as compared to the blurred observation.

An example of the restoration of a naturally blurred and noisy image is presented in figure 2. The blurred observation is an electron microscope picture of a metallic crystal which has been digitized for computer processing. The impulse response of the electron microscope was estimated from measurements of micron diameter plastic pellets, and modelled as a Gaussian shape with L = 11 and $\sigma_b = 2$. The restoration obtained exhibits significantly improved detail.

SUMMARY

A new class of pseudoinverse image restoration computational algorithms has been introduced. These algorithms provide an inherent means of avoiding numerical instability effects; restoration of noisy blurred images is possible even when the blur matrix operator is highly ill-conditioned. The computational algorithms require a two dimensional Fourier transform of the blurred image, followed by scalar weighting of the transform coefficients, followed again by an inverse two dimensional Fourier transform to produce a restored image. With these algorithms computational savings by factors of several thousand are possible, as compared to conventional matrix pseudoinversion. Even greater savings are potentially achievable by utilizing parallel Fourier transformation by electronic or electro-optical means.

(a) blurred observation
electron microscope image

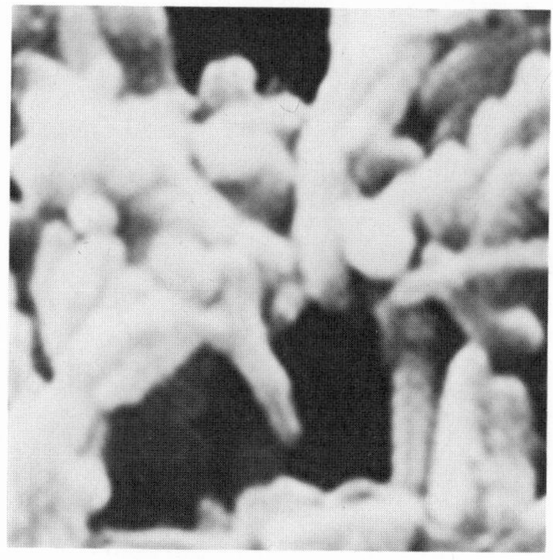

(b) restoration, T = 35

Figure 2. Example of SVD pseudoinverse restoration for Gaussian blur with $\sigma_b = 2$, L = 11, J = 256 employing inverse quadratic attenuation of high order eigen images.

ACKNOWLEDGEMENTS

The assistance of Mr. Ikram Abdou of the University of Southern California Image Processing Institute in the experimental portions of this paper is gratefully acknowledged. This research was supported by the Advanced Research Projects Agency of the Department of Defense and was monitored by the Wright Patterson Air Force Base under Contract No. F-33615-76-C-1203.

REFERENCES

1. M. M. Sondhi, "Image Restoration: The Removal of Spatially Invariant Degradations," Proceedings of the IEEE, Vol. 60, No. 7, July 1972, pp. 842-853.

2. W.K. Pratt and F. Davarian, "Fast Computational Techniques for Pseudoinverse and Wiener Image Restoration," IEEE Transactions on Computers (to be published).

3. H.C. Andrews and C.L. Patterson, "Outer Product Expansions and Their Uses in Digital Image Processing," American Mathematical Monthly, Vol. 1, No. 82, January 1975, pp. 1-13.

4. H.C. Andrews and C.L. Patterson, "Outer Product Expansions and Their Uses in Digital Image Processing," IEEE Transactions on Computers, Vol. C-25, No. 2, February 1976, pp. 140-148.

5. H.C. Andrews and C.L. Patterson, "Singular Value Decompositions and Digital Image Processing," IEEE Transactions on Acoustics, Speech, and Signal Processing," Vol. ASSP-24, No. 1, February 1976, pp. 26-53.

6. W. K. Pratt, Digital Image Processing, Wiley, New York, 1977.

ABOUT NUMERICAL SOLUTIONS OF A MULTIEXTREME PROBLEM OF ALL-PASS FILTER SYNTHESIS

R.D. Baghlay

Institute of Automation and Electrometry
Novosibirsk, USSR

ABSTRACT

Three methods of approximate solution of a multiextreme problem of all-pass filter synthesis (kinoforms) have been considered. The results of phase filter numerical image reconstruction have been given, and a number of important, for applications, characters of approximated solutions have been revealed.

1.

Let us consider the functional

$$\mathcal{J}^{\Omega}(\varphi)=\left\| f-\left|\widecheck{We^{i\varphi}}\right| \right\|_{L_2(S)}, \qquad (*)$$

where $f(x,y)\in L_2(S)$, $W(u,v)\in L_2(\Omega)$ are finite positive functions present, respectively, in the domain of originals and Fourier images or images, which are relative to them; $\check{\zeta}$ is an inverse transformation of the Fourier function ζ .

The problem of all-pass filter synthesis $\left(W=\begin{vmatrix} c,(u,v)\in\Omega \\ 0,(u,v)\notin\Omega \end{vmatrix}, c-const\right)$, intended to reconstruct a fixed module of wave front, or otherwisc, a fixed object image, by optical facilities leads to the statement (*). Problems of radiolocation signal synthesis over the indefinite or correlative function [1], antenna synthesis problems electronic filter problems, and etc. lead to analogous statement, as well. Their precise mathematical solution is unknown. The difficulties are caused by the module of integral in (*). The lack of analytical method, which allows to find the function $\varphi(u,v)$, providing a global minimum to the functional $J(\varphi)$, stimulated a search for approximate solution methods.

1. The following experiment is described in [2]. The phase function $\varphi(u,v,z_0)$ of the wave front $A(u,v,z_0)e^{i\varphi(u,v,z_0)}$ was calculated in the plane uv removed from the object by the distance z_0, when supposing the object to be illuminated with monochromatic diffuse light. Then an all-pass filter (kinoform), realizing the function φ, was created. An image similar to that of an initial object appeared, when illuminating the kinoform with a plane wave, and carrying out the inverse Fourier operation with the help of a lens.

In this experiment the function φ resulting after the direct Fourier transformation $fe^{i\psi_1}=Ae^{i\varphi}$ where ψ_1 is realization of a random function, being represented physically by diffuse light (diffusor), is seen to be taken as a minimizing one.

A mathematical expectation operation (M) was applied to an ensemble of reconstructed images (by the method mentioned above) of the same object with various realizations $\{\psi_n\}_0^\infty$ of the diffuse light Ψ, where the random function Ψ was supposed to be a stationary one $M\Psi=0$, and function $e^{i\psi}$ to be noncorrelated [3]. An average of 78% of wave energy has been shown to be used to form an object image, and 22% of energy to be expended in distortions, being image autoconvolutions. The conclusion is based on the fact that the correlation function of the noise itself $Be^{i\xi}$ [4]

can be determined from the phase correlation function $e^{i\zeta}$ of the narrowband stationary noise $Be^{i\zeta}$.

Instead of the functional (*) one should consider its mathematical expectation, or the expression

$$M\left\| f-\frac{\sum_{k=0}^{n}\left|We^{i\varphi_k}\right|}{n}\right\|^2 ,$$

under conditions of application of many realizations and reconstruction of an average image.

2. To appreciate the meaning of the given example another method, based on a clear mathematical fact, will be recalled

$$\min_{\varphi} = \mathcal{J}(\varphi) = \min_{\psi,\varphi} \left\| fe^{i\psi} - We^{i\varphi}\right\|^2 = \min_{\psi,\varphi} \tilde{\mathcal{J}}(\psi,\varphi),$$

From this it follows that the functional $\mathcal{J}(\varphi)$ local minimum can be found through minimization of the functional $\tilde{\mathcal{J}}(\psi,\varphi)$ in succession over each variable. The order of actions in this method of solution is following

- f is multiplied by $e^{i\psi_0}$, where ψ_0 is an arbitrary initial phase function;
- the Fourier direct transformation is made $\widehat{fe^{i\psi_0}} = \eta_1 e^{i\varphi_1}$;
- W is multiplied by $e^{i\varphi_1}$;
- the Fourier inverse transformation is made $\widecheck{We^{i\varphi_1}} = f_1 e^{i\psi_1}$;
- the norm $\|f - f_1\|$ is calculated.

The first iterative cycle is completed by this. If a norm is great the process will be continued, that is f is multiplied by $e^{i\psi_1}$, and the direct transformation is made $\widehat{fe^{i\psi_1}} = \eta_2 e^{i\varphi_2}$; W is multiplied by $e^{i\varphi_2}$, and the inverse transformation is made $We^{i\varphi_2} = f_2 e^{i\psi_2}$; the norms $\|f - f_2\|$ is calculated, and etc. Iteration is performed until the distance between f and f_k, $k=1,2,\ldots$, is decreasing. If obtained norm value $\|\cdot\|$ is great, another initial phase ψ_0^1 is chosen, and the process is repeated.

Multiplication of f and W by $e^{i\psi_k}$ and $e^{i\varphi_k}$, respectively, in the domain of originals and images is allowable, as f and W are preset by modules, and their phases are arbitrary. It can readily seen that a mathematical operation of orthogonal projecting corresponds to the mentioned above multiplication. Single result of each operation of this kind is ensured by the fact that the functions $f_k e^{i\psi_k}$ and $\eta_k^{i\varphi_k}$ are whole.

Due to the fact that the problem is multiextrem a successful application of this method intirely depends on the initial phase ψ_0 . For various ψ_0 the iterative method can lead to different local minima, and therefore to different reconstruction errors. It should be noted that the process given above does not converge. But as far as applications are concerned it is of little importance, as iteration allows to reduce the magnitude of the initial approximation error, in any case it doesn't increase it.

By comparison we see that a synthesis method used in [2], coinsides with the first iterative cycle of the method given below, when a random function realization is taken for ψ_0 in it. Further we shall se that the fact of accidental nature is of no importance herein, as well. A new step has been done in paper [3], where a quantitative value of a plane wave energy used for formation of an averaged image has been given, and the distortion nature revealed. Practical significance of this step is not great, yet with regard to it the first method assumes its independent meaning. Mechanism of reconstructed image formation and the problem of choice of optimal, as to minimum, reconstruction error of the random function Ψ have not been sufficiently studied in it.

3. Let us consider the third method of approximated solution, based on change of module of integral in the functional (*) for the formula of asymptotic approximation [5,6,7]. At first let us investigate the case

of a function with a single variable $f(x), -X \leqslant x \leqslant X$ and $W(u)$, $-U \leqslant u \leqslant U$, with a unit norm.

Minimization of the functional $\mathcal{J}(\varphi)$ is equivalent to maximization of scalar product $(f, |\widecheck{We^{i\varphi}}|)$. If we suppose that W changes slowly, we shall search for the maximum on the class of strict convexity functions $\{\varphi(u)\}$ from C^2 of such kind, that $\varphi'(u)$ changes from $-X$ to X, while u changes from $-U$ to U (values $\varphi'(u)$ have physical meaning of a group shift at frequency u). On this assumption and assume, for example, $\varphi''(u_0)>0$, we can write:

$$\frac{1}{2\pi}\int_{-U}^{U} W(u)e^{i(\varphi(u)-ux)}\,du =$$

$$= W(u_0)(2\pi\varphi''(u_0))^{-\frac{1}{2}}\, e^{i(\varphi(u_0)-u_0x+\frac{\pi}{4})} + O(x)\frac{1}{XU},$$

where the stationary point u_0 is determined from the equation

$$\varphi'(u_0)=x,$$

$O(x)$ is a function slightly depending on x. When limitations on the function W are not important for practice, the correction term $O(x)\frac{1}{XU}$ can be arbitrarily decreased at the expense of the increase of U for the fixed value X. In particular, this fact is used in Fourier spectrum theory of frequency modulated signals, at great indices of modulation [7,8].

Taking into account the triangle rule, we can write

$$|\widecheck{We^{i\varphi}}| = W(u_0)(2n\varphi''(u_0))^{-\frac{1}{2}} + O\left(\frac{1}{UX}\right).$$

Then

$$(f, |\widecheck{We^{i\varphi}}|) \simeq \int_{-X}^{X} f(x)W(u_0)(2\pi\varphi''(u_0))^{-\frac{1}{2}}\,dx = I$$

and changing x for $\varphi'(u_0)$, we have

$$I=(2\pi)^{-\frac{1}{2}}\int_{-U=\varphi'^{-1}(-X)}^{U=\varphi'^{-1}(X)} f(\varphi'(u))W(u)\left(\frac{d\varphi'(u)}{du}\right)^{\frac{1}{2}} du .$$

According to Schwarz inequality

$$I^2\leqslant\int_{-X}^{X} f^2(\varphi')d\varphi'(2\pi)^{-1}\int_{-U}^{U} W(u)du$$

for the upper limit of scalar product we'll have an equation

$$f^2(\varphi')d\varphi'=(2\pi)^{-1}W^2(u)du.$$

At boundary condition $\varphi'(U)=X$ its solution gives the requred phase φ to an accuracy of the arbitrary constant.

In case of functions of two independent variables the condition $\varphi''(u)>0$ is changed for the following

$$\varphi''_{u,u}\varphi''_{v,v}-(\varphi''_{u,v})^2>0,$$

and the stationary point (u_0,v_0) is determined from the system

$$\varphi'_u(u_0,v_0)=x$$

$$\varphi'_v(u_0,v_0)=y.$$

If we change the integral module $|\widehat{We^{i\varphi}}|$ for the formula of asymptotic approximation [6,7], from Schwarz inequality for the upper limit of a scalar product we have for φ non-linear elliptical equation of the Monge-Ampere type:

$$f(u,v)(\varphi''_{u,u}\varphi''_{v,v}-(\varphi''_{u,v})^2)=(2\pi)^{-2}W^2(u,v). \quad (1)$$

Monographs [9,10] are devoted to investigation of equations of this type. Practical application of the results of these investigations to the problem being under consideration is difficult, because of the lack of the developed method of numerical solution and constractive estimates of approximation. Yet, to solve problems (1) is not difficult, when synthesizing all-pass filters, and under assumption that f is the product $f(x,y)=f(x)f(y)$. As these functions play an important role when solving optical problems, we gave numerical experiments with functions of this kind only, while using the third method.

From described above, it is seen that the results of numerical experiments can be of practical use in the problem under consideration.

2.

Experimental conditions are the following: images were preset through values at points on a regular grid of 64 x 64; a diffuse light effect was simulated with the use of a random numbers (noise) generator with regular distribution; Fourier transformation was performed with a FFT (Fast Fourier transformation);modules of initial and reconstructed images were dumped by six gradations.

Principle aims of the experiment: using some examples, let us condsider dependence of the part of plane wave energy, forming a reconstructed image, on the magnitude of noise dispersion, and see if the fact of a random change of a phase function, when reconstructing through a single iterative cycle, is important or not; we shall illustrate the nature of distortions arrising while noise dispersion is reducing to zero, get an averaged image, and give the results of iterative procedure application; we shall investigate reconstruction efficiency according to image size at different values of noise dispersion; we shall give examples of image reconstruction when selecting the phase function from the equation (1); we shall compare typical characters of reconstructed images with the use of kinoforms and diffuse holograms.

In the first experiment the initial function

$f(n,m)$, $-32 \leq n,m \leq 32$, was preset by units on the grid of 24 x 24, and by zeroes at the rest of the points of the domain mentioned (Fig. 1a). Later on functions of this kind will be called binary functions. Applied interest to binary function reconstruction is conditioned, in particular, by the fact that they can represent a memory page which, later on, is coded in the form of kinoforms or holograms. The function reconstructed $|We^{i\varphi_1}|$, where $W(n,m)=c$, $-32 \leq n,m \leq 32$ $\varphi_1 = \arg f e^{i\psi_0}$ and ψ_0 is realization of the random function Ψ with regular distribution at $M\Psi = 0$ and $|\Psi_{max}| = \pi$, is shown in Fig. 1b. We notice that it is easy to follow its contour visually, but the values at different points, embraced by the contour, are "scattered", to a great extent. Ratio of energy at the points of the domain where the initial function is different from zero to total energy in the field of 64 x 64 points was 90 %, accuracy to the unit of the last decimal place.Nothing has practically changed when realization of binary noise of the amplitude was used as ψ_0. And nothing has changed, as well, when a regular function, e.g. a sign--variable function increasing linearly in both arguments, was used as ψ_0. At the same time, the Fourier function spectrum ψ_0 is important to be rich in higher harmonics, i.e. to be close to a uniform one.

The results of the experiment with the same function f, at $|\Psi_{max}| = \frac{\pi}{4}; \frac{\pi}{16}; \frac{\pi}{64}; 0$ are given in Fig. 1c,d,e,f respectively. At the same time, the following energy magnitudes have been obtained - 84%, 79%, 62% and 39%.

For the function $f(n,m)$, preset by units on the grid 12 x 12, and by zeroes, at the rest of the points in the same order of values $|\Psi_{max}|$, the following energy magnitudes have been obtained - 85%, 82%, 70%, 49%, and 31%. These experiments and those analogous to them with other binary functions showed that the value $|\Psi_{max}|$ changes, within the $\pi \div \frac{\pi}{2}$ range, have little effect on the energy magnitude, which gets to a number of points, embraced by the contour of the image reconstructed.

Reconstructed realization and the result of averaging over twenty realizations,are given in Fig.2a,b. One-

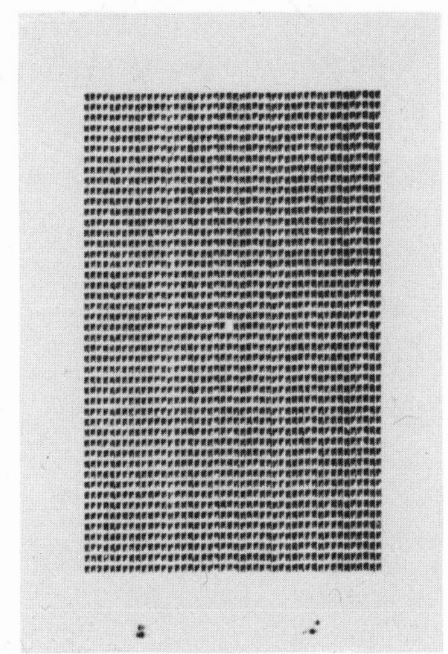

Fig. 1a

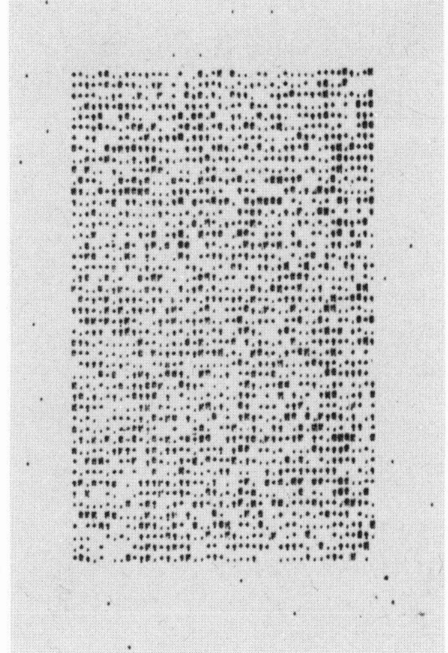

Fig. 1b

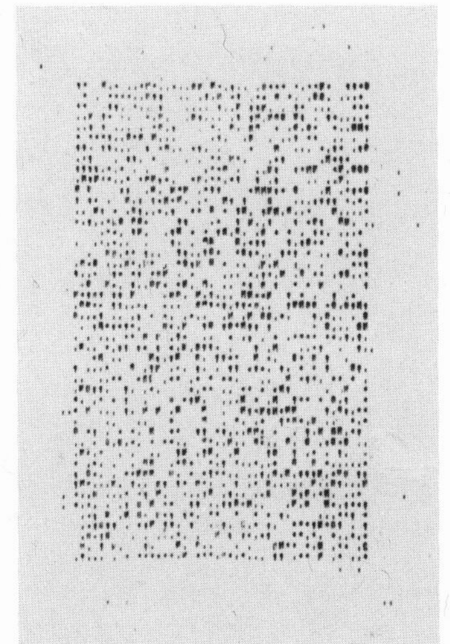

Fig. 1c

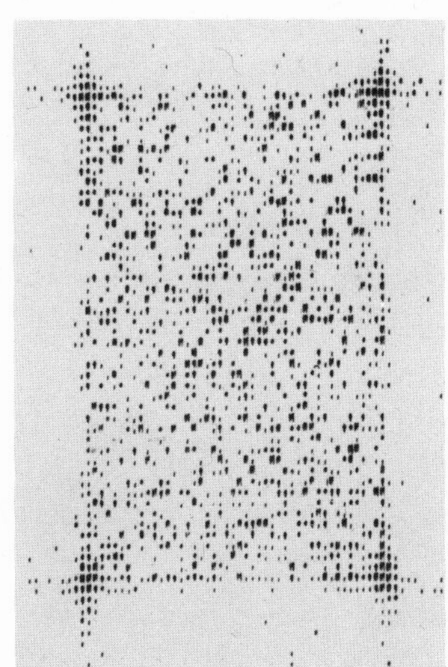

Fig.1d

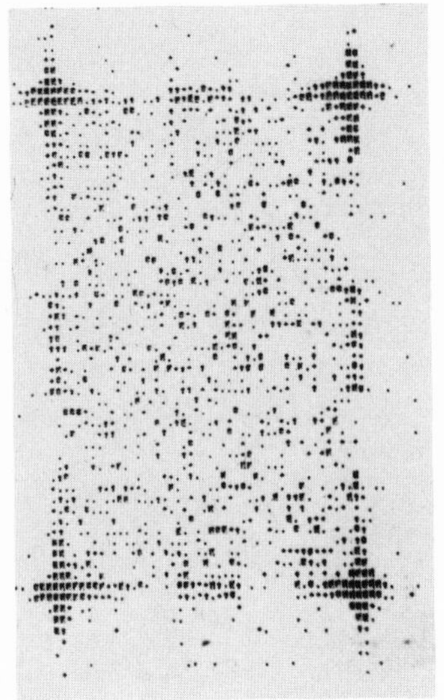

Fig. 1e

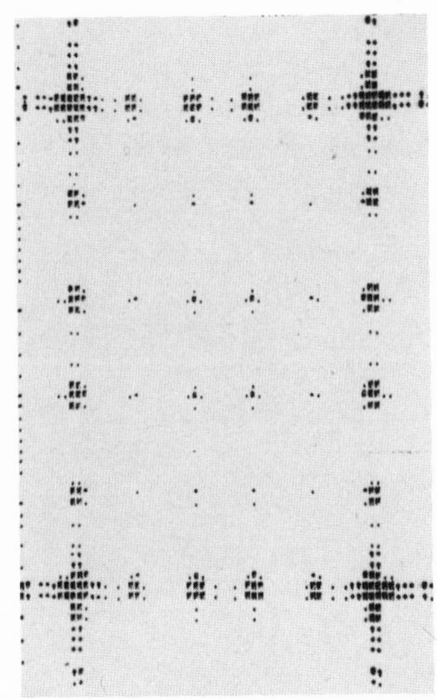

Fig. 1f

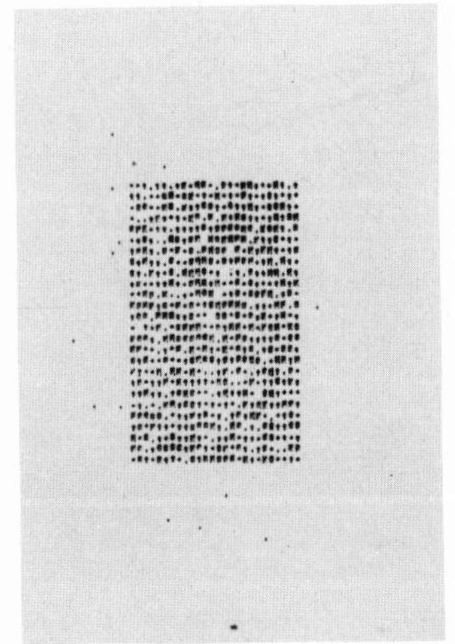

Fig. 2a

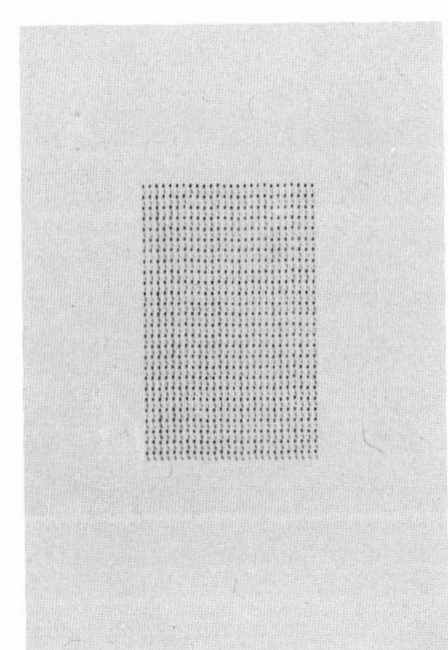

Fig. 2b

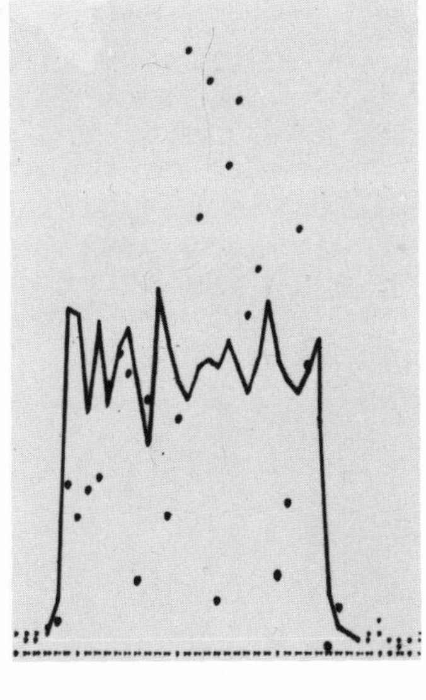

Fig. 2c

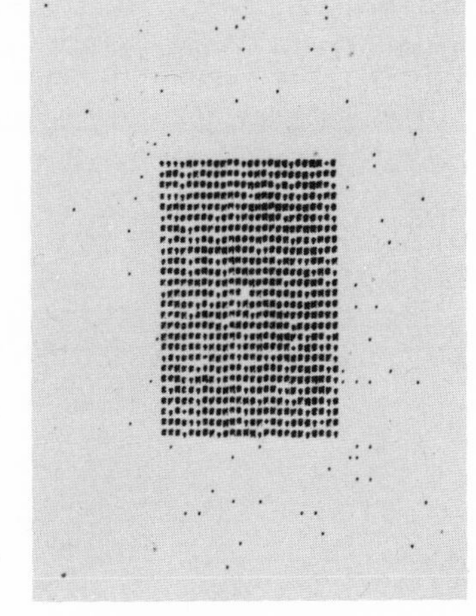

Fig. 2d

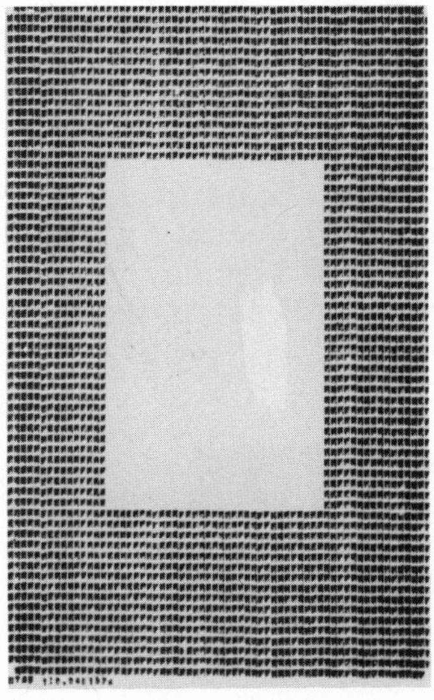

Fig. 3a

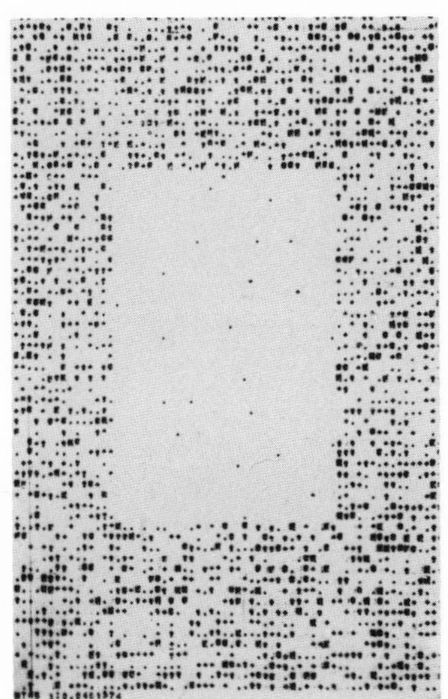

Fig. 3b

-dimensional functions being cross profiles along the same line of reconstructed image, before averaging (a broken line) and after averaging (a continuous line), are shown in Fig. 2c. When comparing them, averaging is seen to give a good result. But it is difficult to use it in a physical experiment. The same function, obtained after five iterative cycles, is given in Fig. 2d. When comparing it with the result obtained after one cycle (Fig. 2a), we see that the reconstructed image has been considerably improved. Practically, it appeared to be the same as the averaged one (Fig. 2b). Yet, iteration was not a great success, when applied to another binary function Fig. 3a; after the fifth cycle (Fig. 3b) the result remained the same as after the first one.

Initial functions which are not binary, and the examples of their reconstruction at $|\Psi_{max}| = \pi$ are given in Fig. 4a,b.

For comparison we adduce here the results of image reconstruction with a complete diffuse hologram and its parts. Complete image Fresnel diffuse hologram of figure 6, and two conjugated images, obtained with its help, are shown, respectively, in Fig. 5a,b (a component, caused by its background was not dumped). Different parts of the same hologram and images resulting from their use are shown in Fig. 5c,d, and 5e,f. We see, when using a part of a hologram, the situation, analogous to the previous one, that is contours of a reconstructed image are well seen, but the values at its different points are greatly scattered. This circumstance prevents a lot of hologram and Kinoform applications.

As the experiments have shown, successful use of Kinoforms depends on the size of the initial object. Fig. 6a,b,c represents the images of the word "SOS", different in size, reconstructed at $|\Psi_{max}| = \pi$. In the last case the solution "ruined". Fig. 7a,b, represents an unsuccessful attempt to obtain a figure 6 image, small in size. Though the same picture is readily reconstructed with a diffuse hologram (Fig. 8a), including multiplication (Fig. 8b). However, when reducing the magnitude $|\Psi_{max}|$ up to $\pi/2$, we obtain Fig. 6d from Kinoform. Thus, the problem of compromise selecting of the magnitude $|\Psi_{max}|$ arises, when synthesizing kinoforms for objects are small in size.

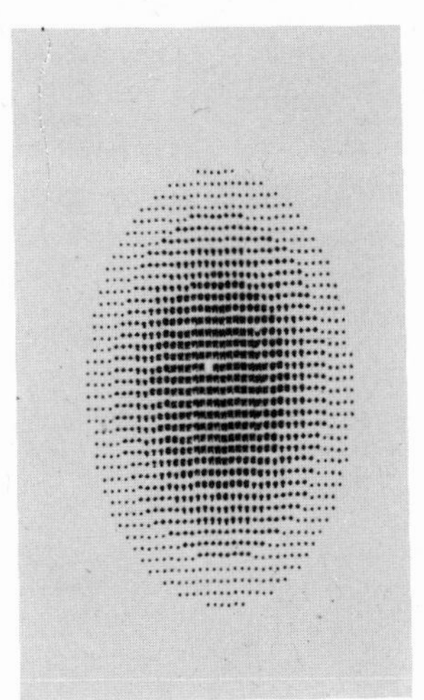

Fig. 4a

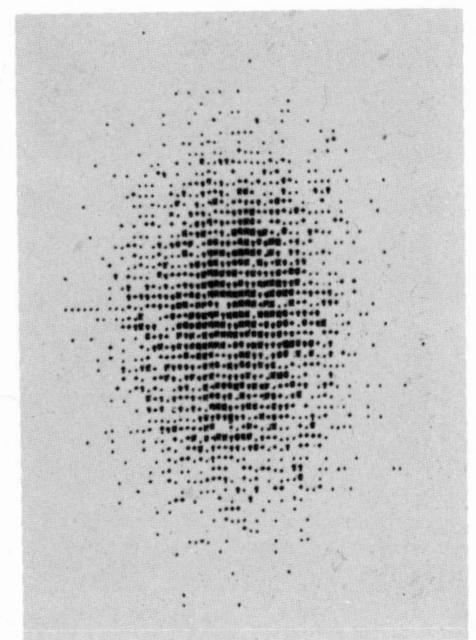

Fig. 4b

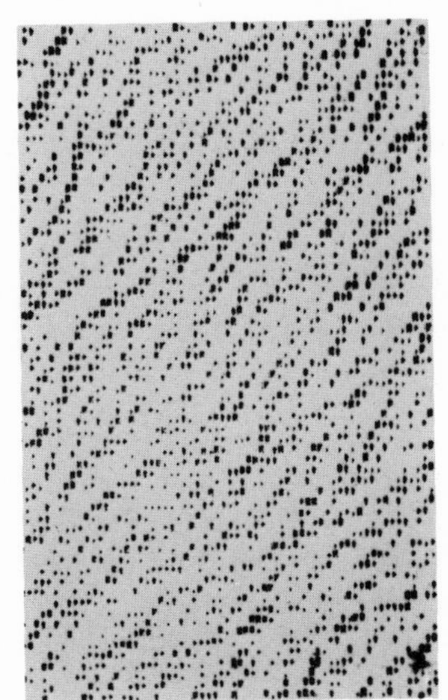

Fig. 5a

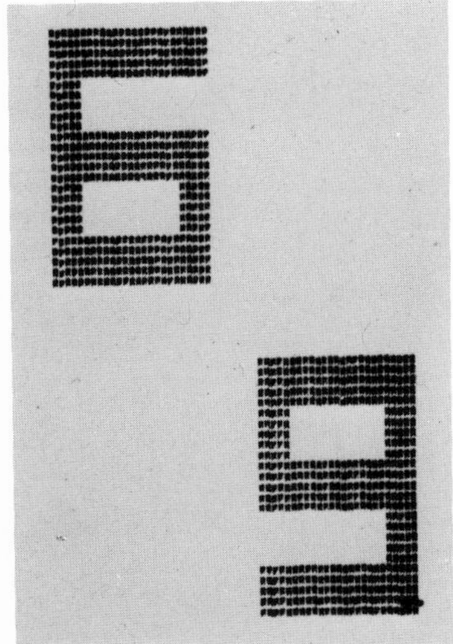

Fig. 5b

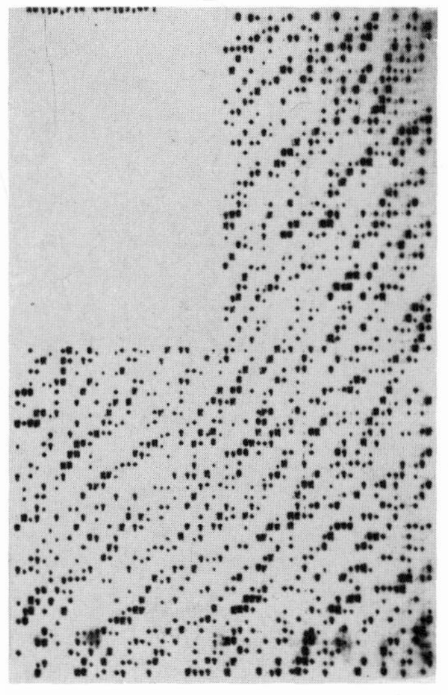

Fig. 5c

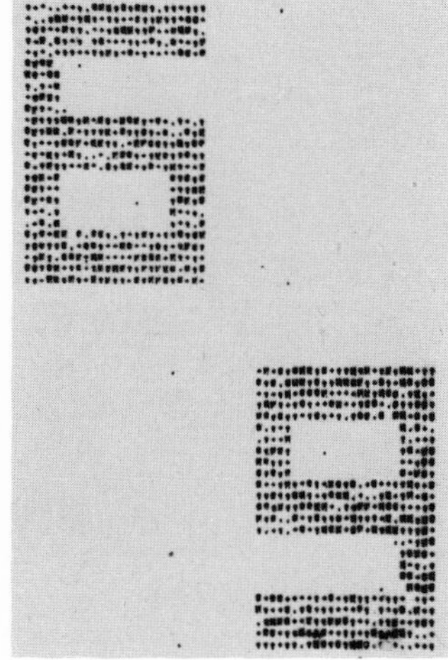

Fig. 5d

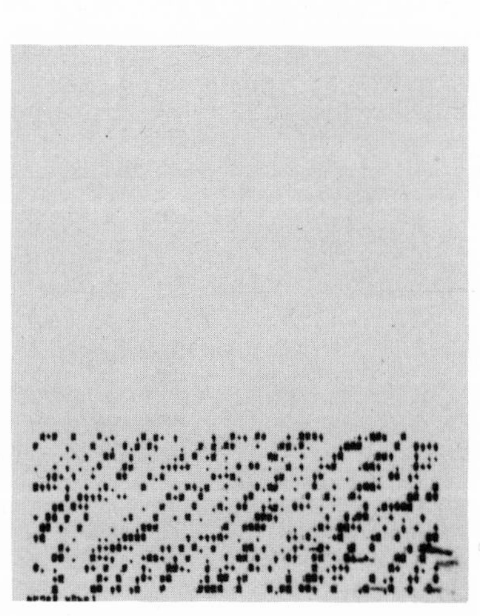

Fig. 5e

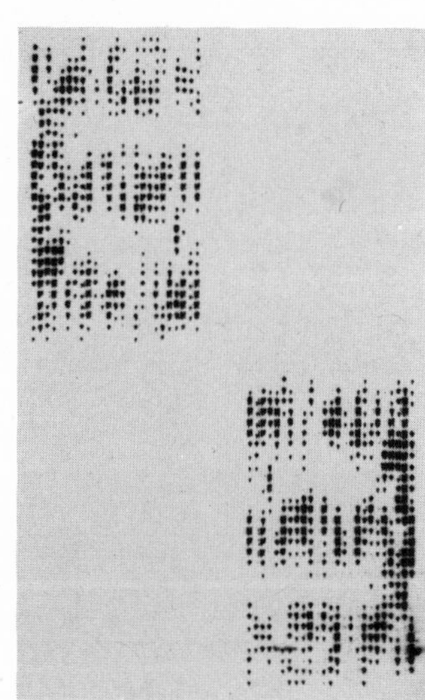

Fig. 5f

Fig. 6a

Fig. 6b

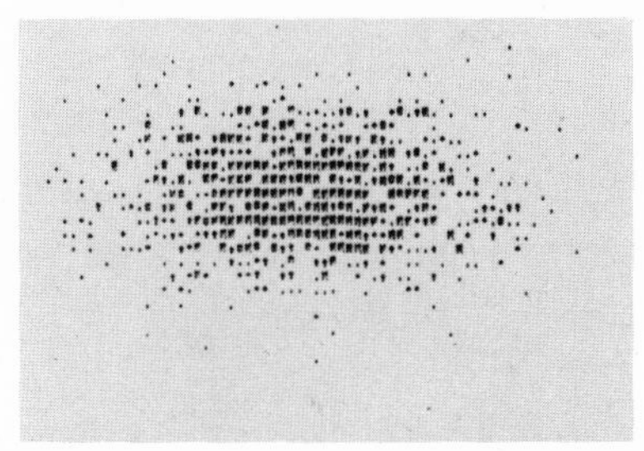

Fig. 6c

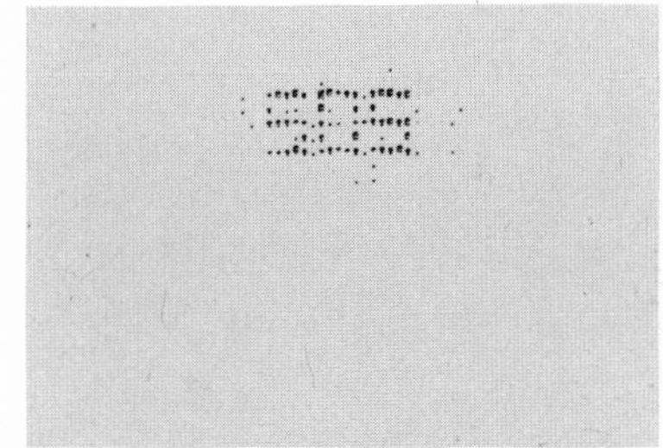

Fig. 6d

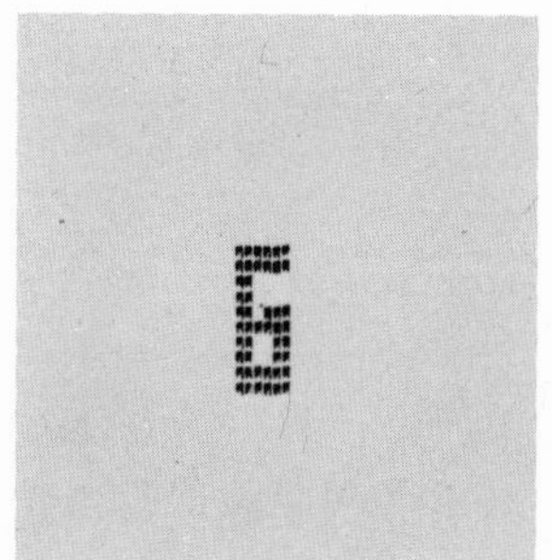

Fig. 7a

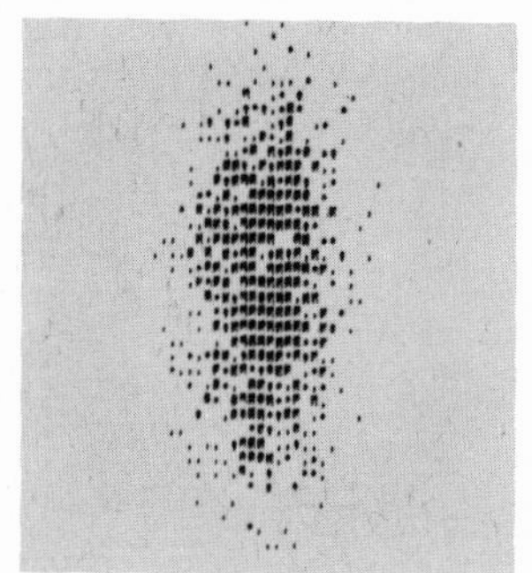

Fig. 7b

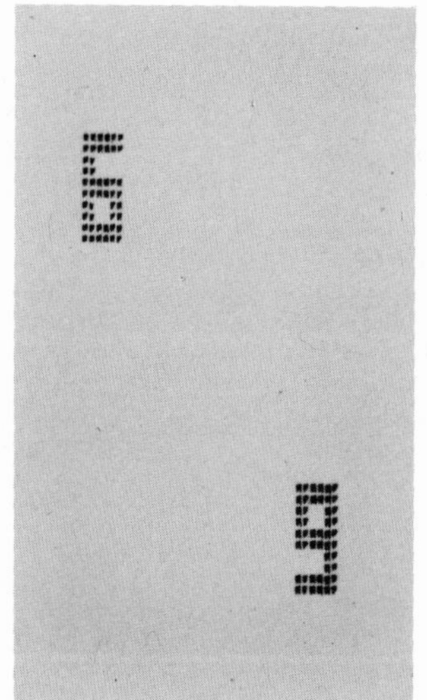

Fig. 8a

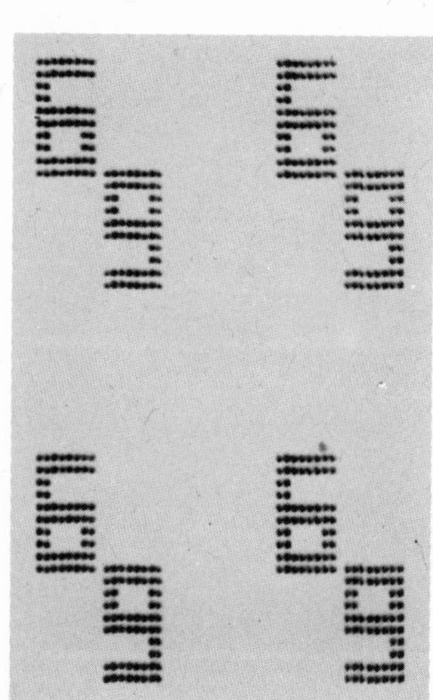

Fig. 8b

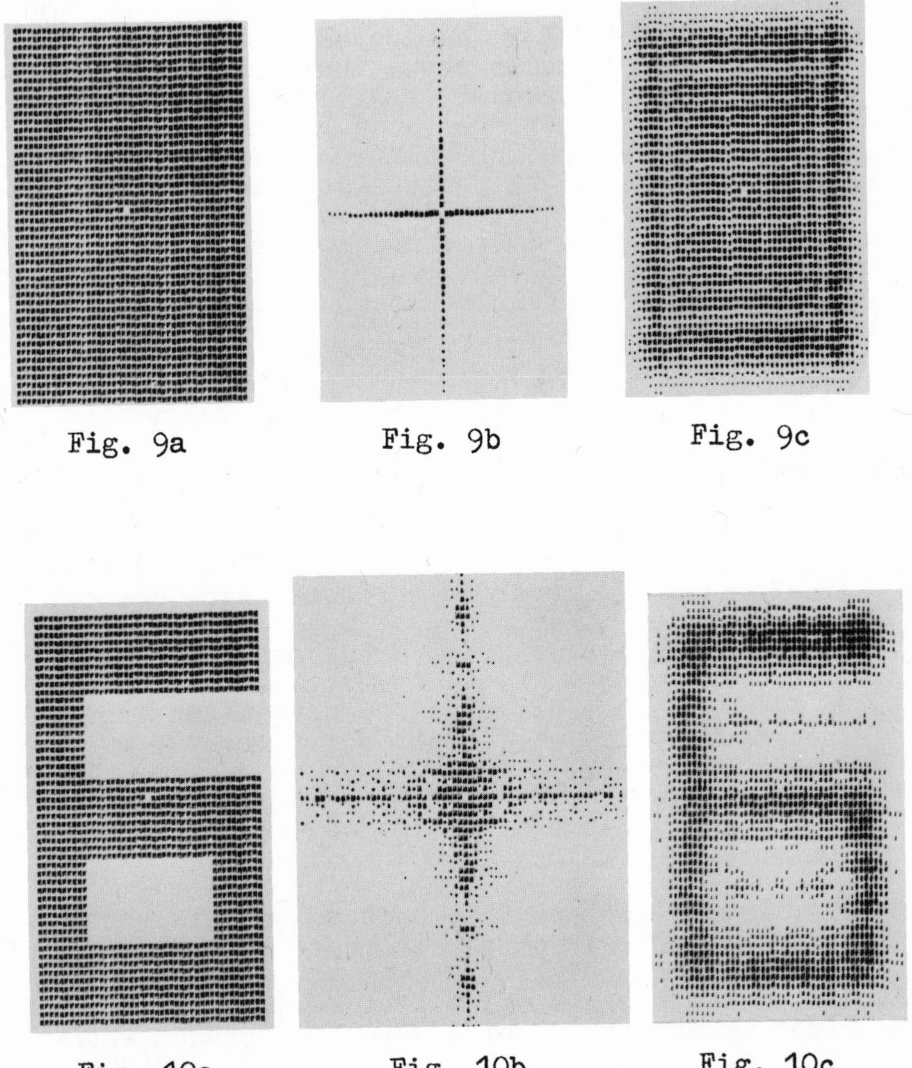

Fig. 9a Fig. 9b Fig. 9c

Fig. 10a Fig. 10b Fig. 10c

The results, obtained by using the third method of solution, i.e. when the required phase function φ was determined from equation (1), are given in Fig. 9c and 10c. In doing so, binary functions (Fig. 9a, 10a) were preset in the domain of Fourier images. Their inverse transformations at $\varphi = 0$ are given in Fig. 9b, 10b, respectively, and at φ selected from equation (1) - in Fig. 9c, 10c. Fig. 9c shows that characteristic distortions caused by Fresnel oscillations, which amplitude increases to the edges of a reconstructed image, where f and W are irregular, arise here. The mean--square error caused by these distortions is reduced as Ω domain increases, and becomes negligible, in the limit. The result of reconstruction can be considerably improved for reasonable Ω magnitudes, if the function W, preset in the Fourier image domain, will not have sharp overfalls.

REFERENCES

1. D.E. V a k m a n, P.M. S e d l e t s k i i. Voprosi sinteza signalov. - M., "Sovetskoe radio", 1973.

2. L.B. L i s e m, P.M. H i r s h, J.A. J o r d a n. The kinoform: a new maveform reconstruction device.-IBM Res.Dev., 1969, March, pp. 150-154 ("Zarubezhnaya radioelectronika"), 1969, N 12, str. 41-49.

3. D. K e r m i s c h. Image reconstruction from phase information only. - J. Opt. Soc.Amer., v. 60, 1970, Januar, pp.15-17.

4. D. M i d d l e t o n. An Introduction to Statistical Communication.Theory. 1960.

5. M.A. Lavrentyev and B.V. Shabat. Metodi teorii functzii komplexnogo peremennogo. M., "Nauka",1965.

6. M.I. K o n t o r o v i t c h, Yu.K. M u r a v y e v. Vivod zakona otrazheniya geometritcheskoi optiki na osnove asumptotitcheskoi traktovki zadachi difraktsii. Zh. TF, XXII, 3, 1952, stp.394-407.

7. D.E. V a k m a n. Asimptotitcheskie metodi v lineinoi radotekhnike. M., "Sovetskoe radio", 1962.

8. A.A. H a r k e v i t c h. Spektri i analis. Fizmatgiz, M., 1962.

9. A.V. P o g o r e l o v. Ob uravneniyakh Monge-Ampere elliptitcheskogo tipa. Harkov, Izd-vo HGU 1960.

10. I.Ya. B e k e l m a n. Geometritcheskie metodi reshenia elliptitcheskikh uravnenii. M., "Nauka", 1965.

AN INTRODUCTION TO INTEGRATED OPTICS*

Herwig Kogelnik
Bell Laboratories
Holmdel, New Jersey 07733

I. INTRODUCTION

It is now five years since the name "integrated optics" was coined [1], and there are now research activities in this field in an increasing number of industrial, governmental, and university laboratories. For the interested reader there is already available a good collection of survey articles [2]-[5] which review both the history and the fundamentals of this new discipline. The purpose of this paper is to give the reader an introduction to integrated optics and an illustration of present day activities. While no attempt is made to present a complete survey, we will try to collect the major thoughts and arguments which have motivated and stimulated work in this field.

Clearly, one impetus stems for the promise of optical-fiber transmission systems [6] which is rooted in the recent achievement of low transmission losses in optical fibers. Integrated optics may one day provide circuits and devices for the repeaters of these systems, in particular, when higher transmission speeds are envisaged, or it may offer possibilities for wavelength multiplexing or switching of optical signals. The spectral range of interest is, therefore, in the visible and near infrared where fiber losses are low. There is also activity at 10.6 μm, the wavelength of the CO_2 laser where device improvements are needed for communications and the processing of laser radar signals.

The area defined by the name "integrated optics" has expanded gradually, and it now includes all exploration towards the use of guided-wave techniques to fabricate new or improved optical devices.

*Appeared originally in IEEE Transactions on Microwave Theory and Techniques (MTT), Vol. MTT-23, No. 1, pp. 2-16, Jan. 1975. Reprinted by permission.

With this, one associates compact and miniaturized devices, and hopes that small size will lead to better reliability, better mechanical and thermal stability, and to lower power consumption and drive voltages in active devices. Of course, there is also the promise of integration, i.e., that one will be able to combine several guided-wave devices and form more complicated circuits on a common substrate or chip. However, some of the new guided-wave devices, lasers or modulators, for example, may well be able to compete on their individual merits with their bulk-optical counterparts.

The waveguides used in integrated optics are dielectric waveguides, usually in the form of a planar film or strip with a refractive index higher than that of the substrate. The guiding mechanism is total internal reflection. Typical refractive index differences are 10^{-3} - 10^{-2}, and guide losses of less than 1 dB/cm are desirable.

The fabrication techniques are reminiscent of the technology of electronic integrated circuits. Planar structures predominate and photolithographic techniques are employed. Naturally there are new requirements, a particularly stringent one being the need for an edge roughness smaller than 500 Å to keep light scattering losses to a tolerable level.

The devices we talk about are the counterparts of familiar microwave and optical devices. They are couplers, junctions, directional couplers, filters, wavelength multiplexers, and active devices such as modulators, switches, light deflectors, and lasers. Later sections will illustrate some of the recent device work in more detail, but first we will give a brief survey of the characteristics of dielectric waveguides.

II. DIELECTRIC WAVEGUIDES

The properties of dielectric waveguides are discussed in great detail in the cited review articles [2]-[5] as well as in recent textbooks on this subject [7]-[9]. The planar slab waveguide is one of the simplest waveguide structures and it is also one of the most commonly used in integrated optics. We shall use it in the following to illuminate the principal characteristics of dielectric guides.

A. The Asymmetric Slab Waveguide

The planar asymmetric slab guide structure is illustrated in Fig. 1. We have a film of thickness f (usually a fraction of a micron) and of refractive index n_f, and a cover and substrate material of lower index n_c and n_s respectively, so that

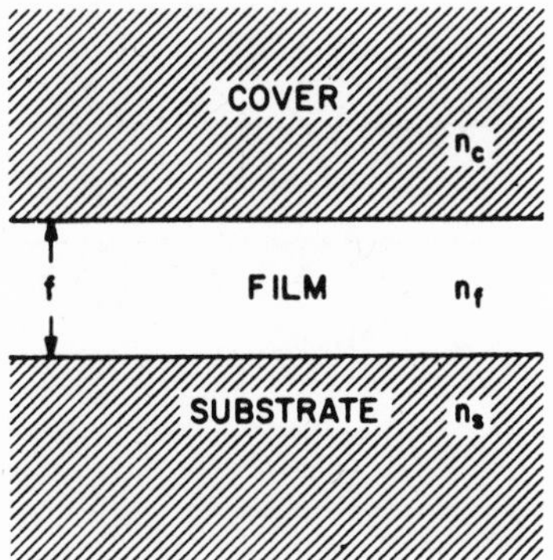

Figure 1. Schematic of asymmetric dielectric slab waveguide.

$$n_f > n_s \geq n_c. \quad (1)$$

The structure is called asymmetric when $n_s \neq n_c$ and symmetric when $n_s = n_c$.

In dielectric guides, we distinguish between two types of electromagnetic modes, guided modes and radiation modes. Following Tien [4] we can use ray arguments to distinguish between these types. This is illustrated in Fig. 2, where we have shown light rays incident on the structure with increasing angles of incidence. In Fig. 2(a) the incident ray penetrates through the structure. On its passage it is refracted according to Snell's law:

$$n_f \sin \Theta_f = n_s \sin \Theta_s = n_c \sin \Theta_c \quad (2)$$

The light escapes again from the structure in this situation which we associate with a radiation mode. This is also the case for the situation depicted in Fig. 2(b) where the light is incident at a somewhat larger angle Θ_f. We have, still, refraction at the substrate-film interface, but at the film-cover interface we surpass the critical angle and the light is totally reflected and escapes through the substrate. In Fig. 2(c) the angle $\Theta \equiv \Theta_f$ is still larger, surpassing the critical angle both at the film-substrate and film-cover interfaces resulting in total internal reflection at both locations. The light is trapped in the film which corresponds to a guided mode of the structure.

According to this zigzag model a guided mode is represented by plane waves traveling in a zigzag path through the film. The model fields propagate like exp $(-j\beta z)$ in the z direction with a propagation constant β related to the zigzag angle θ by

$$\beta = kn_f \sin \theta, \quad (3)$$

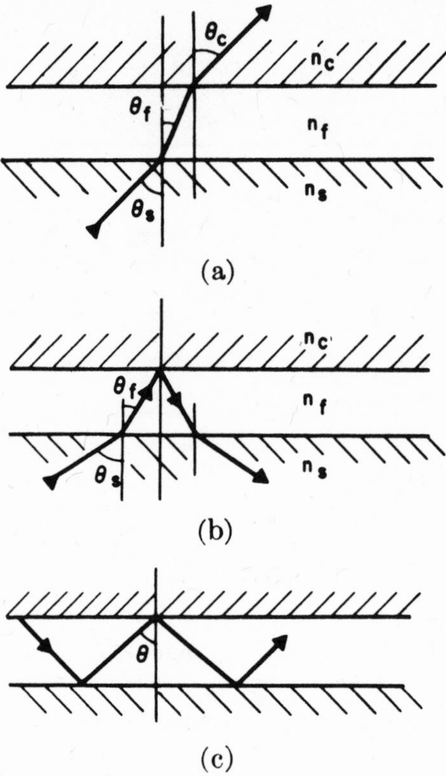

Figure 2. Ray pictures illustrating radiation modes (a), substrate modes (b), and guided modes (c) in s dielectric slab waveguide.

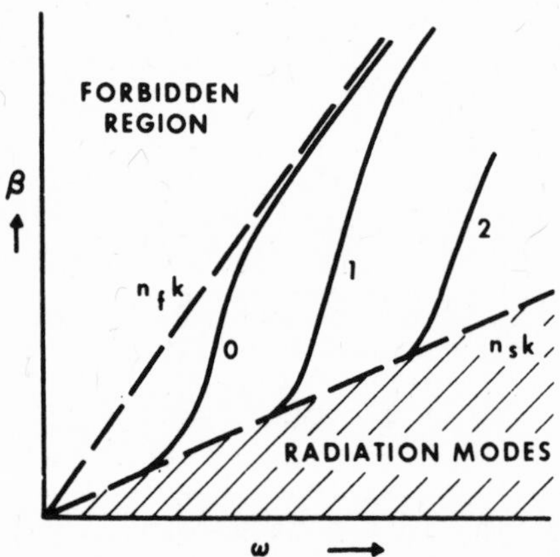

Figure 3. Typical ω-β diagram for dielectric slab waveguides. Three discrete guided modes are shown.

where $k = 2\pi/\lambda = \omega/c$, λ is the free-space wavelength, ω the angular frequency of the light, and c the velocity of light. Equation (3) and the critical angle produce bounds for β which are

$$kn_s < \beta < kn_f. \tag{4}$$

The lower bound is reached at cutoff, i.e., at the critical angle and the upper bound is approached for glancing angles of incidence θ. Figure 3 shows a typical ω-β diagram for the first three modes (labeled 0, 1, and 2) which indicates the variation of β with frequency ω (or relative film thickness). Typically, we have a discrete spectrum of guided modes and a continuous spectrum of radiation modes.

In discussing dielectric waveguides it is often convenient to define an effective guide index N which is related to the propagation constant β by

$$N = \beta/k = n_f \sin\theta. \tag{5}$$

Its values are bounded by the substrate and film indices

$$n_s < N < n_f. \tag{6}$$

We will mention one use of this effective index later in connection with strip guides.

While the field of a guided mode can be viewed as a zigzagging plane wave in the film, we have evanescent fields in the substrate and the cover which decay as exp $(-\gamma x)$. The expressions for the decay constants γ_s and γ_c follow from Maxwell's equations as

$$\gamma_s^2 = k^2(N^2 - n_s^2)$$

$$\gamma_c^2 = k^2(N^2 - n_c^2) \tag{7}$$

where the subscripts s and c refer to substrate and cover, respectively.

We can use the zigzag model to determine the propagation constant of a guided mode if we bear in mind the phase shift that occurs on total reflection of plane waves. For phase shifts $2\phi_s$ and $2\phi_c$ at the film-substrate and film-cover interfaces we have the formulas [4]

$$\tan^2\phi_s = (N^2 - n_s^2)/(n_f^2 - N^2)$$

$$\tan^2\phi_c = (N^2 - n_c^2)/(n_f^2 - N^2) \tag{8}$$

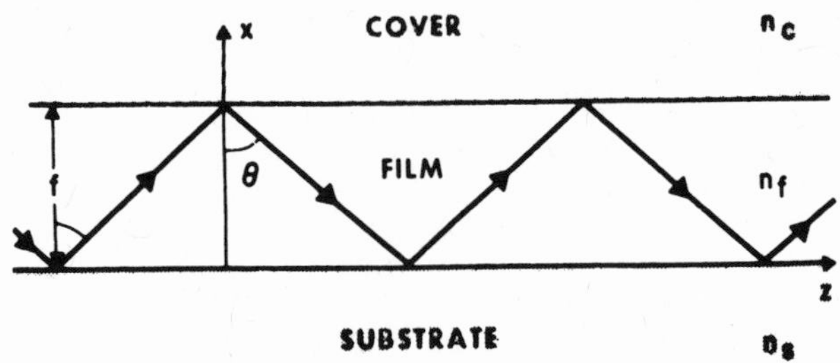

Figure 4. Zigzag wave picture for slab waveguides.

for waves with TE polarization, i.e., for electric fields perpendicular to the plane of incidence.

Figure 4 shows the zigzag model and our choice of the coordinate system. To obtain the self-consistent solution corresponding to a waveguide mode we require that the phase shifts accumulated during one complete zigzag add up to a multiple of 2π. The phase shift corresponding to a single transversal of the film (in the z direction) is $kn_f f \cos\theta$. For a complete zigzag we have

$$2kn_f f \cos\theta - 2\phi_s - 2\phi_c = 2m\pi. \tag{9}$$

This is the dispersion relation which determines $\beta(\omega)$. Approximate solutions for this are available [10], but for the general case, we have to resort to a numerical evaluation. To make the numerical results broadly applicable it is useful to introduce a series of normalizations for the guide parameters. The first is the normalized film thickness V defined as

$$V = k \cdot f \cdot (n_f^2 - n_s^2)^{1/2} \tag{10}$$

and the second is a measure of the asymmetry of the structure [11] given by

$$a = (n_s^2 - n_c^2)/(n_f^2 - n_s^2). \tag{11}$$

This measure applies to the TE modes and ranges in value from zero for perfect symmetry ($n_s = n_c$) to infinity for strong asymmetry ($n_s \neq n_c$ and $n_s \to n_f$). As an illustration we list in Table 1 the refractive indices of some practical guide structures together with the asymmetry measure (a_E). The third is the definition of a "normalized guide index" b related to the effective index N by

$$b = (N^2 - n_s^2)/(n_f^2 - n_s^2) \tag{12}$$

which takes on zero value at cutoff and has a maximum value of unity. Figure 5 shows a plot of the index b as a function of the

Table 1. Asymmetry measures for TE modes (a_E) and TM modes (a_M).

Waveguide	n_s	n_f	n_c	a_E	a_M
GaAlAs, double heterostructure	3.55	3.6	3.55	0	0
Sputtered glass	1.515	1.62	1	3.9	27.1
Outdiffused $LiNbO_3$	2.214	2.215	1	881	21,206

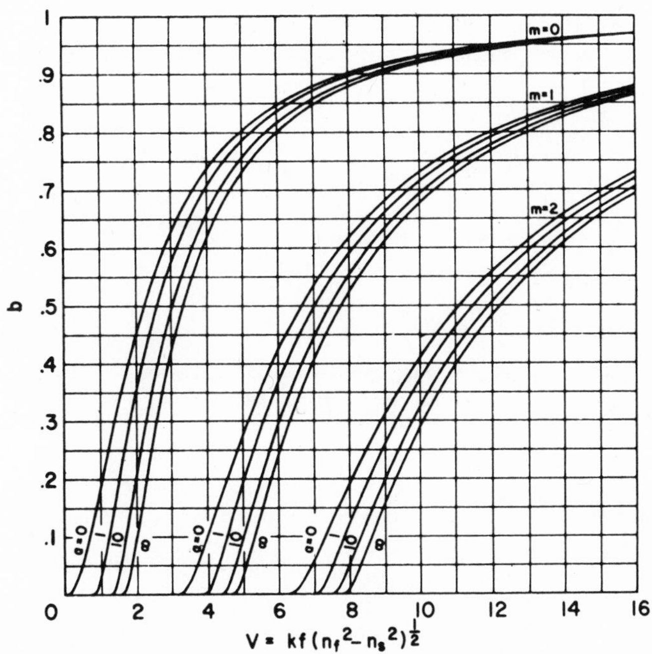

Figure 5. Guide index b as a function of normalized frequency V for slab waveguides with various degrees of asymmetry. (From [11]).

normalized film thickness V (which is also proportional to ω) for various degrees of asymmetry in the guide structure. Values for the first three TE modes are plotted. This plot is a close relative of the ω-β diagram, in particular as we have a simple linear relation

$$N \approx n_s + b(n_f - n_s) \tag{13}$$

when the difference between the film and substrate indices is small ($n_f \approx n_s$). In this latter case the chart of Fig. 5 can also be applied to TM modes as long as we define the asymmetry measure a in a somewhat different manner, namely by [11]

$$a = \frac{n_f^4}{n_c^4} \cdot \frac{n_s^2 - n_c^2}{n_f^2 - n_s^2} . \tag{14}$$

Table I includes values for this measure under a_M.

Figure 5 reflects the fact that the fundamental mode of a symmetric guide has no cutoff. The cutoff value V_0 for the fundamental mode of an asymmetric guide is obtained from the dispersion equation as

$$V_0 = \tan^{-1} a^{1/2}. \tag{15}$$

We can write this in terms of film thickness and wavelength in the form

$$(f/\lambda)_0 = \frac{1}{2\pi(n_f^2 - n_s^2)^{1/2}} \tan^{-1} a^{1/2}. \tag{16}$$

In an oversized waveguide many modes are allowed to propagate, with the number m of allowed guided modes given by

$$m = \frac{2f}{\lambda} (n_f^2 - n_s^2)^{1/2} \tag{17}$$

which is independent of the degree of asymmetry.

B. Effective Guide Thickness

In Fig. 4 we have associated the drawn arrows with the wave normals of zigzagging plane waves. When we investigate phenomena in the waveguide which involve the exchange or flow of energy we have to go one step further and consider the behavior of wave packets or rays. These are indicated in Fig. 6. Burke [12] has pointed out that this ray picture should include the Goos-Haenchen shifts which occur on total reflection from the film-substrate and film-cover boundaries. These ray shifts of $2z_s$ and $2z_c$ are included in the figure. One can show [13] that this shift of the reflected wave packet is related to a derivative of the phase shifts ϕ_s and ϕ_c:

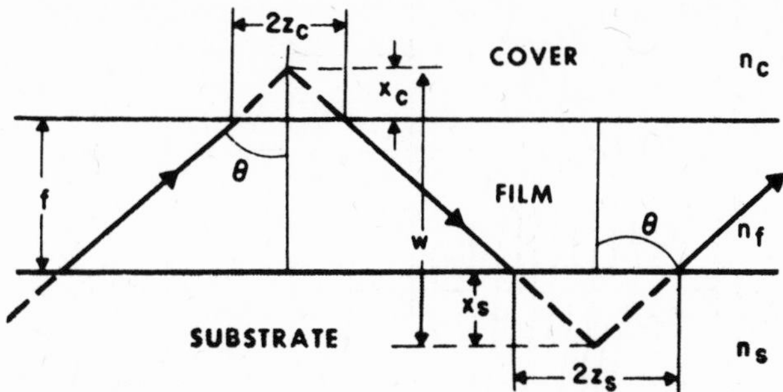

Figure 6. Zigzag wave picture for slab waveguides with Goos-Haenchen shifts included.

$$z_s = d\phi_s/d\beta = \frac{1}{\gamma_s}\tan\theta$$

$$z_c = d\phi_c/d\beta = \frac{1}{\gamma_c}\tan\theta \qquad (18)$$

where the expressions given on the right-hand side apply to the TE modes. This is the spatial analog of the $d\omega/d\beta$ relation for the group velocity which follows from consideration of timelike wave packets. As a consequence of this shift the ray appears to penetrate into substrate and cover as indicated in the figure. We calculate the apparent penetration depths x_s and x_c from (18) as

$$x_s = 1/\gamma_s$$

$$x_c = 1/\gamma_c \quad . \qquad (19)$$

As the appearance of the decay constants γ suggests, there is a close relation between this ray penetration and the existence of evanescent fields in substrate and cover.

As a consequence of the ray penetration the waveguide appears to possess an effective thickness w

$$w = f + 1/\gamma_s + 1/\gamma_c \qquad (20)$$

which is larger than its actual thickness f and which always turns up as a characteristic parameter when questions of energy exchange are involved.

Figure 7 gives a plot for the normalized guide thickness

$$W = kw(n_f^2 - n_s^2)^{1/2} \qquad (21)$$

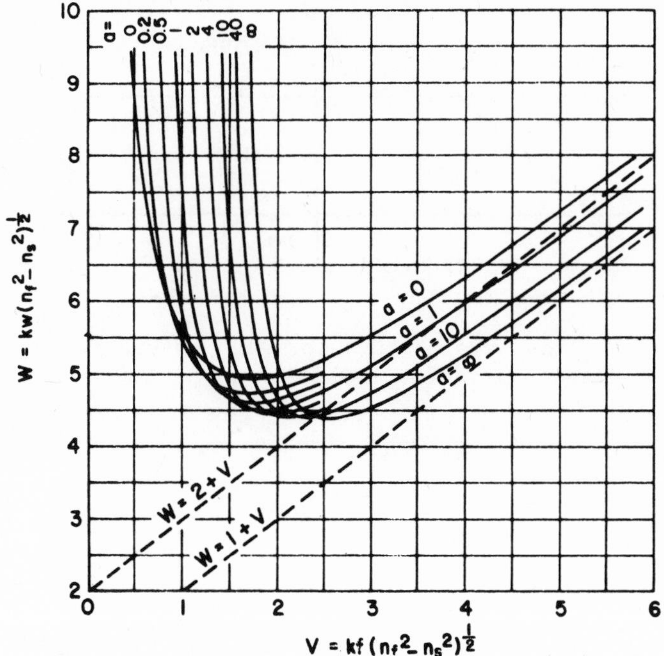

Figure 7. Normalized effective guide thickness W as a function of normalized frequency V for slab guides with various degrees of asymmetry. (From [11]).

for the TE_0 modes of asymmetric slab guides for various degrees of asymmetry [11]. This figure also gives an idea of the degree of light confinement that we can achieve with a dielectric waveguide. On the chart we observe a broad minimum of $W = 4.4$ at $V = 2.55$ occurring for $a = \infty$. This implies a minimum achievable effective width w_{min} of

$$w_{min}/\lambda = 0.7\cdot(n_f^2 - n_s^2)^{-1/2} \tag{22}$$

which is a function of the film-substrate index difference. For $n_s = 1.5$ and $n_f = 1.6$ we get $w_{min} \approx 1.3\lambda$.

C. Strip Guides

A planar slab guide provides no confinement of the light in the film plane. For some devices this confinement is not necessary, and for others it is not even desired. Examples for the latter are planar devices such as the acoustooptic light deflectors in thin-film form [14], and the planar film lenses and prisms proposed for optical data processing [15], [16]. Strip guides provide confinement in the film plane. In some active devices such as lasers and modulators this additional confinement is very desirable as it can lead to a reduction in drive voltage and power consumption. A considerable number of devices can, and some have been, made in both the planar and the strip version. An example are directional couplers which one can make by fabricating two strip guides with a close spacing, or by separating two planar slab guides by a thin layer of lower index material [17]. Generally, it appears that the planar device versions are easier to fabricate, while the strip-guide versions provide more compactness and flexibility.

Strip guides can be made in a variety of ways. Figure 8 shows four examples of possible strip-guide (x-y) cross sections which have a typical width of a few micrometers. Figure 8(a) shows a raised strip guide, which can be fabricated from a slab guide by masking the strip and removing the surrounding film by reverse sputtering, ion-beam etching, or chemical etching. The imbedded strip in Fig. 8(b) can be made by ion implantation through a mask. The ridge guide of Fig. 8(c) is fabricated by the same process as the raised strip but with incomplete removal of the surrounding film. The strip-loaded guide of Fig. 8(d) is formed by depositing a strip of material with lower index $n_c < n_f$ onto a slab guide. For reasons of simplicity the sketches of the figure show abrupt changes of refractive index or film thickness. However, some fabrication methods, such as diffusion through a mask lead to guides with graded transitions.

The ridge guide is, of course, a close relative of the recently reported single material fiber [18]. Both the ridge (or "rib")

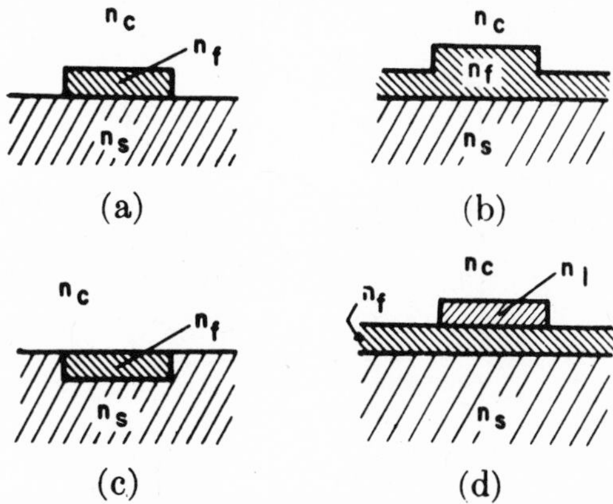

Figure 8. Cross sections of various strip guide configurations. (a) Raised strip. (b) Embedded strip. (c) Ridge guide. (d) Strip-loaded guide.

guide [19], [20] and the strip-loaded guide [21], [22] have been proposed because of their promise in relaxing the stringent fabrication requirements for resolution and edge smoothness. Both use a propagating surround, i.e., a surrounding slab guide allowing at least one guided mode.

No exact analytic solutions for the modes of strip guides are available. Numerical calculations have been made for rectangular dielectric guides embedded in a uniform surround [23], [24], and Marcatili [10], [25] has given approximate solutions applicable to a large class of strip guides operating sufficiently away from cutoff.

The effective index method is another simple way to get some fairly good predictions for the behavior of strip guides such as the ridge and the strip-loaded guides [22], giving good agreement with experimental results. In Fig. 9 we have sketched this method for the ridge guides. Corresponding to the ridge region on the one hand and the surround on the other we consider two unbounded slab guides of thickness l and f and normalized thickness V_l and V_f, respectively. We use these parameters to determine the corresponding guide indices b_l and b_f and the effective indices N_l and N_f, for example, via Fig. 5. To predict the guide characteristics in the y direction we consider a symmetric slab guide with a thickness equal to the ridge width a, and with substrate index N_f and film index N_l as indicated in the figure. To this "equivalent" slab guide we can apply all the results available for slab guides and arrive at estimates for quantities of experimental interest; e.g., we can determine the number of guided modes m_y allowed in the y dimension from (17), which becomes

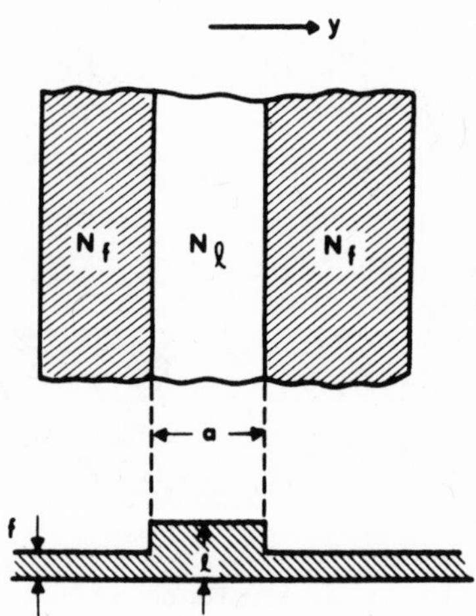

Figure 9. Schematic illustrating the application of the effective index method to a ridge guide.

$$m_y = \frac{2a}{\lambda}(N_1^2 - N_f^2)^{1/2} = \frac{2a}{\lambda}[(n_f^2 - n_s^2)(b_1 - b_f)]^{1/2} \quad (23)$$

or we can use (21) in combination with Fig. 7 to determine the minimum effective width $w_{y\ min}$ in the y direction from

$$\begin{aligned} W_{y\ min} &= kw_{y\ min}(N_1^2 - N_f^2)^{1/2} \\ &= kw_{y\ min}[(n_f^2 - n_s^2)(b_1 - b_f)]^{1/2} \approx 5. \end{aligned} \quad (24)$$

III. THE USE OF COUPLED WAVES AND PERIODIC STRUCTURES

Many phenomena in physics and engineering can be viewed as coupled-wave phenomena. To these belong the diffraction of x-rays in crystals [26], the scattering of light by acoustic waves [27] or by hologram gratings [28], the directional coupling of microwaves [29], [30], and the energy exchange between electron beams and slow-wave structures in microwave tubes [30], [31]. In integrated optics we encounter a great variety of coupled-wave phenomena, and Yariv [32] has recently given a summary of these. They include the conversion of TE to TM modes in anisotropic waveguides [33], [34] or in magnetooptic modulators [35], or the backward scattering in distributed feedback structures of lasers [36].

Dealing with coupled-wave phenomena we usually consider two waves or modes (R and S) of a guiding structure which propagate freely and uncoupled as long as the structure is not perturbed. A perturbation of the original structure, e.g., an induced change of the refractive index of a film guide, will lead to a coupling of the two waves and an exchange of energy between them. As the interaction over distances of a wavelength is usually weak, we talk of a slow exchange of energy and a slow change of the complex amplitudes R(z) and S(z) of the coupled waves. Typical interaction lengths useful for integrated optics devices are between 100 μm and 1 cm. An important requirement for a significant interaction is the synchronism or "phase matching" between the two coupled waves, which, in the simplest case, is the requirement for the equality

$$\beta_R = \beta_S \quad (25)$$

of the propagation constants β_R and β_S of the two waves.

In the following discussion we will distinguish between the coupling of two guided modes and the coupling to radiation modes. Another distinction is between waves with codirectional and waves with contradirectional flow of energy.

A. Codirectional Coupling of Guided Modes

When the group velocities of R and S point in the same direction, we talk of codirectional coupling or forward scattering (or the Laue case). This is indicated in Fig. 10(a) where we have sketched the change in wave amplitudes. The usual boundary condition in this case is

$$R(o) = R_0$$
$$S(o) = 0 \qquad (26)$$

i.e., a reference wave R with an incident amplitude R_0 and a scattered wave S starting with zero amplitude. For synchronism one gets a periodic interchange of energy described by amplitudes [26]-[32]

$$S(L) = -jR_0 \sin(\kappa L)$$
$$R(L) = R_0 \cos(\kappa L) \qquad (27)$$

where κ is called the coupling constant and L is the interaction length. For this case we can get a complete transfer of power from R to S if we choose an interaction length of $\pi/2\kappa$.

The coupling coefficients depend on the particular device structure. Snyder [37] and Marcuse [9] have applied a coupled mode theory to perturbed dielectric waveguides and found that κ is proportional to an overlap integral containing the (normalized) electric field distributions $E_R(x,y)$ and $E_S(x,y)$ of the two coupled modes. For TE modes this is of the form

$$\kappa \propto \iint dx\, dy (n^2 - n_0{}^2) E_R \cdot E_S{}^* \qquad (28)$$

where $n_0(x,y)$ is the index profile of the unperturbed waveguide and $n(x,y,z)$ is the index profile of the perturbed structure. Reference [32] gives the coefficients for nonlinear interactions and for electrooptic, magnetooptic, and photoelastic modulation in dielectric waveguides.

B. Contradirectional Coupling of Guided Modes

Contradirectional coupling or backward scattering occurs when the group velocities of R and S point in opposite directions as indicated in Fig. 10(b). This is also referred to as the Bragg case. Here the usual boundary conditions are

$$R(o) = R_0$$
$$S(L) = 0 \qquad (29)$$

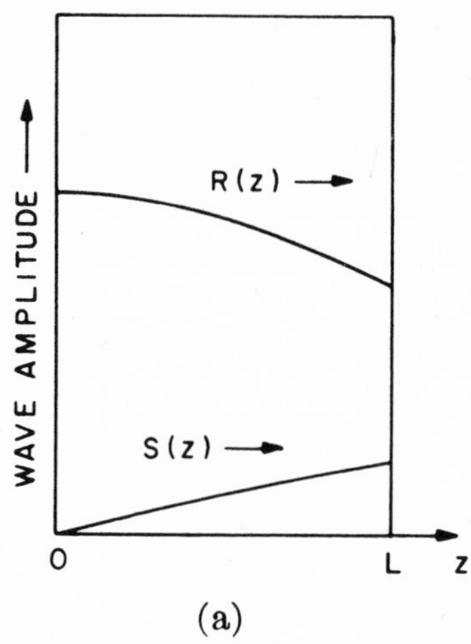

(a)

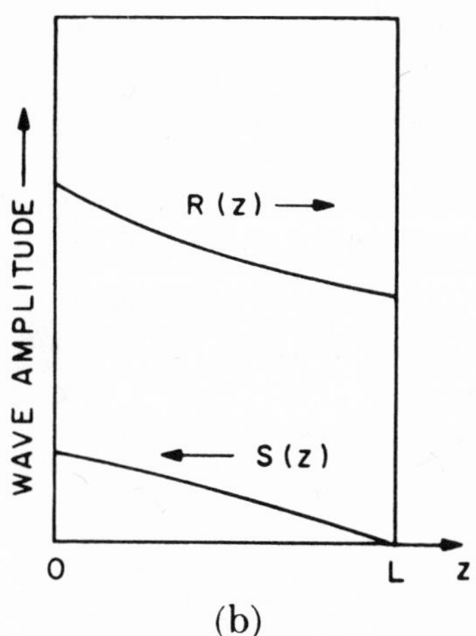

(b)

Figure 10. Variation of coupled wave amplitudes with diatance. (a) Forward scattering. (b) Backward scattering.

with the scattered wave S starting with zero amplitude at the end of the interaction region. The resulting energy interchange at synchronism is described by amplitudes [26]-[28], [31], [32]

$$S(o) = jR_0 \tanh(\kappa L)$$

$$R(L) = R_0/\cosh(\kappa L). \tag{30}$$

The scattered wave appears on reflection from the structure.

C. Periodic Structures

Periodic structures are used in integrated optics in several forms, including films with surface corrugations and index perturbations produced by an acoustic wave [14], and for several different purposes, including light deflection [14], filtering [32], [38], [39], and for the spectral narrowing of laser output [36]. Figure 11 shows the example of a film with a surface corrugation. The scattering of light by a periodic structure can be viewed as a coupled wave process. We can associate a grating vector K

$$K = 2\pi/\Lambda \tag{31}$$

with the structure which is related to its period Λ. The periodicity establishes synchronism between two modes with different propagation constants β_1 and β_2 if

$$\beta_2 = \beta_1 + K. \tag{32}$$

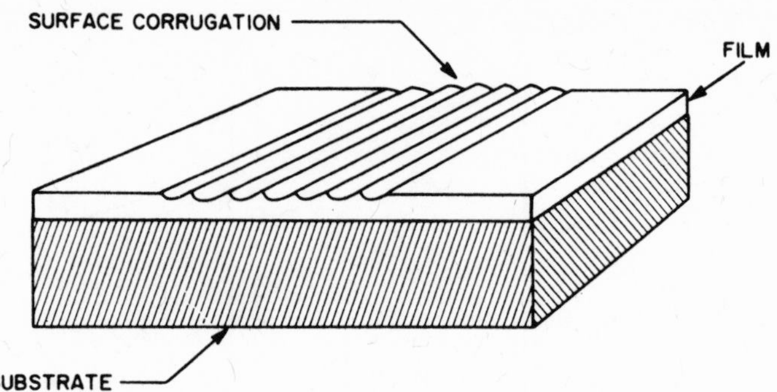

Figure 11. Sketch of thin-film waveguide with a surface corrugation.

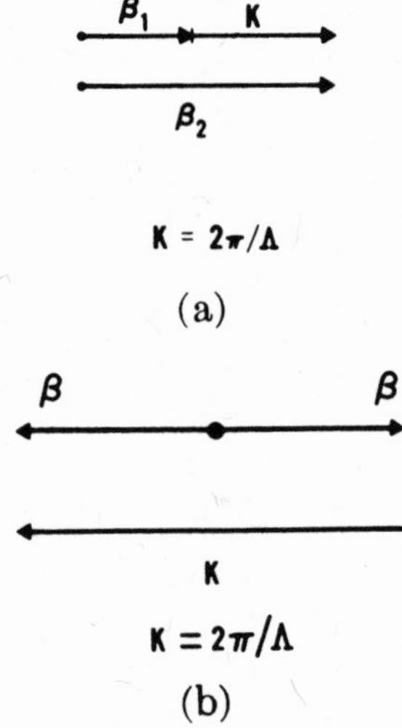

Figure 12. Wave vector diagrams. (a) For mode matching forward scattering. (b) Backward Bragg scattering of guided modes (β) by a periodic structure (K).

This leads to forward scattering as sketched in Fig. 12(a). The magnetooptic modulators of [35] and the mode conversion by means of acoustic waves [40] are two experiments where this method was used. The structures here can be relatively coarse with periods Λ of many micrometers.

Figure 12(b) shows the case of backward scattering where a mode of propagation constant β is coupled to a mode traveling in the opposite direction with −β. This interaction is strongest if

$$K = 2\beta \cdot \tag{33}$$

This implies the need for very small periods Λ of the order of 1000-2000 Å. These have been used for filters [38] and distributed feedback lasers [41]-[44].

A derivation of the coupling coefficient κ for the corrugated film structure of Fig. 11 is given in [9]. For backward scattering of TE modes it can be written in the form [42]

$$\kappa = \frac{\pi h}{\lambda} \frac{(n_f^2 - N^2)}{2wN} \tag{34}$$

which is accurate to first order in the corrugation height h. N and w are the effective index and the effective guide thickness defined earlier. The dependence of κ on the film thickness f is shown in Fig. 13, which is taken from [44] and applies to a GaAlAs waveguide at λ = 8300 Å with n_f = 3.59, and n_s = 3.414. The scale of the normalized quantity $\bar{\kappa} = \lambda\kappa/\pi h$ is used for the ordinate. Results are shown for the three lowest order modes (m = 0,1,2). The solid lines refer to a cover index of n_c = 3.294 ($Al_{0.5}Ga_{0.5}As$). We note that the maximum value for κ occurs fairly close to cutoff, and that a change in the value of n_c has relatively little influence on this value.

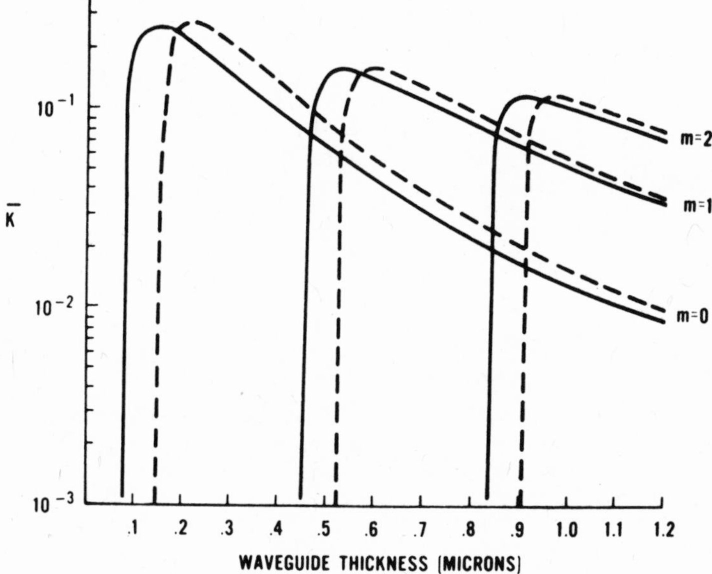

Figure 13. Normalized coupling constant as a function of waveguide thickness for n_s = 3.414, n_f = 3.59, n_c = 1 (solid curves), and n_c = 3.294 (dashed curves), and λ = 0.83 μm. (From [44].)

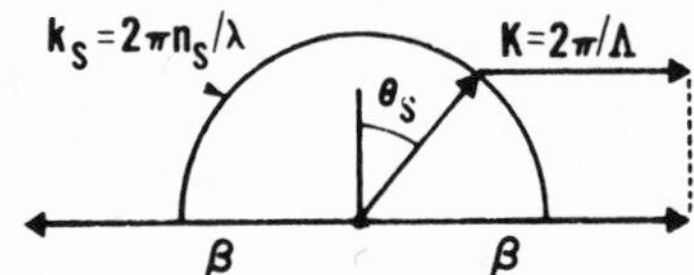

Figure 14. Wave vector diagram indicating the grating coupling of a guided mode (β) by a periodicity in the waveguide (*K*).

D. Coupling to Radiation Modes

When we choose a grating vector K for a periodic waveguide such that

$$\beta = K + n_s k \sin \Theta \tag{35}$$

then we couple a guided mode to a radiation mode. This results in light leaking from the guide into the substrate at an angle Θ with respect to the film normal. We have drawn the wave vector diagram for this case in Fig. 14, where the circle is the locus of all wave vectors allowed in the substrate. Due to the leakage the guided wave amplitude R decreases exponentially with distance

$$R(z) = R_0 \exp(-z/l) \tag{36}$$

where l is called the leakage length. Coupling to radiation modes is the mechanism by which waveguide inhomogeneities can cause scattering losses [8], [9], and it is also the basis for the grating couplers [45], [46] which can feed laser beams into guided film modes.

IV. FABRICATION TECHNIQUES

The past few years have seen the exploration of a great variety of techniques for the fabrication of guiding films with low loss, the precise delineation of circuit and device patterns, and for the machining of these patterns with high resolution. The following examples were selected to illustrate this variety.

A. Film Fabrication

To form a planar slab guide we need to fabricate a uniform layer of higher index on a planar substrate. One of the earlier methods to achieve this employed ion exchange in glass [47]. A recent workhorse for the fabrication of devices has been a film of Corning 7059 glass produced by RF sputtering on microscope slides [48]. This technique yields single mode guides with losses below 1 dB/cm and indices of about $n_f = 1.62$ and $n_s = 1.52$. The use of ion implantation is in its early stages of exploration. Waveguides have been made by implanting Li in fused silica [49] and protons in GaP [50] and GaAs [51] substrates.

Solid-state diffusion is another promising technique which can also be used to form waveguides in crystal substrates. Naturally the result is a guide with a graded refractive index profile. Waveguides have been made by diffusion of Se into CdS crystals [52], of Cd and Se into ZnS [53], and of Cd into ZnSe [53] substrates. Outdiffusion [54] of Li has been used to produce waveguides in $LiNbO_3$ and $LiTaO_3$ both modulator materials of great interest. The confinement obtainable with these guides is limited to about 12 μm. Recently, two promising techniques have been reported which can produce low-loss single-mode waveguide layers as thin as 1 μ. These employ the diffusion of Nb into $LiTaO^3$ [55], and the diffusion of Ti into $LiNbO_3$ [56]. Interesting because of their high index are films of Ta_2O_5 prepared by oxidizing a sputtered Ta film [57].

The epitaxial growth of single-crystal films has become a key element in the production of electrooptic, magnetooptic, or laser-active guiding layers. To achieve the required higher index these layers are usually of a different material composition than the substrate and the matching of lattice constants becomes an important consideration. The epitaxial GaAlSs heterostructure layers made for junction lasers [58]-[60] are a prime example for the use of this method. Other examples are the electrooptic epitaxial ZnO layers on sapphire [61], the epitaxial ADP-KDP mixed-crystal films on KDP substrates [62], and the epitaxial layers of $LiNbO_3$ grown on $LiTaO_3$ substrates [63], [64]. Epitaxial magnetooptic waveguides have been prepared in a variety of garnets [65] which are relatives of the magnetic bubble materials.

The exploration of organic films as waveguides has included the use of epoxy films [66], [67], of organosilicon films [68] showing extremely low losses, and of layers of photoresist [69], [70] or electron resist [71].

B. Circuit Patterns by Scanning Electron Microscopy

Strip guides and more complicated integrated optics circuit and device structures are made by a planar technology which is very similar to the electronic integrated circuit technology and to that used for surface acoustic wave devices. Figure 15 illustrates a typical fabrication sequence [72] for a strip guide. In the first step (a) one starts with a planar slab guide (glass) on which one spin-coats a suitable photoresist or electronresist layer (polymethyl methacrylate). The resist is exposed to an image of the device pattern which is produced optically or by a scanning electron beam. After development, the resist masks the film areas to be etched

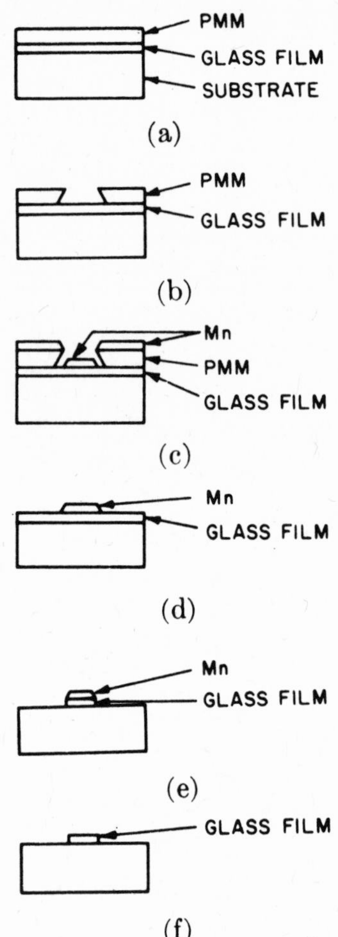

Figure 15. Typical sequence for fabricating strip guide patterns. (From [72].)

later (b). In the third step one deposits a metal layer (Mn) over mask and film (c). Subsequently, the resist is removed (with acetone), and one is left with a metal strip masking the film (d). Now the surrounding film material is etched away, e.g., by RF back sputtering, leaving the structure shown in (e). After removal of the metal (with hydrochloric acid), the desired strip guide pattern is obtained.

For integrated optics work we usually require patterns with a high resolution (guide separation of 1 μm are typical) and waveguides with an edge roughness better than 500 Å. Conventional photolithographic techniques are not good enough for this job. Satisfactory results are expected from scanning electron beam techniques. These are being explored by several groups and some devices have already been fabricated that way [71]-[74].

When we think about the economics of scanning electron beam techniques we should recall that high-resolution pattern replication techniques are under development for the mass fabrication of electronic integrated circuits. In one example [75], soft X-rays are employed to reuse a mask with submicrometer resolution many times. In another technique [76] a master mask is used which, upon exposure to UV, emits electrons in the desired pattern. These techniques are also of interest for integrated optics.

C. Grating Patterns by Laser Interference

The masks for periodic structures of high resolution can be written by a technique borrowed from holography. Here a photoresist layer (e.g., Shippley AZ 1350) is exposed to the interference pattern produced by two coherent laser beams incident at an angle to each other [42], [77], [78]. The angle of incidence and the wavelength of the laser light control the fringe spacing. To achieve ultrafine grating periods ($\Lambda \approx 1000$ Å) the UV line at 3250 Å of the CW helium-cadmium laser has been used.

It appears that the laser-interference technique is more naturally suited to the production of high-resolution gratings than the scanning electron-beam techniques, yielding good uniformity of the grating period with greater ease.

D. High-Resolution Etching Techniques

The previously mentioned sputter etching technique [72] is not the only way to produce strip guides or more complicated patterns. Strip guides have also been made by masked diffusion [52], [53] and by ion implantation through a mask [79], [80]. Other techniques under exploration are the embossing [81] and the photolocking [82] of strip guide patterns written directly with a laser beam. A technique capable of very high resolution is ion-beam etching using photoresist as a mask material [42], [77], [78]. Typically, the etching is done with a beam of argon ions at a few kilovolts, and the substrate remains relatively cool during the process. Figure 16 shows a surface corrugation on a glass waveguide etched that way to make a filter device [38]. The period shown there is 1900 Å, but the technique is capable of producing corrugations with periods as small as 1000 Å, and corrugation depths of about 500 Å [78], [43].

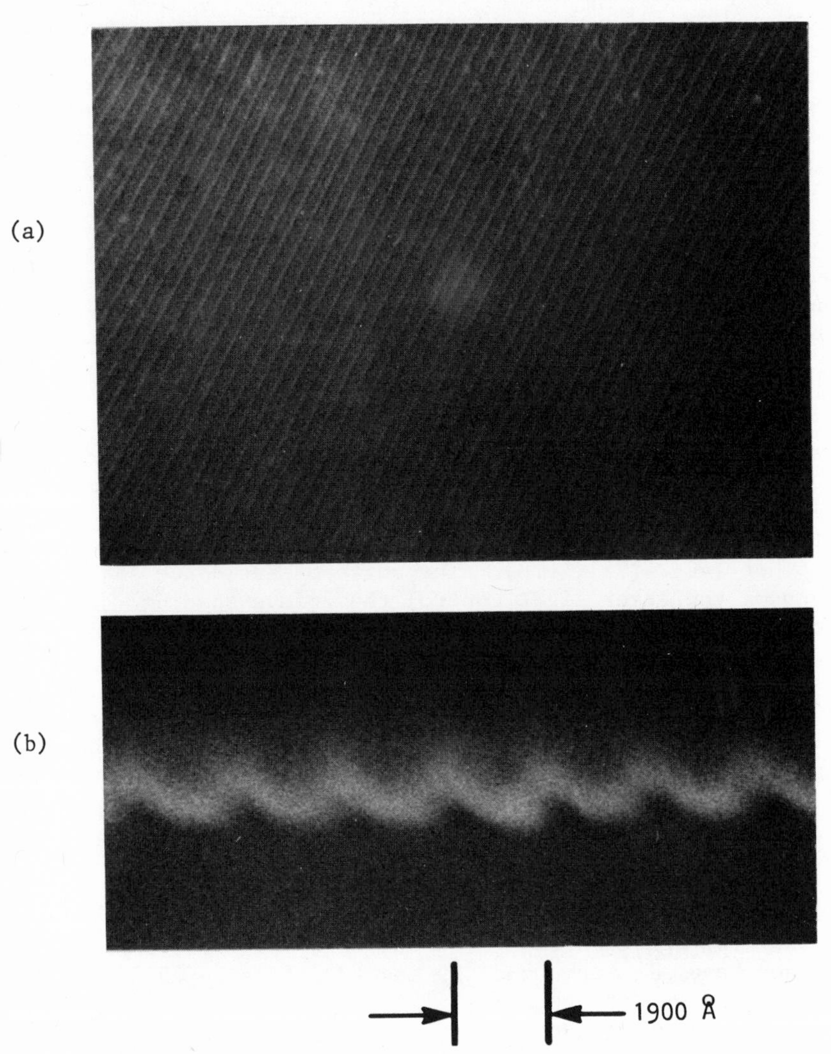

Figure 16. Ion beam etched gratings. Scanning electron micrographs of corrugation etched into a glass waveguide. (a) Top view. (b) End view.

V. DEVICES

In the past few years a great variety of devices have been explored experimentally, including directional couplers, filters, modulators, light deflectors, and the application of guided-wave techniques to lasers. In this section we will try to illustrate this work. Because of space limitations, this illustration has to remain quite selective. For more complete and up-to-date reviews the reader should consult the articles in this issue by Kaminow [83] on modulators, and by Panish [84] on GaAs injection lasers.

A. Beam to Film Couplers

Beam-to-film couplers are devices that allow us to feed the light from a laser beam into a film waveguide. Figure 17 shows two coupler types used in the laboratory, the prism coupler [85], [86] and the grating coupler [45], [46]. In the first method [Fig. 17(a)] a prism of high index n_p

$$n_p \gtrsim n_f \qquad (37)$$

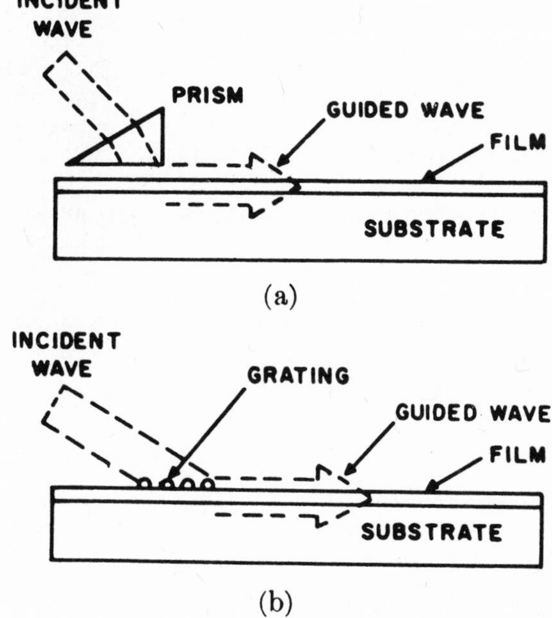

Figure 17. (a) Schematic of prism coupler. (b) Schematic of grating coupler. (From [5].)

is brought close enough to the film guide (typically 1000 Å) so that it can interact with the evanescent field of a guided mode. This makes the structure leaky and coupling is possible. In order to achieve coupling we also have to provide synchronism (phase matching) and satisfy

$$\beta = kn_p \sin \Theta_p \tag{38}$$

by choosing the correct injection angle Θ_p in the prism relative to the film normal.

To make a grating coupler [Fig. 17(b)] we introduce a periodicity into the guide, e.g., by depositing a photoresist grating. This will lead to coupling of a guided mode to radiation modes as discussed earlier in Section III.

The overall behavior of prism and grating couplers is rather similar. For uniform coupling, i.e., uniform prism film gaps or uniform grating strength, the theoretically predicted maximum coupling efficiency is about 81 percent for both uniform or Gaussian beam cross section [85]-[89]. This is achieved when the width of w_B of the incident laser beam matches the leakage length l of the coupler [see (36)], i.e., if

$$w_B \approx l. \tag{39}$$

To approach this value with grating couplers, care must be taken that unwanted grating orders are suppressed. This was done in a recently reported coupler [87] using a photoresist grating, reverse coupling [46], and light incident through the substrate. Efficiencies exceeding 70 percent were obtained.

The theoretical coupler efficiencies can approach 100 percent when the coupling is tapered to match the input beam to the leakage field of the guide [88], [89].

A third coupler type is described in [90]. Here the film thickness is tapered down at the edge of the guide resulting in coupling of light into the substrate.

B. Directional Couplers

The directional coupler is a familiar microwave device. The strip-guide version of its integrated optics analogue is shown in Fig. 18, where we have two strip guides approaching each other, running close and parallel over the interaction distance, and then separating again. When the propagation constants of the two guides are matched we can get codirectional coupling and exchange of power between the guides. The coupling constant κ (and coupling length) between the parallel guides is determined by their separation c and the decay constant γ_y in the film plane

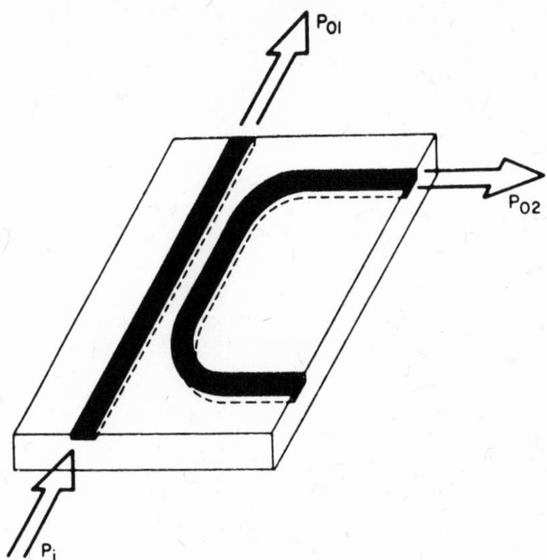

Figure 18. Sketch of a strip-guide configuration forming an optical directional coupler.

$$\kappa \propto \exp\,(-\gamma_y c) \tag{40}$$

where the exponential dependence indicates a great sensitivity of the coupling length to variations in the gap spacing.

In the first experimental work on directional couplers, the exchange and complete transfer of power [see (27)] between parallel strip guides in GaAs has been observed [19], and couplers of the geometry shown in Fig. 18 have been made by scanning electron beam techniques [72], [74]. Figure 19 shows the smooth edges of the glass guides in the interaction region of the coupler of [72], where the guide spacing c was about 1 μm.

C. Filters

The filter characteristics of periodic waveguides have also been seen experimentally [38], [39]. A periodic guide of length L provides a band rejection filter with a fractional bandwidth of approximately

$$\Delta\lambda/\lambda \approx \Lambda/L \tag{41}$$

centered at the Bragg wavelength $2N\Lambda$ [see (33)].

Figure 20 shows the filter response of a corrugated glass waveguide at 0.57 μm reported in [38]. The filter was fabricated by

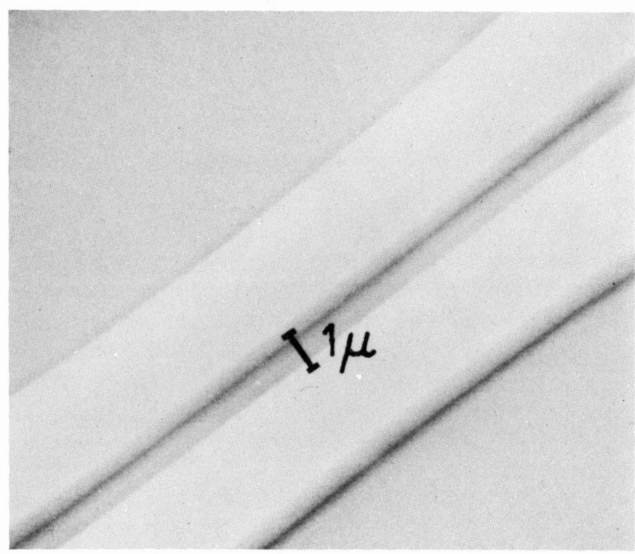

Figure 19. Scanning electron micrograph of the interaction region of a directional coupler formed by two strip guides. (From [72].)

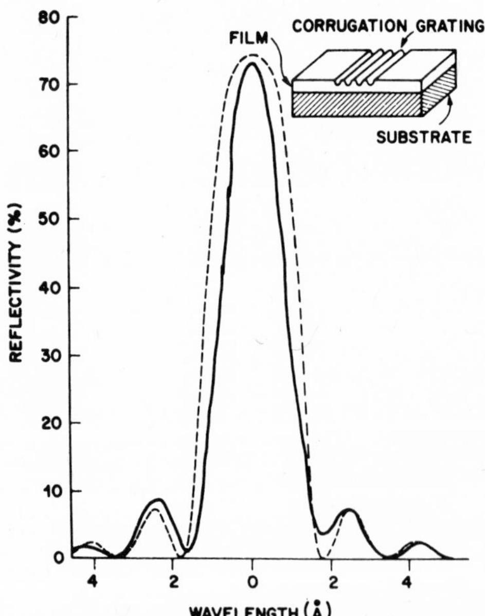

Figure 20. Frequency response of a corrugated waveguide filter. The dashed curve shows the theoretical prediction. (From [38].)

means of UV laser exposure and ion-beam etching. It was L = 0.57 mm long, and had corrugations with a period of about Λ = 2000 Å and a depth of 460 Å. The response was measured with a tunable dye laser showing a bandwidth of 4 Å and good agreement with the theory of contradirectional coupling.

In recent experiments [91] filter bandwidths as low as 0.1 Å were obtained. This requires longer filter lengths and careful attention to tolerances imposed on the film thickness. The achieved bandwidth indicates that the fabricated grating period Λ had deviations from uniformity smaller than 10^{-5} over the length of the filter.

D. Modulators

Guided-wave modulators are, perhaps, the integrated optics devices which have developed most rapidly in the recent past. Here, apart from compactness and stability, the guided-wave approach promises low drive powers and low drive voltages. The reason for this is that no diffraction spreading occurs and the light can be contained within long structures having small cross sections with diameters of the order of a wavelength. However, when one contemplates the use of these modulators as separate devices, one should bear in mind two potential drawbacks. One is the need for coupling to such a device, and the other is a possible limitation in the guided optical power to prevent damage or breakdown due to the high power densities associated with small guide cross sections.

Several modulation effects and device structures have been explored and [83] gives a detailed review. We will mention here electrooptic, magnetooptic, and electroabsorption modulators. An important measure of device performance is the drive power dissipated per unit bandwidth of the signal for a given modulation index. Others are the drive voltage, the speed, and the optical insertion loss. If the light is confined in a strip guide with an effective guide cross section $w_x \cdot w_y$ and length L then we have a drive power P per bandwidth $\Delta\nu$ which is proportional to

$$P/\Delta\nu \propto w_x w_y / L. \tag{42}$$

Here we have assumed that the applied modulating field is confined to the same volume as that occupied by the light. This means that an electrode configuration is required that matches the applied field distribution to that of the guided light.

Electrooptic modulators use materials in which a change of the refractive index can be induced by an applied electric field. This results in a phase modulation of the light passing through. Promising electrooptic materials are $LiNbO_3$ and $LiTaO_3$ which are transparent for wavelengths from 0.4 to 4 μm. Outdiffused waveguides in

$LiNbO_3$ have been used to make high-speed guided-wave phase modulators [92], [93]. Figure 21 shows a sketch of a recent device [93] where a ridge guide is employed to confine the light within the film plane as well. The electrodes were evaporated along the ridge which is about 20 μm wide. This modulator requires a drive voltage of 1.2 V to produce a phase modulation of 1 rad and uses a power of 20 μW/MHz of bandwidth. The method of diffraction by electrooptically induced gratings has been proposed to obtain amplitude modulation of the light [94] and modulators in ZnO waveguides have been constructed that way [95]. Figure 22 shows the electrode structure used to induce such a grating in a thin $LiNbO_3$ crystal [96]. Grating modulation has also been reported with the new Nb-indiffused waveguides in $LiTaO_3$ [55]. Amplitude modulation can also be achieved with only two electrodes if one uses the induced prism effect due to the inhomogeneity of the applied field. This was demonstrated in work on Schottky-barrier GaAs modulators operating at 10.6 μm [97].

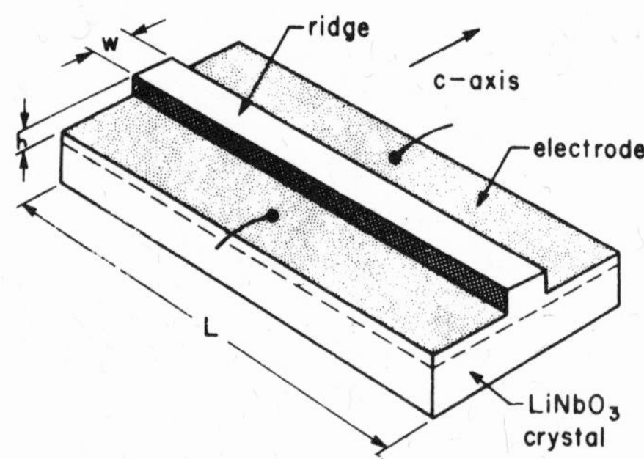

Figure 21. Sketch of $LiNbO_3$ phase modulator with a ridge guide on an outdiffused waveguide layer. (From [93].)

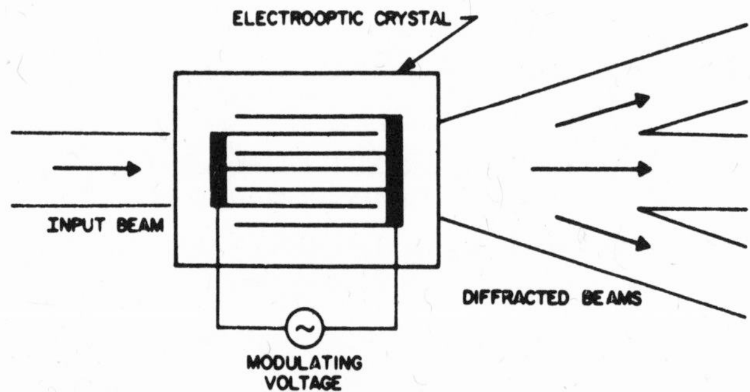

Figure 22. Interdigital electrode structure used to induce an electrooptic grating. (From [96].)

Electrooptic effects are also obtained in the depletion layer of reverse-biased semiconductor diodes which can also serve as a waveguide. Such guided-wave modulators have been made using the p-n junctions in GaP [98] and Schottky barriers in GaAs [99]. Recently, a GaAlAs double heterostructure was used to construct a modulator requiring 10 V and 0.1 mW/MHz of bandwidth for phase modulation of 1 rad at 1.15 μm [100]. This device has a potential bandwidth of 4 GHz.

In magnetooptic modulators one uses the Faraday rotation to achieve mode conversion and, thereby, amplitude modulation. Guided-wave modulators of this type have been built with epitaxial films of iron garnets [35]. They have operated at frequencies up to 80 MHz at a wavelength of 1.15 μm. These devices are very sensitive and require applied magnetic fields of only 1/100 Oe to switch the output light on or off.

Electroabsorption, or the Franz-Keldysh effect, is another way to achieve direct amplitude modulation. These devices use an electric-field induced shift of the band edge (and, therefore, operate near the band edge). An electroabsorption modulator operating at 0.9 μm was reported in [101]. It uses GaAlAs waveguides and requires 0.2 mW/MHz for an amplitude modulation of 90 percent.

E. Light Deflectors

The technology of bulk acoustooptic devices for the deflection and modulation of light is already fairly well developed. In the integrated optics analogues of these devices guided optical waves are scattered from surface acoustic waves [14]. This approach allows a close confinement and overlap of the optical and acoustical fields resulting in a reduction of the required acoustic drive power. The surface acoustooptic interaction is similar to the scattering of light sketched in Fig. 22 with the difference that the gratings are induced acoustooptically. The figure shows Raman-Nath scattering with several diffracted orders as it may be used for modulators. For deflectors one uses a codirectional coupled wave interaction. with light incident at the Bragg angle and only one significant diffracted order.

In order to launch the surface waves we need surfaceacoustic transducers. Comparing these to bulk transducers we find that, at present, it is more difficult to obtain good efficiencies over large bandwidth, but there are also advantages: surface-wave transducers can be fabricated with a planar technology, and complicated electrode patterns for steering the acoustic column can be made with relative ease [102]. Such a pattern is sketched in Fig. 23(b) while Fig. 23(a) shows the hard-to-fabricate stepped surface required for a comparable transducer array for bulk waves.

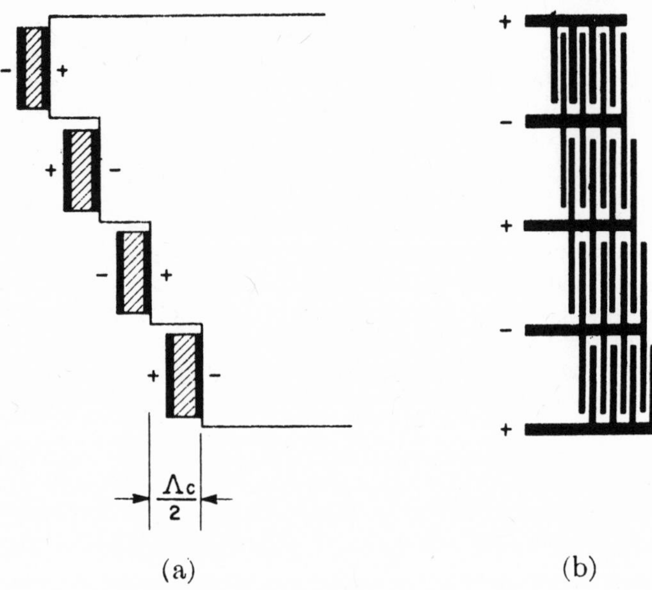

Figure 23. Sketch of phased arrays of acoustic bulk (a) and surface wave (b) transducers.

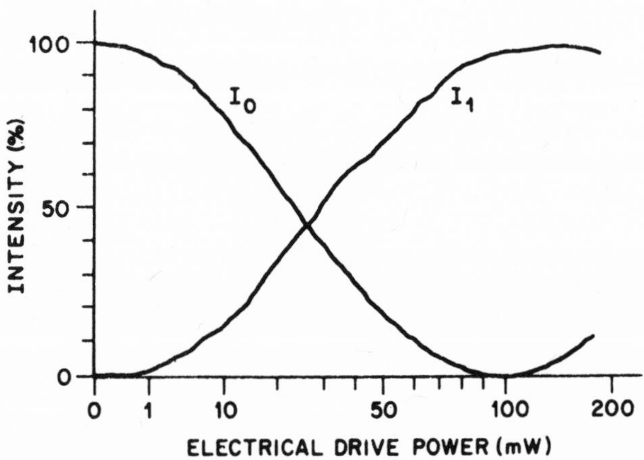

Figure 24. Deflected (I_1) and transmitted (I_0) light intensity as a function of acoustic drive power for a thin-film light deflector using titanium-diffused waveguides on $LiNbO_3$. (From [105].)

Recently, optical guides have been fabricated on materials such as $LiNbO_3$ which have a better acoustooptic figure of merit than the materials used for earlier devices (e.g., quartz). Experiments with light deflectors have been made with As_2Se_3 guides [103] on $LiNbO_3$, with outdiffused $LiNbO_3$ guides [104] and with titanium-diffused guides [105] on $LiNbO_3$. Figure 24 shows the results achieved with the recent titanium-diffused waveguides at an acoustic frequency of 170 MHz [105]. The Bragg diffracted light intensity I_1 is shown as a function of acoustic power; about 8 mW are needed to diffract 70 percent of the incident light over a bandwidth of 30 MHz of the acoustic frequency. Note also that the measured data agree well with the coupled-wave behavior predicted by (27). We should recall here that $LiNbO_3$ is not the ultimate in acoustooptics and still better materials exist. An example is paratellurite (TeO_2) which has been used for high-performance light deflectors in bulk form [106]. No optical waveguides on these materials are available as yet. Experimental studies were made of a variety of other acousto-optic interaction geometries different from those described above. These include the inducing of a grating by launching acoustic shear waves normal to the film guide [107], and the scattering of light from the guide into the substrate induced by a surface-acoustic wave [108] which acts like a grating coupler.

F. Lasers

Along with the promise of small size, compactness, and compatibility with other waveguide devices, the application of guided-wave techniques to lasers promises stability of output, low threshold, low power consumption, and low heat dissipation. The techniques under exploration include the confinement of the light to a thin film, the use of structures such as stripe contacts or ridge guides for confinement in the film plane and for transverse mode control, and the use of periodic structures to provide (distributed) feedback, control of the longitudinal modes, and a narrow output spectrum. Guided-wave techniques are in various phases of their development for semiconductor lasers, solid-state lasers, gas lasers, and organic dye lasers. While the use of guiding films in semiconductor lasers is almost as old as the semiconductor laser itself, the use of wave-guides in dye lasers has, so far, been confined to the laboratory for the purpose of exploring new guided-wave structures. For thin-film solid-state lasers, optical pumping with light-emitting diodes or other lasers has been considered [109]. Here problems of pump inefficiencies and the associated lower overall efficiency have to be overcome. The present standard of comparison for these devices are the electrically pumped semiconductor lasers which can offer relatively high efficiencies.

Recently, $Al_xGa_{1-x}As$ heterostructure junction lasers have become available which are capable of long-lived CW operation at room temperature. These devices are of great interest for optical communications. For a detailed review of these lasers we refer the reader to [84], [110], [111], or [112]. The refractive index of the $Al_xGa_{1-x}As$ material decreases with the Al concentration x following very nearly the rule

$$n(x) = n(o) - 0.45x. \quad (43)$$

A heterostructure waveguide can thus be formed by sandwiching an epitaxial layer of GaAs with lower index $Al_xGa_{1-x}As$ material. A typical concentration value is $x = 0.3$. While heteroepitaxy is now the standard for junction lasers, the exploration of other fabrication techniques is being continued. An example is the recent production of junction lasers formed by implantation of Zn^+ ions in n-type GaAs doped with tellurium [113]. Pulsed lasers operation of these devices was observed at low temperature (77K).

For the optically pumped thin-film solid-state lasers a high concentration of active ions is preferred. Concentration quenching of the fluorescence, however, calls for compromise solutions [109]. Initial attempts with thin-film solid-state lasers have already shown some success. Lasers with Ho^{3+} doped aluminum garnet films [114] have operated at 2.1 μm, and lasers with similar films doped with Nd^{3+} have operated at 1.06 μm [115]. Both film types were epitaxially grown on YAG substrates. Relatively large index differences were achieved with Nd^{3+} doped epitaxial films of YAG (n_f = 1.818) and $CaWO_4$ (n_f = 1.89) grown on sapphire (n_s = 1.77) substrates [116]. Thin-film lasers have also been made of $CdS_{1-x}Se_x$ layers grown on CdS substrates [117].

GaAs junction lasers with a stripe geometry have been developed to provide lasers operating in a stable and pure transverse mode [112]. Stripe contacts and mesa stripes are two key examples for stripe-geometry structures. These structures usually confine both the light and the charge carriers. In principle, separated optical confinement can be achieved by the use of strip waveguides such as those discussed in Section II-C. Recently, ridge waveguides have been investigated for this purpose [118], and confinement of light and pure modes were observed in passive structures. The ridges were formed by anodization etching of GaAs films on $Al_xGa_{1-x}As$ substrates. Figure 25 shows such a ridge guide with a width of approximately 10 μm and a film thickness of about 0.8 μm.

Distributed feedback structures promise to provide compact low-loss optical cavities for thin-film lasers which allow longitudinal mode control and frequency selection. Essentially, these are periodic waveguide structures superimposed on the gain medium. The feedback mechanism is backward Bragg scattering. While the first studies of the distributed feedback mechanism were done with dye

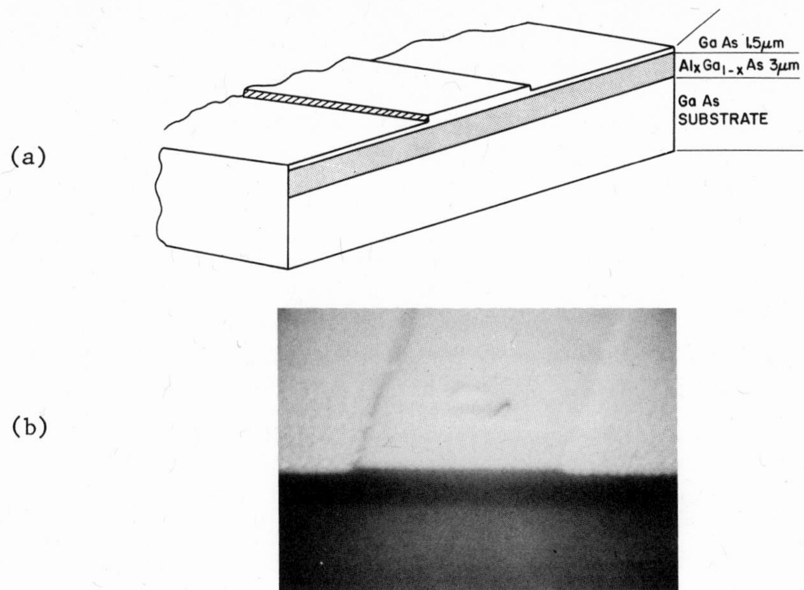

Figure 25. Sketch (a) and scanning electron micrograph (b) of a GaAs ridge guide. (From [118].)

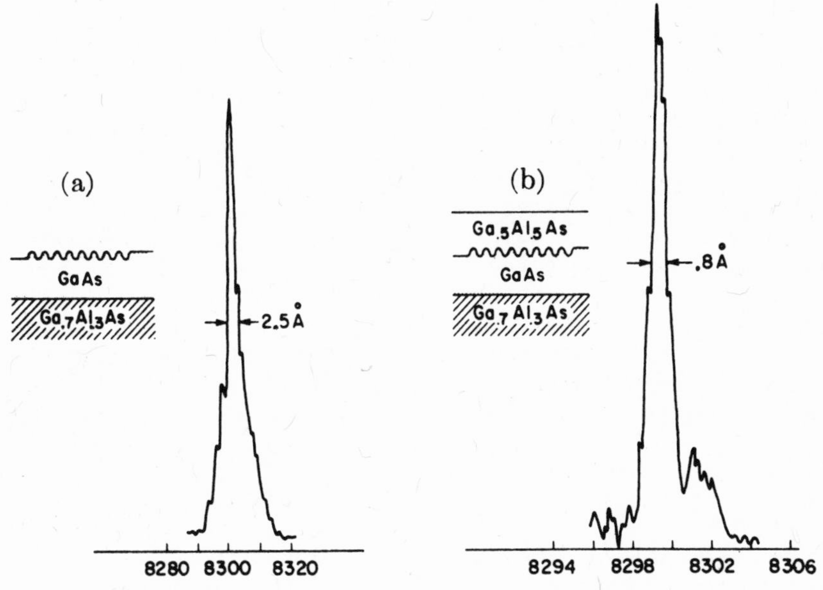

Figure 26. Output spectra of optically pumped AlGaAs lasers with periodic heterostructures. In case (a) the cover material was air, and in case (b) regrown AlGaAs. (From [44].)

lasers, the recent work has focused on the materials and fabrication problems of providing distributed feedback for AlGaAs lasers [43], [44], [119], [120]. Here the structures of interest take the form of a surface corrugation of the GaAs layer, as indicated in Fig. 26. Problems under study include the fabrication of the ultrafine grating-like structure where periods of about 1000 Å are needed, and the regrowing of AlGaAs crystal layers upon the corrugated GaAs [44], [119], [120]. Figure 26 shows the narrowing of the linewidth achieved with the periodic heterostructure lasers of [44], where the corrugation was fabricated by UV-laser exposure and ion-beam etching techniques and regrowth was accomplished with a dummy-seed-crystal technique preventing back-dissolving of the corrugation. These lasers were pumped optically with a pulsed dye laser and output spectra less than 1 Å in width were observed at 77K. Very recently [120], distributed feedback operation was demonstrated in electrically pumped single-heterojunction GaAs lasers at low temperatures (77K). Here corrugations were used which were etched 1300 Å deep and had a period of 3500 Å designed to utilize third-order scattering.

VI. CONCLUSIONS

We have tried to collect here the chief principles, underlying thoughts and promises of the field of integrated optics, and to illustrate recent experimental work on guided-wave devices. Whereas the field is still in its infancy it has attracted considerable interest and stimulated an extensive exploratory effort. This is reflected in the over 100 references we have cited, and these are only a fraction of the literature that is already available. Indeed, since this paper was written there have appeared two additional papers reviewing integrated optics [121], [122]. Integrated optics has drawn on many other disciplines including microwave theory and techniques, integrated circuit technology, solid-state physics, and optics and new devices, materials, and fabrication techniques have emerged. The field is now in a state of great flux and there are still many problems awaiting a solution, a main problem being the lack of a simple and efficient method of coupling optical fibers to thin-film or strip-guide devices.

REFERENCES

[1] S. E. Miller, "Integrated optics: An introduction," Bell Syst. Tech. J., Vol. 48, p. 2059, Sept. 1969.

[2] J. E. Goell, R. D. Standley, and T. Li, Electronics, Vol. 20, pp. 60-67, Aug. 1970.

[3] J. E. Goell and R. D. Standley, "Integrated optical circuits," Proc. IEEE (Special Issue on Optical Communications), Vol. 58, pp. 1504-1512, Oct. 1970.

[4] P. K. Tien, "Light waves in thin films and integrated optics," Appl. Opt., Vol. 10, p. 2395, Nov. 1971.

[5] S. E. Miller, "A survey of integrated optics," IEEE J. Quantum Electron. (Special Issue on 1971 IEEE/OSA Conference on Laser Engineering and Application, Part II of Two Parts), Vol. QE-8, pp. 199-205, Feb. 1972.

[6] S. E. Miller, E. A. J. Marcatili, and T. Li, "Research toward optical-fiber transmission systems (Invited paper)," Proc. IEEE, Vol. 61, pp. 1703-1753, Dec. 1973.

[7] N. S. Kapany and J. J. Burke, Optical Waveguides. New York: Academic, 1972.

[8] D. Marcuse, Light Transmission Optics. New York: Van Nostrand Reinhold, 1972.

[9] _____, Theory of Dielectric Optical Waveguides. New York: Academic, 1974.

[10] E. A. J. Marcatili, Bell Syst. Tech. J., Vol. 48, p. 2071, Sept. 1969.

[11] H. Kogelnik and V. Ramaswamy, Appl. Opt., Vol. 13, pp. 1857-1862, Aug. 1974.

[12] J. J. Burke, Opt. Sci. Newsletter (Univ. Arizona), Vol. 5, p. 31, 1971; also Opt. Sci. Newsletter (Univ. Arizona), Vol. 5, p. 6, 1971.

[13] H. Kogelnik and H. P. Weber, J. Opt. Soc. Amer., Vol. 64, pp. 174-185, Feb. 1974.

[14] L. Kuhn, M. L. Dakss, P. F. Heidrich, and B. A. Scott, Appl. Phys. Lett., Vol. 17, p. 265, Sept. 1970.

[15] R. Shubert and J. H. Harris, "Optical surface waves on thin films and their application to integrated data processors," IEEE Trans. Microwave Theory Tech. (1968 Symposium Issue), Vol. MTT-16, pp. 1048-1054, Dec. 1968.

[16] R. Ulrich and R. J. Martin, Appl. Opt., Vol. 10, p. 2077, Sept. 1971.

[17] F. Zernike, Appl. Phys. Lett., Vol. 24, p. 285, Mar. 1974.

[18] P. Kaiser, E. A. J. Marcatili, and S. E. Miller, Bell Syst. Tech. J., Vol. 52, p. 265, 1973.

[19] S. Somekh et al., Appl. Opt., Vol. 13, p. 327, Feb. 1974.

[20] J. E. Goell, Appl. Opt., Vol. 12, p. 2797, Dec. 1973.

[21] H. Furuta, H. Noda, and A. Ihaya, Appl. Opt., Vol. 13, p. 322, Feb. 1974.

[22] V. Ramaswamy, Bell Syst. Tech. J., Vol. 4, p. 697, Apr. 1974.

[23] W. Schlosser and H. G. Unger in Advances in Microwaves. New York: Academic, 1966, p. 319.

[24] J. E. Goell, Bell Syst. Tech. J., Vol. 48, p. 2133, Sept. 1969.

[25] E. A. J. Marcatili, Bell Syst. Tech. J., Vol. 4, p. 645, Apr. 1974.

[26] W. H. Zachariasen, Theory of X-Ray Diffraction in Crystals. New York: Wiley, 1945.

[27] C. F. Quate, C. D. Wilkinson, and D. K. Winslow, "Interaction of light and microwave sound," Proc. IEEE (Special Issue on Ultrasonics), Vol. 53, pp. 1604-1623, Oct. 1965.

[28] H. Kogelnik, Bell Syst. Tech. J., Vol. 48, p. 2909, Nov 1969.

[29] S. E. Miller, Bell Syst. Tech. J., Vol. 33, p. 661, May 1954.

[30] J. R. Pierce, J. Appl. Phys., Vol. 25, p. 179, Feb. 1954.

[31] W. H. Louisell, Coupled Mode and Parametric Electronics. New York: Wiley, 1960.

[32] A. Yariv, "Coupled-mode theory for guided-wave optics," IEEE J. Quantum Electron., Vol. QE-9, pp. 919-934, Sept. 1973.

[33] S. Wang, M. L. Shah, and J. D. Crow, Appl. Phys. Lett., Vol. 19, p. 187, 1971.

[34] T. P. Sosnowski and H. P. Weber, Opt. Commun., Vol. 7, p. 47, Jan. 1973.

[35] P. K. Tien et al., Appl. Phys. Lett., Vol. 21, p. 394, Oct. 1972.

[36] H. Kogelnik and C. V. Shank, J. Appl. Phys., Vol. 43, p. 2327, May 1972.

[37] A. W. Snyder, J. Opt. Soc. Amer., Vol. 62, p. 1267, Nov 1972.

[38] D. C. Flanders, H. Kogelnik, R. V. Schmidt, and C. V. Shank, Appl. PHys. Lett., Vol. 24, p. 194, Feb. 1974.

[39] F. W. Dabby, M. A. Saifi, and A. Kestenbaum, Appl. Phys. Lett., Vol. 22, p. 190, Feb. 1973.

[40] L. Kuhn, P. F. Heidrich, and E. G. Lean, Appl. Phys. Lett., Vol. 19, p. 428, Nov. 1971.

[41] H. Kogelnik and C. V. Shank, Appl. Phys. Lett., Vol. 18, p. 152, Feb. 1971.

[42] D. P. Schinke, R. G. Smith, E. G. Spencer, and M. F. Galvin, Appl. Phys. Lett., Vol. 21, p. 494, Nov. 1972.

[43] H. W. Yen et al., Opt. Commun., Vol. 9, p. 35, Sept. 1973.

[44] C. V. Shank, R. V. Schmidt, and B. I. Miller, Appl. Phys. Lett., Vol. 25, pp. 200-201, Aug. 1974.

[45] M. L. Dakss, L. Kuhn, P. F. Heidrich, and B. A. Scott, Appl. Phys. Lett., Vol. 16, p. 523, June 1970.

[46] H. Kogelnik and T. P. Sosnowski, Bell Syst. Tech. J., Vol. 49, p. 1602, Sept. 1970.

[47] H. Osterberg and L. W. Smith, J. Opt. Soc. Amer., Vol. 58, p. 1078, Sept. 1964.

[48] J. E. Goell and R. D. Standley, Bell Syst. Tech. J., Vol. 48, p. 3445, Dec. 1969.

[49] R. D. Standley, W. M. Gibson, and J. W. Rodgers, Appl. Opt., Vol. 11, p. 1313, 1972.

[50] M. K. Barnoski, R. G. Hunsperger, R. G. Wilson, and G. Tangonan, J. Appl. Phys., Vol. 44, p. 1925, 1973.

[51] E. Garmire, H. Stoll, A. Yariv, and R. G. Hunsperger, Appl. Phys. Lett., Vol. 21, p. 87, 1972.

[52] H. F. Taylor, W. E. Martin, D. B. Hall, and V. N. Smiley, Appl. Phys. Lett., Vol. 21, p. 95, Aug. 1972.

[53] W. E. Martin and D. B. Hall, Appl. Phys. Lett., Vol. 21, p. 325, Oct. 1972.

[54] I. P. Kaminow and J. R. Carruthers, Appl. Phys. Lett., Vol. 22, p. 326, Apr. 1973.

[55] J. M. Hammer and W. Phillips, Appl. Phys. Lett., Vol. 24, p. 545, June 1974.

[56] R. V. Schmidt and I. P. Kaminow, Appl. Phys. Lett., Vol. 25, pp. 458-460, Oct. 1974.

[57] D. H. Hensler, J. D. Cuthbert, R. J. Martin, and P. K. Tien, Appl. Opt., Vol. 10, p. 1937, May 1971.

[58] Z. I. Alferov et al., Sov. Phys.--Semicond., Vol. 2, p. 1289, Apr. 1969.

[59] I. Hayashi, M. B. Panish, and P. W. Foy, "Low-threshold room-temperature injection laser," IEEE J. Quantum Electron (Corresp.), Vol. QE-5, pp. 211-212, Apr. 1969.

[60] H. Kressel and H. Nelson, RCA Rev., Vol. 30, p. 106, Mar. 1969.

[61] J. M. Hammer, D. J. Channin, M. T. Duffy, and J. P. Wittke, Appl. Phys. Lett., Vol. 21, p. 358, Oct. 1972.

[62] V. Ramaswamy, Appl. Phys. Lett., Vol. 21, p. 183, Sept. 1972.

[63] S. Miyazawa, Appl. Phys. Lett., Vol. 23, p. 198, 1973.

[64] P. K. Tien et al., Appl. Phys. Lett., Vol. 24, p. 503, May 1974.

[65] ____, Appl. Phys. Lett., Vol. 21, p. 207, Sept. 1972.

[66] J. H. Harris, R. Shubert, and J. N. Polky, J. Opt. Soc. Amer., Vol. 60, p. 1007, Aug. 1970.

[67] R. Ulrich and H. P. Weber, Appl. Opt., Vol. 11, p. 428, Feb. 1972.

[68] P. K. Tien, G. Smolinsky, and R. J. Martin, Appl. Opt., Vol. 11, p. 637, Mar. 1972.

[69] D. B. Ostrowsky and A. Jacques, Appl. Phys. Lett., Vol. 18, p. 556, June 1971.

[70] H. W. Weber, R. Ulrich, E. A. Chandross, and W. J. Tomlinson, Appl. PHys. Lett., Vol. 20, p. 143, Feb. 1972.

[71] J. C. Dubois, M. Gazard, and D. B. Ostrowsky, Opt. Commun., Vol. 7, p. 237, Mar. 1973.

[72] J. E. Goell, Appl. Opt., Vol. 12, p. 729, Apr. 1973.

[73] J. J. Turner et al., Appl. Phys. Lett., Vol. 23, p. 333, Sept. 1973.

[74] D. B. Ostrowsky, M. Papuchon, A. M. Roy, and J. Trotel, Appl. Opt., Vol. 13, p. 636, Mar. 1974.

[75] D. L. Spears and H. I. Smith, Electron. Lett., Vol. 8, p. 102, Feb. 1972.

[76] B. Fay, D. B. Ostrowsky, A. M. Roy, and J. Trotel, Opt. Commun., Vol. 9, p. 424, Dec. 1973.

[77] H. L. Garvin et al., Appl. Opt., Vol. 12, p. 455, Mar. 1973.

[78] C. V. Shank and R. V. Schmidt, Appl. Phys. Lett., Vol. 23, p. 154, Aug. 1973.

[79] J. E. Goell, R. D. Standley, W. M. Gibson, and J. W. Rodgers, Appl. Phys. Lett., Vol. 21, p. 72, July 1972.

[80] S. Somekh et al., Appl. Phys. Lett., Vol. 22, p. 46, Jan. 1973.

[81] R. Ulrich et al., Appl. Phys. Lett., Vol. 20, p. 213, Mar. 1972.

[82] E. A. Chandross, C. A. Pryde, W. J. Tomlinson, and H. P. Weber, Appl. Phys. Lett., Vol. 24, p. 72, Jan. 1974.

[83] I. P. Kaminow, "Optical waveguide modulators," this issue pp. 57-70.

[84] M. B. Panish, "Heterostructure injection lasers," this issue, pp. 20-30.

[85] P. K. Tien, R. Ulrich, and R. J. Martin, Appl. Phys. Lett., Vol. 14, p. 291, May 1969.

[86] J. H. Harris and R. Shubert, in International Scientific Radio Union Spring Meeting Conf. Abstracts, Apr. 1969, p. 71.

[87] D. Dalgoutte, Opt. Commun., Vol. 8, p. 124, June 1973.

[88] R. Ulrich, J. Opt. Soc. Amer., Vol. 61, p. 1467, Nov. 1971.

[89] J. H. Harris and R. Shubert, "Variable tunneling excitation of optical surface waves," IEEE Trans. Microwave Theory Tech., Vol. MTT-19, pp. 269-276, Mar. 1971.

[90] P. K. Tien and R. J. Martin, Appl. Phys. Lett., Vol. 18, p. 398, May 1971.

[91] R. V. Schmidt, D. C. Flanders, C. V. Shank, and R. D. Standley, Appl. Phys. Lett., Vol. 25, p. 651, Dec. 1974.

[92] I. P. Kaminow, J. R. Carruthers, E. H. Turner, and L. W. Stulz, Appl. Phys. Lett., Vol. 22, p. 540, May 1973.

[93] I. P. Kaminow, V. Ramaswamy, R. V. Schmidt, and E. H. Turner, Appl. Phys. Lett., Vol. 24, p. 622, June 1974.

[94] J. F. S. Ledger and E. A. Ash, Electron. Lett., Vol. 4, p. 99, March. 1968.

[95] J. M. Hammer, D. J. Channin, and M. T. Duffy, Appl. Phys. Lett., Vol. 23, p. 176, Aug. 1973.

[96] M. A. deBarros and M. G. Wilson, Proc. Inst. Elec. Eng., Vol. 119, p. 807, July 1972.

[97] P. K. Cheo, Appl. Phys. Lett., Vol. 22, p. 241, Mar. 1971.

[98] F. K. Reinhart, J. Appl. Phys., Vol. 39, p. 3426, June 1968.

[99] D. Hall, A. Yariv, and E. Garmire, Opt. Commun., Vol. 1, p. 403, Apr. 1970.

[100] F. K. Reinhart and B. I. Miller, Appl. Phys. Lett., Vol. 20, p. 36, Jan. 1972.

[101] F. K. Reinhart, Appl. Phys. Lett., Vol. 22, p. 372, Apr. 1973.

[102] R. M. De La Rue, C. Stewart, C. D. Wilkinson, and I. R. Williamson, Electron. Lett., Vol. 9, p. 326, July 1973.

[103] Y. Omachi, J. Appl. Phys., Vol. 44, p. 3928, Sept. 1973.

[104] R. V. Schmidt, I. P. Kaminow, and J. R. Carruthers, Appl. Phys. Lett., Vol. 23, p. 417, Oct. 1973.

[105] R. V. Schmidt and I. P. Kaminow, IEEE J. Quantum Electron. (Corresp.), Vol. QE-11, pp. 57-59, Jan. 1975.

[106] A. W. Warner, D. L. White and W. A. Bonner, J. Appl. Phys., Vol. 43, p. 4489, Nov. 1972.

[107] M. L. Shah, Appl. Phys. Lett., Vol. 23, p. 75, July 1973.

[108] F. R. Gfeller and C. W. Pitt, Electron. Lett., Vol. 8, p. 549, Nov. 1972.

[109] J. P. Wittke, RCA Rev., Vol. 33, p. 674, Dec. 1972.

[110] M. P. Panish and I. Hayashi, Applied Solid State Science. New York: Academic, 1974.

[111] H. Kressel, in Lasers, A. K. Levine and A. deMaria, Ed. New York: Dekker, 1971.

[112] L. A. D'Asaro, J. Luminescience, Vol. 7, pp. 310-337, 1973.

[113] M. K. Barnoski, R. G. Hunsperger, and A. Lee, Appl. Phys. Lett., Vol. 24, pp. 627-628, June 1974.

[114] J. P. van der Ziel, W. A. Bonner, L. Kopf, and L. G. Van Uitert, Phys. Lett., Vol. 42A, p. 105, 1972.

[115] J. P. van der Ziel et al., Appl. Phys. Lett., Vol. 22, p. 656, June 1973.

[116] J. G. Grabmaier et al., Phys. Lett., Vol. 43A, p. 219, Mar. 1973.

[117] M. Kawabe, H. Kotani, K. Masuda, and S. Namba, to be published.

[118] F. K. Reinhart, R. A. Logan, and T. P. Lee, Appl. Phys. Lett., Vol. 24, p. 270, Mar. 1974.

[119] M. Nakamura et al., Appl. Phys. Lett., Vol. 24, p. 466, May 1974.

[120] D. R. Scifres, R. D. Burnham, and W. Streifer, Appl. Phys. Lett., Vol. 25, pp. 203-206, Aug. 1974.

[121] W. S. Chang, M. W. Muller, and F. J. Rosenbaum, "Integrated optics," in Laser Applications, Vol. 2. New York: Academic, 1974.

[122] H. F. Taylor and A. Yariv, "Guided wave optics," Proc. IEEE, Vol. 62, pp. 1044-1060, Aug. 1974.

OPTO-ELECTRONIC SYSTEM AND AUTOMATIZATION OF RESEARCH

Yu.E. Nesterikhin

Institute of Automation and Electrometry
USSR Ac.Sc., Siberian Branch, Novosibirsk

The paper is a brief report on the work of the Institute of Automation and Electrometry USSR Ac.Sc., Siberian Branch in the field of automatization of scientific researches (ASR).

The main problems the Institute is engaged in within the bounds of creating the standard automated systems is to provide various physical-technical and biological experiments of the Siberian Branch institutes with a uniform ideology and means.

Such systems should be realized not in "each laboratory separately", but within the frames of the all-institute ASR complex on the uniform constructive, system-technical, and program basis. The first and the most complicated problem was the choice of hardware and software architecture, which allows to solve the problems set successfully.

The first step in the way of designing unified structure was standardization of equipment occupying intermediate position between an experiment and a computer (i.e. orientation for a wide use of the CAMAC systems /1-3/, development of the program of creating series of CAMAC modularsof common use and specialized modulars,as well, necessary for automatization of experimental investigation of high-speed processes).

But even more delicate and principal appeared to be the problem of defining the "global" structure of the complex, i.e. the method of providing interaction of basic functional subsystems of data acquisition and control over experiments; prior experimental data processing; computing center of the Institute. Based on the analysis fulfilled and on acquaintance with the ideology of designs which exist now (see, for example, /4/) the main requirements to organization of the communication subsystem of the ASR complex were formulated: it must possess greater symmetry as compared to that of teleprocessing systems. The subsystem must be realized, it must be machine-independent.

We shall restrict our description to a concrete system, which matches the given requirements (Fig. 1).

1. A communication subsystem of the ASR complex developed at the Institute - it is a unified exchange bus system (UEBS) /5, 6/, i.e. multicrate modular bus system managed according to the CAMAC principles, intended to provide interaction between "users": a computer, specialized peripheral devices, capture and control systems, and etc. Each user is presented in UEBS by a unified user controller (all of the user controllers are similar) with which it is connected via user interface, reflecting specific features of the device being connected with the UEBS. Information interchange between the users is carried out with the help of corresponding user controllers under control of hardware communication processors. Information is transmitted through a 24-digital channel with a clock frequency 1 megc (inside the crate). The UEBS operates in the mode "Question-Answer" so that time-sharing mode is possible in the channel. The UEBS is a system with a distributed control that is control functions are realized both by hardware (communication processors) and by software computers.

2. Due to UEBS application as a communication subsystem of the complex a number of aims of a system character has been achieved. First note that due to "symmetry" of the system, i.e. availability of any user for another one we managed to increase efficiency of

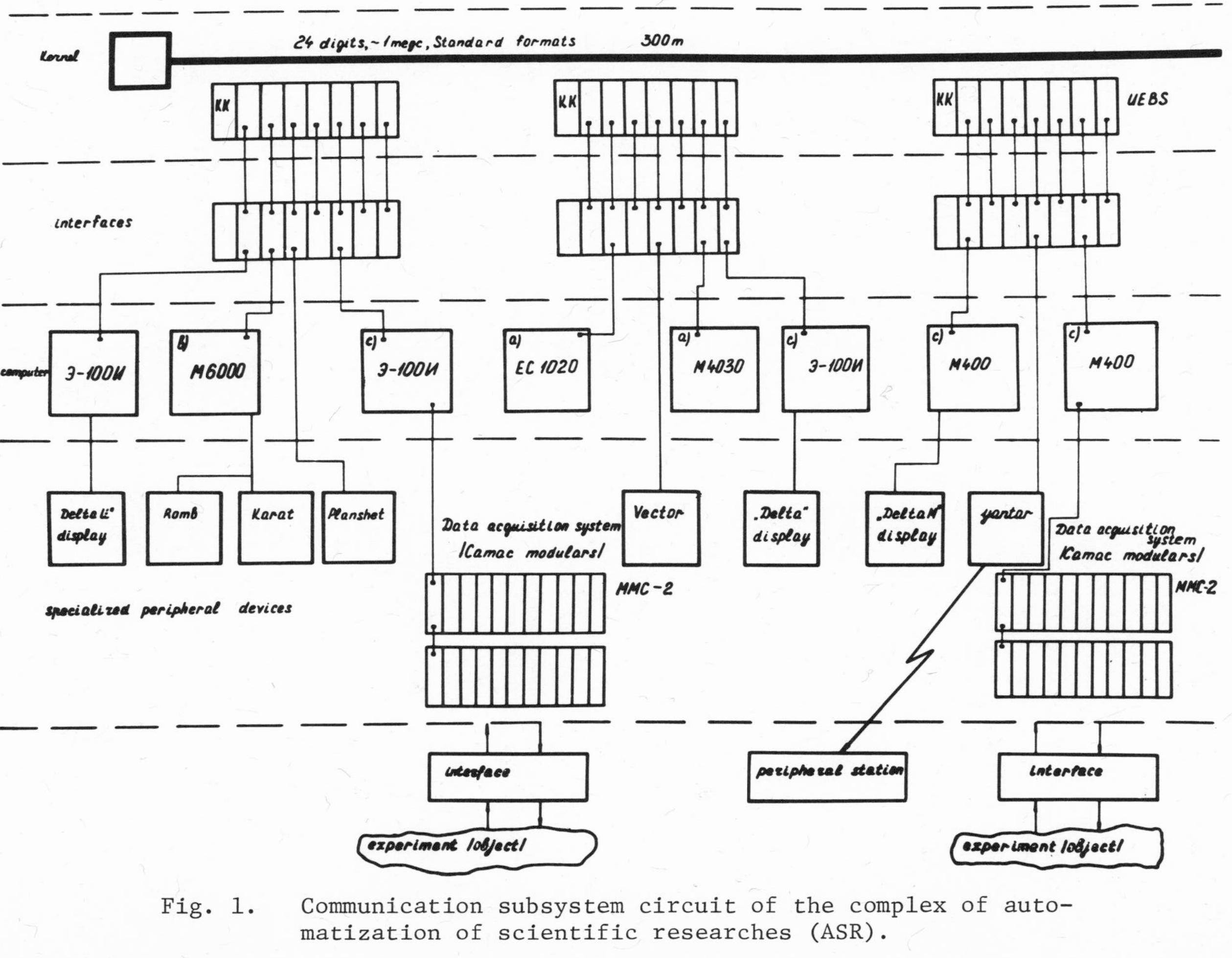

Fig. 1. Communication subsystem circuit of the complex of automatization of scientific researches (ASR).

using specialized peripheral devices to which "multiple access" is provided. An extremely important circumstance is the easiness of the complex expansion: as far as the hardware is concerned it is caused by the fact that a new user connection to the UEBS is reduced to creating a single user interface, i.e. a linear dependence of the interface equipment volume on the number of users takes place; as far as the software is concerned it is caused by the fact that for a computer-user the whole complex is a single not standard peripheral device.

Experimental usage of the complex began in 1973 and it confirmed rationality of the principles forming the basis.

3. As far as specialized peripheral devices are concerned the work of the Institute is directed to development of its own devices and first of all of optical image input/output devices and graphical interactive systems among them.

For example a number of graphical displays /7, 8/ were developed at the Institute; "Delta" is one of them, and in the Institute ASR complex it is the basis when solving the problems of data processing and simulating in the interactive mode /9, 10/. Besides a series of plotters with step motors "Vector" /11/ a high-speed plotter-coder with analog driver "Planshet" /12/ was created in which optimal high-speed algorithms of plotting are realized. A x-ray tube device for outputing graphical and alphanumerical information on a microfilm /13/ is widely used.

All these devices have been developed on the basis of a minicomputer Э-100И with the help of which they are connected with the system (Fig. 2) /14/.

3a. A peripheral device "Zenit" should be specially marked in this instance which is intended for active processing of photographic images of microbiological samples ; it is included in the ASR complex via its own computer; complicating of image processing problems leads to use of a special computer and, respectively, increase of the role of mathematical methods when analyzing the results of images.

Fig. 2. "Zenit" device for active photographic image processing.

The image is considered as a numerical array encoded in a particular way. The manner of useful information packing in the picture is not the best one in the majority of cases from the mathematician wiew point.

An image, e.g. a photographic one, comprises nothing but a two-dimensional function of optical density, given, naturally, in an analogue form. Converting this function into a digital form requires to make measurements of coordinates and optical density. At the same time the difficulty connected with the image redundancy arises immediately. The experimental array redundancy is defined in a final analysis by a priori data which one manages to formalize and use in processing. It is these data that play the part of the addressed system making possible to carry out search for useful information in the experimental data array.

Most often useful information localizes at the final number of separate image fragments (shots of a star sky, bubble tracks). In case of smooth images,e.g. interferograms, nabulae, and etc. useful information localizes at the set of parameters of analytical description given in accordance with the available a priori data. In this case the possibility of obtaining the required accuracy of evaluation of parameters by optimum arrangement of some minimal number of points on the image at which the density is measured appears. Usually an investigator is able to preset more or less easy formalized character according to which informative fragments of the image can be discovered, or point to the procedure which determines the count density destribution on the image analysed. Development of these principles leads to solving the problem of image processing when each measurement provides in the average the maximum quantity of information for achievement of the final result.

The question is in the adaptive input of the image into a computer which can be efficiently fulfilled if consider the problem of scanning together with the whole problem of processing.

Practical realization of this principle demands a special approach to designing devices which must provide comparatively quick measuring the optical density (or transmission coefficient) of a shot in any extremely small area which coordinates within the shot limits are given on a computer command. The result of measurement must come directly into the computer processor to make the following calculations. The size of the shot was chosen 300 x 300 mm^2 in accordance with the standard for photograph plates produced for scientific-technical aims. For accurate measuring the coordinates laser interferometers with the count discreteness equal to 0.32 μ were used as coordinate sensors. The device is used as a "writing" one. That is why it admits a change of a reading unit for a writing one and when supplied with proper assemblies and carriers it should be used as an optical memory of the archive type. To obtain high fast response a cathode-ray tube was used in a reading unit, the light spot in it can occupy any of 4096 x 4096 positions on the computer command in a time not more than 30 sec. The light from the CRT screen is focussed on the shot plane with the help of a relay optical system. The light which passed a photographic material is registered by a photomultiplier, simultaneously measurement of the light intensity radiating by the screen is made with the help of an additional lightreceiver. The result expressing the coefficient of transmission of the image under analysis is fed to the computer. According to the type of objectives being used the CRT can serve a square fragment of the size from 3 x 3 to 1 x 1 mm^2 and less on the shot. The operating diameter of the light spot in the focusing plane is about 2 μ . Focusing is obtained with a pneumatic servo system of arranging the objective. Operation through the whole area is provided by joint action of CRT and the system of electromechanical displacement of the carriage. Displacements of the carriage for a distance of 150 μ and more occur in acceleration-braking mode with an acceleration of the order of 3 m/s^2. Such mode is used for rapid transit from one image fragment to another, the carriage velocity may be as much as 1 m/s^2. Displacement for the distance less than 150 μ is carried out at small acceleration with the help of nonlinear feedback circuit realizing a transfering process close to the optimum one. The carriage may occupy the given position with discreteness of 0.32 μ . Displacement for the maximum distance and stopping at the given position occurs for the time of the order of 1 second. All calculations necessary for motor control are performed with a specialized device and the computer presets the required coordinates of the carriage. Carriage movement along any curve is possible, as well. As it takes place current indications of laser sensors are fed to the computer. The minimum speed of uniform displacement is 5μ/s.The device comprises commutator direct-current motors with a small moment of inertia and clearanceless ribbon transmission. This allowed to use a single electromechanical system of transmission for all modes of movement.

Joint action of electronoptical and mechanical scanning systems provides certain practical conveniences in spite of the presence of two reference systems and the necessity of introducing corresponding transformations. A special program enables to calculate automatically the elements of a conversion matrix, to specify scale coefficients and to take into account the angles between the axise of coordinates, to measure the optical system distortion, to evaluate the diameter of a scanning light spot and light distribution within the spot. Besides, using the proper algorithm one can measure the curve of the carriage mechanical guides and the angle between them. Scanning and construction of images can be performed when the carriage is in motion. In so doing there is no necessity to drive the carriage with high accuracy. It is sufficient to provide coming the required point of a photoplate into the view of the optical system, the resedual misalignment is eliminated by the corresponding light spot addressing. Due to this reason the device appears to be insensitive to mechanical low-frequency vibrations of the carriage.

Practical use of the system requires to develop a special software. Computer commands providing a necessary connection of the computer with all basic functional units have been defined. A sufficient number of reserved commands to control additional units and devices has been allowed for. Programming is carried out using the FORTRAN language. There are more than 50 programs at present. They include the programs carrying out the error analysis, subroutines of testing electronic and electromechanical assemblies, the programs giving different modes of carriage motion.

The program file for analysing the shots of the star sky has been developed. It includes the programs of measuring haze level and its statistical characte-

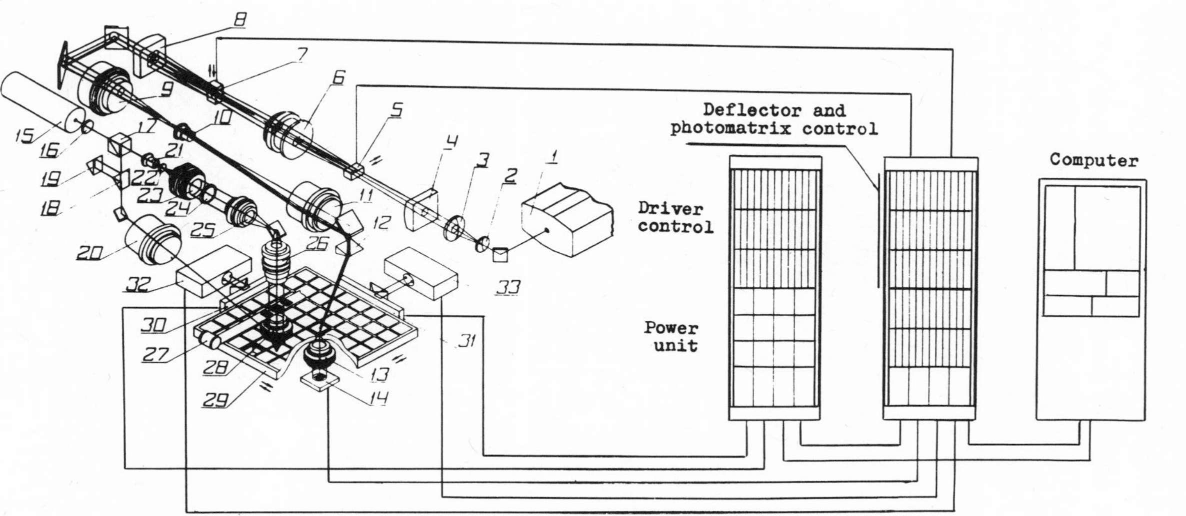

Fig. 3. Archive holographic memory with a movable carrier.

Reading system		Writing system		Elements of displacement	
1	laser	15	laser	29	case of memory modulators
2,3	collimator	16	shutter	30	mechanism of displacements X
4,5,6, 7,8,9	deflector	17	image divider	31	meter of displacements Y
10,11	telescopic system	18,19	aligning system	32	meter of displacements X
12	wedge	20	reference beam objective	33	meter of displacements Y
13	reconstructive objective	21,22,23	collimator		
14	photomatrix	24	phase mask		
		25,26	phase mask projection system		
		27	changeable phototransparencies		
		28	Fourier transformation objective		

ristics, program of retrieval of images of stars on the shot, the program of automatic identifying images of stars with the data of the star catalogue, the programs of estimating the coordinates of the centres of stars and star magnitudes, the programs of recognizing star images, twin-stars, galaxies, and nabulae.

To evaluate "Zenit" possibilities the programs for measurement and retrieval of defects on the photomasks of integrated microcircuits have been created. An adaptive method of scanning in combination with a priori information on the nature of various defects is widely used here.

Application of special optics allowed to turn from the analysis of photographic images directly to scanning microbiological objects. The device allows to distinguish with confidence the details inside the cell nuclear and to make necessary measurements.

The same device provided additionaly with a gas laser as a writing device is used to construct synthesized phase holograms on "photo" materials. The laser radiation intensity is controlled with the computer which makes all computations necessary for construction including correction of nonlinearity of the light modulator and the photomaterial sensitivity.

Two new versions of mechanical systems for the carriage displacement have been designed. In one of them aerostatic suspension of all mobile elements is used which provides an increased accuracy of positioning with the same characteristics at the expense of absence of mechanical "hysteresis" of the mobile system.

In another version linear electromotors matched with the guides along which the carriage moves are used. This version has an enlarged size of the operating field 450 x 450 mm^2.

We consider the chosen structure of the complex to be a successful one for development of experiments on creating hibrid optoelectronic data processing systems. To day, practically, besides a device "Zenit", including (via a buffer minicomputor) an archive holographic memory (Fig. 3) and a number of specialized optical processors is being realized, as the UEBS users.

REFERENCES

1. FUR 4100e. CAMAC. A Modular Instrumentation System for Data Handling. Revised Description and Specification. ESONE Committee, 1972.

2. FUR 4600e. CAMAC. Organization of Multi-Crate Systems Specification of the Branch Highway and CAMAC Crate Controller Type A. ESONE Committee, 1972.

3. FUR S100e. CAMAC. A Modular Instrumentation System for Data Handling. Specification of Amplitude Analogue Signals. ECONE Committee, 1972.

4. Computer Communications Proceeding of the IEEE.v.60. N11. November 1972. Special Issue.

5. Yu.E. Nesterikhin, A.N. Ginzburg, Yu.N. Zolotukhin, A.M. Iskoldski, Z.A. Lifshitz, Yu.K. Postoenko. Organizatsia sistem avtomatizatsii nauchnikh issledovanii (problemi, metodi, perspektivi). - Avtometria, 1974, N 4,

6. V.D. Bobko, Yu.N. Zolotukhin. Yu.M. Krendel, Z.A. Lifshits, A.P. Yan. Magistralnaya sistema obmena informatsiei. - Avtometria, 1974, N 4.

7. B.S. Dolgovesov, A.M. Kovalev, V.N. Kotov, A.A. Zubkov, Yu.E. Nesterikhin, K.F. Obertishev, A.S. Tokarev. Sistema "Ekran" dlya grafitcheskogo vzaimodeistvia cheloveka s EVM. - Avtometria, 1971, N 4.

8. A.M. Kovalev, V.N. Kotov, A.A. Lubkov, A.S. Tokarev. Grafitcheskii displei "Delta". - Avtometria, 1974, N 4.

9. E.G. Babat, B.S. Dolgovesov, F.M. Izrailev. Ispolzovanie dialogovogo grafitcheskogo terminala "Ekran" dlya reshenia prikladnikh zadatch. - Avtometria, 1971, N 6.

10. Yu.G. Bokovikov, Yu.I. Rodionov. Ispolzovanie razgovornoi mashinnoi grafiki v zadatche analiza aerodinamitcheskikh kharakteristik krilyev. - Avtometria, 1974, N 4.

11. A.N. Ginzburg, E.L. Emelyanov, V.V. Ivanov, K.P. Kasheev, K.V. Kotelnikov, V.I. Kuznetsov, A.V. Loginov, A.P. Lukashenko, Grafopostroitelnaya sistema "Vektor". - Upravlyaushie sistemi i mashini. 1974, N 5.

12. V.M. Aleksandrov, G.I. Gromilin, I.S. Karlson, N.N. Karlson, L.B. Kastorski, A.S. Kuznetsov, V.I. Litvintsev, M.M. Lyapunov, N.I. Pokrovski. "Planshet" - ustroistvo vvoda/vivoda grafitcheskoi informatsii. - Avtometria, 1976, N 1.

13. V.S. Avdeev, S.T. Vaskov, G.M. Mamontov, Yu.V.Obidin, A.K. Potashnikov, S.E. Tkatch. "Karat" - ustroistvo dlya vivoda grafitcheskoi informatsii na mikrofilm. - Avtometria, 1976, N 1.

14. L.V. Burii, V.P. Koronkevitch, Yu.E. Nesterikhin, A.A. Nesterov, B.M. Pushnoi, S.E. Tkatch, A.M. Scherbatchenko. Pretsizioznii fotogrammetritcheskii avtomat. - Avtometria, 1974, N 4.

A SURVEY OF OPTICAL INFORMATION PROCESSING IN SOME ADVANCED DATA PROCESSING APPLICATIONS

E. S. Barrekette

IBM Thomas J. Watson Research Center
P.O. Box 218
Yorktown Heights, New York 10598

The state-of-the-art within the information processing industry is reviewed and a comparison made with that of optical approaches to define those areas where the latter may play a role. As a case in point, the various elements of the automated office are examined from the point-of-view of pinpointing those areas where optical approaches might displace existing technologies. Since these technologies present a moving target, new candidates cannot enter the marketplace, unless they offer significant price and/or performance advantages. It is concluded that optical technologies have at best a negligible chance of entering the storage marketplace in the next decade and then only at the very high end. However, they have a more significant opportunity for success in communications, and they have all but arrived in input and output devices.

I. INTRODUCTION

Since the invention of the laser, almost three decades ago, many attempts have been made [1], utilizing its unique properties as well as the inherent advantages of parallelism offered by optical approaches, to introduce optical information processing systems into the marketplace. These attempts have at best been only marginally successful. Yet, one would expect that at least in some aspects of the processing of visual, graphic, and pictorial information, optical approaches should have a competitive edge. The purpose of this paper is to examine the processing of such information by those means which make some sense, and to determine where optical ones can play a role.

Because optical systems can handle large amounts of information in parallel, any application which would be amenable to their early successful entry should involve the handling of large amounts of information, much of it in "visual" form. A problem area, where electronic means are only beginning to make inroads, but have not as yet been fully established, is the management of the office information explosion. Accordingly, optical approaches to the various elements of an automated office are examined in order to determine if they can offer significant price and/or performance advantages to overtake the moving target presented by existing technologies by a sufficient margin to enable them to succeed in the marketplace. It is concluded that optical technologies have at best a negligible chance of entering the storage marketplace in the next decade and then only at the very high end. However, they have a more significant opportunity for success in communications, and they have all but arrived in input and output devices.

II. THE OFFICE INFORMATION ARENA

The volume of paper handled in modern offices has been growing at an exponential rate. This office information explosion created the problem of managing the cost of not just paper, but also of reproduction, distribution, and storage and retrieval.

A typical filing cabinet may contain from 5×10^3 to 2×10^4 documents, and the information content of these documents can range from 10^4 to 10^9 bits per page. For example, a coded typewritten page contains $1 - 3 \times 10^4$ bits, while a photograph or graphic arts quality page with sixteen levels of gray and color can have more than 3×10^9 bits. Table I gives the information content of a page of non-coded information recorded by means of various conventional printing technologies.

Table I. Information Content of a Non-Coded Document

Type	Resolution pel/cm	Binary	16 Gray Levels	16 Gray level and Color
Facsimile	50	1.5×10^6	6×10^6	2×10^7
Electrophotography	200	2.4×10^7	10^8	3×10^8
CRT Composer	300	5.4×10^7	2×10^8	6×10^8
Engraved Type, Photograph	600	2×10^8	10^9	3×10^9

Generally, the mix of documents in a filing cabinet consists primarily (> 90%) of coded (or codable) pages with a small proportion of non-coded ones. But even with only about 10% non-coded binary documents, 1% non-coded documents with gray scale and only a few photographs (< 0.1%) the information content of a five-drawer file can be as high as $10^9 - 10^{10}$ bits.

A medium sized establishment or a department of a large enterprise would have a few hundred individuals with associated active files containing as many as $10^{11} - 10^{12}$ bits of information and backed by archival files of several trillion bits. Even with data compaction (and some error correction and detection codes to improve reliability) to reduce storage and communications bandwidth requirements, a medium sized establishment would have files approaching a trillion bits.

It is this volume of information which one would seek to handle automatically by means of a centralized data base accessed through terminals or by a network of distributed intelligent work stations. A possible configuration is shown in Fig. 1. Currently, conventional technologies are being applied to the solution of the problems raised by the office paper explosion. It is anticipated that substantial cost savings could be had by reducing the flow of paper and automating the storage, retrieval, and distribution of documents. In the following sections, the various elements of an automated information handling system are examined from the point-of-view of determining whether and where optical means can play a role. The main elements of concern being Storage and Memory, Communications, and Input/Output.

III. STORAGE & MEMORY

Although optical candidates have regularly appeared claiming a capability to fill the needs of some storage and/or memory systems, none have had more than very limited success. In this section, the formidable competitive position of the entrenched technologies is reviewed and it is concluded that optical candidates have only a negligible chance of overtaking the continuously moving target of increased performance at rapidly decreasing costs.

A. Magnetic Recording

Today, a single disk pack (Fig. 2) can store the entire active file of a typical principal, and be available on-line at a cost which is equivalent to that of a secretarial word processing station. Similarly, two tape cartridges (Fig. 2) from the IBM 3850 mass storage system have a comparable capacity and provide on-line read/write storage at a lower cost per word than paperback novels. These are but temporary occupants of the front rank in a passing parade of products with ever decreasing costs and continually improving performance.

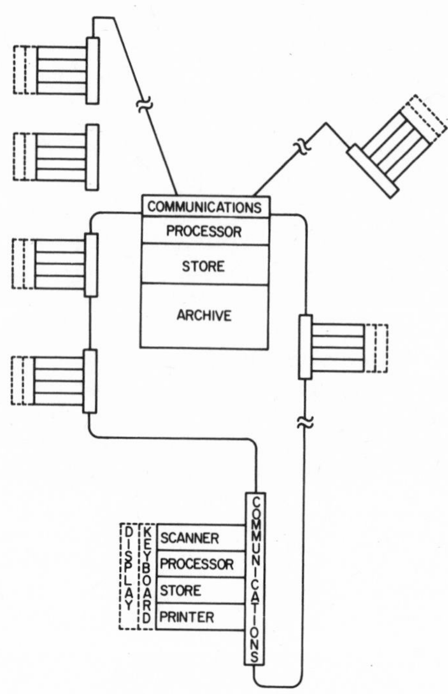

Fig. 1. Schematic of a general purpose system for an automated office/electronic mail.

Fig. 2. Tape cartridges for IBM 3850 Mass Storage System with a disk pack of comparable capacity.

Areal density in magnetic recording is approaching [2] 10^6 bits/cm^2 and the trend (Fig. 3a) which has brought an increase in density of about three orders of magnitude in two decades does not appear to be abating as yet [3]. Such densities have been achieved through a significant reduction in the dimensions (Fig. 3b) of the magnetic head gap, head flying height, and thickness of the magnetic media. These have reached magnitudes comparable to the wavelength of visible light [4,5]. At these densities and dimensions, recording and readout are accomplished at relative velocities approaching 50 m/sec or 200 km/hr.

With increasing densities and other improvements, the price per bit of magnetic recording (Fig. 4) has been dropping at the rate of more than an order of magnitude per decade leading to an acceleration in the consumption of on-line direct access storage (Fig. 3c) at a rate substantially in excess of an order of an magnitude per decade.

It is interesting to note that the main emphasis has been on cost reduction (primarily through substantial increases in storage density). However, access time, one measure of performance, has not improved significantly (Fig. 3d) in the most recent past; this is due to the fact that the constraints of mechanical access can be overcome for a price, but the improvement has not been attractive enough to warrant the investment.

B. Semiconductor Memory

Although quite remarkable, progress in magnetic recording is overshadowed by developments in semiconductor memory which has replaced magnetic cores as the prime technology within the last few years. Here too reduction in cost has been the prime motivator for innovation; its success is vividly illustrated in Fig. 4. The price of semiconductor memory has been dropping at the rate of much more than two orders of magnitude per decade, and that trend is still in progress. This has been accomplished primarily through two complementary concurrent approaches toward increased density: 1) improved device design reducing the area of a bit cell at a given level of lithography and 2) higher resolution lithography.

Although progress in device design [6], (Fig. 5), seems to have brought us to what appears to be a limit: the area of the intersection of two lines, new ideas such as the possibility of multi-level CCD storage may yet lower this limit by allowing several bits to be stored in a single cell. Similarly, although photolithography has been employed to make devices with ever decreasing geometries, the limit posed by the wavelength of light has been circumvented with the advent of electron beam exposure [7,8] which is beginning

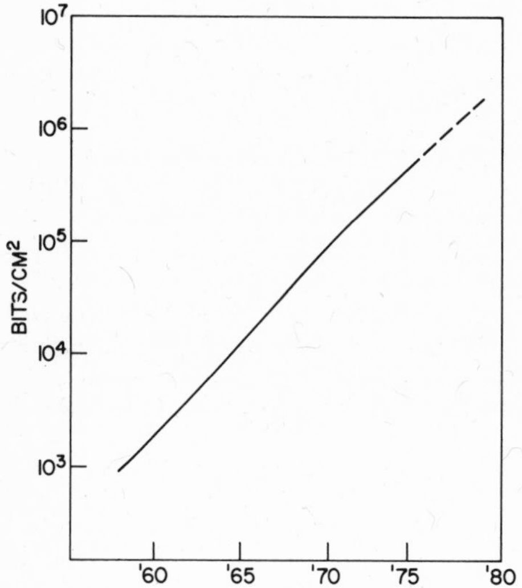

Fig. 3a. Trend in areal bit density of magnetic disk.

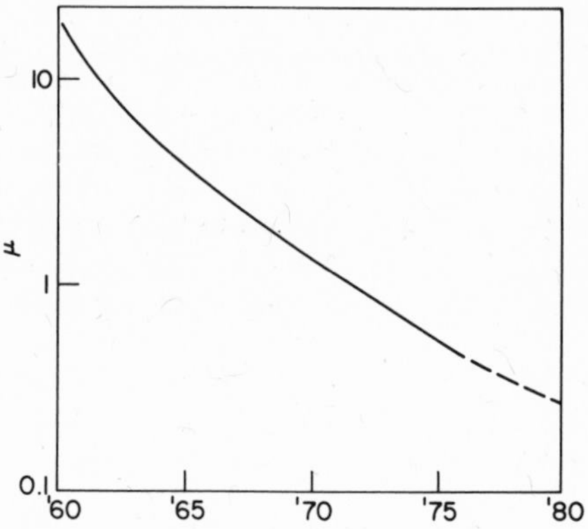

Fig. 3b. Trend in typical dimensions of magnetic disk recording.

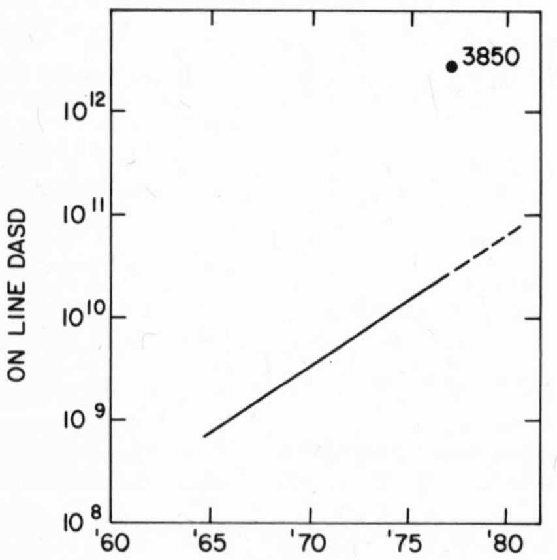

Fig. 3c. Trend in on-line file capacity.

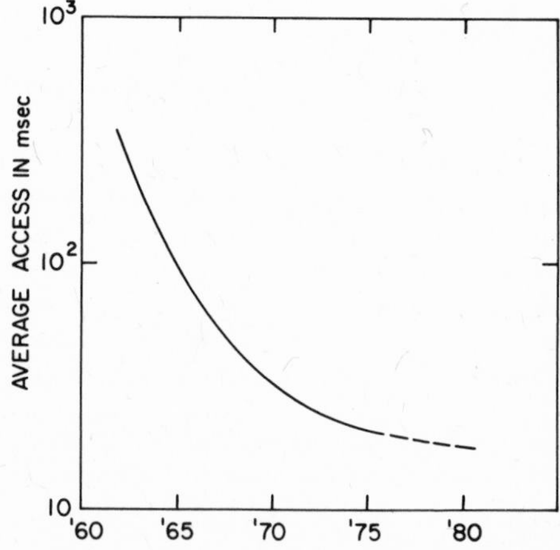

Fig. 3d. Trend in average access time in magnetic disk files.

to make an entry into the manufacturing environment, and x-ray lithography [8] which has shown promise in the laboratory. Both new approaches have the potential of submicron linewidth capability. Figure 6a shows an 8K bit FET chip with 1.2 μm gates [9] which was made using electron beam lithography and Fig. 6a shows a detail from a similar chip, but exposed at double the resolution [9], (i.e., with 0.6 μm gates). Figure 7 shows a bubble structure with μm dimensions fabricated using x-ray exposure [10].

These developments, as well as other progress in semiconductor circuit fabrication, have contributed to the trend toward larger chips with increased resolution. As a result, the number of bits per chip [11] (Fig. 8a) has been increasing at more than two orders of magnitude per decade leading to the spectacular drop in prices mentioned in the foregoing. Consequently, the cost (Fig. 8b) of operating semiconductor memory has been dropping rapidly encouraging substantial increases in memory consumption. On-line memory (Fig. 8c) has grown so that today it is approaching the levels of on-line disk files in the early 60's.

C. Magnetic Bubbles

Magnetic bubbles which have just entered the marketplace (in sample quantities) have cost and performance between magnetic storage and semiconductor memory. Their potential is exemplified by the initial entries [12] (e.g., Texas Instruments) which approach 10^5 bits per chip at tens of millicents per bit, or a fraction of RAM costs, and by development devices from Rockwell [13] with 10^6 bits per chip. Thus, so very early in the learning process, bubbles already exceed the capacities and densities of the much more mature FET memory devices and the somewhat newer CCD chips. Predictions of fantastic densities in the future relying on a) submicron linewidths, and b) new device configurations, are supported by developments (Fig. 7) of the type cited in the foregoing in x-ray lithography and by progress with bubble lattice files [14] (Fig. 9a) and contiguous disk devices [15] (Fig. 9b) both of which allow higher densities of bubbles at a given level of lithography.

D. Optical Storage

Despite the efforts of many workers pushing optical memory and storage for many years [1] no alterable optical memory either bit oriented or holographic has yet been brought to a commercial product stage.

Initially, read-only optical systems were introduced at the high end -- 10^{12} bits. IBM's Photo Digital Storage System [16]

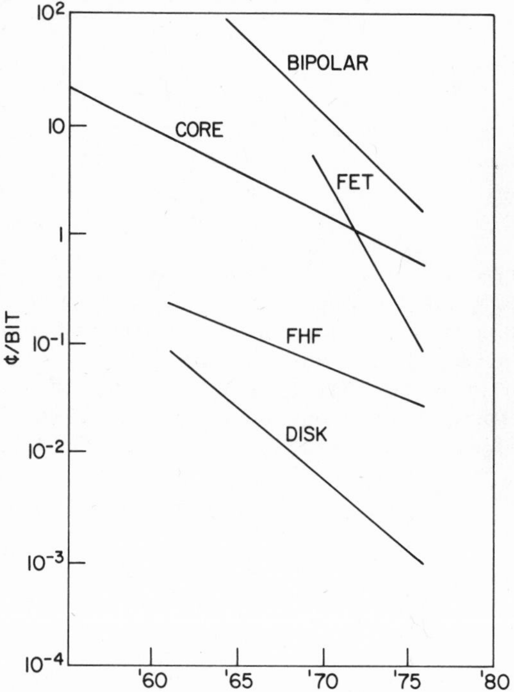

Fig. 4. Trends in the price of storage.

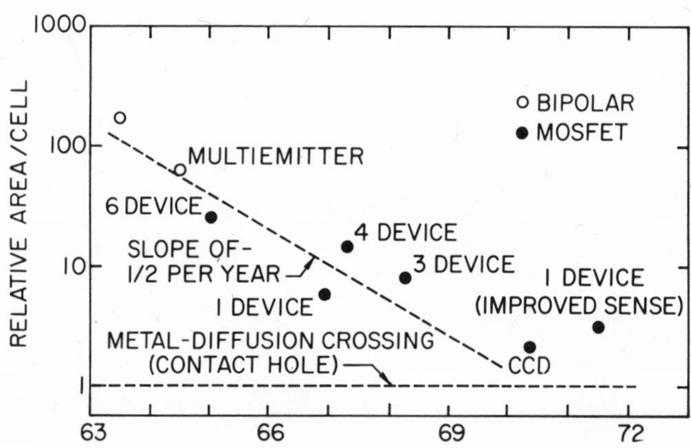

Fig, 5. Memory cell improvement due to device and circuit innovation - fixed lithography rules.

Fig. 6a. Electron beam fabricated 8K bit FET chip with 12 μm gates.

Fig. 6b. Detail of FET structure with 0.6 μm gates.

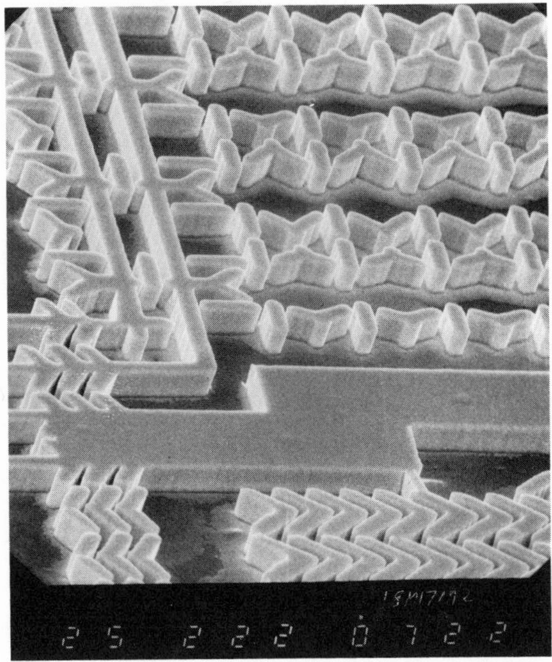

Fig. 7. X-ray fabricated bubble structures with μm dimensions.

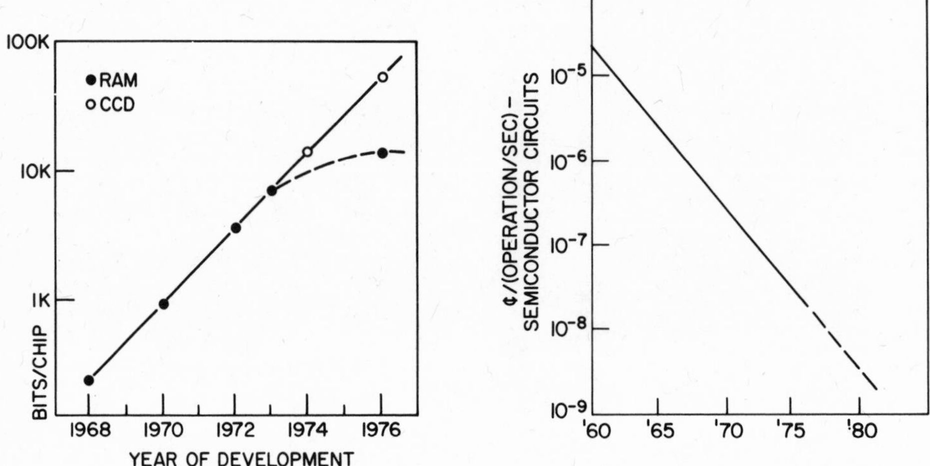

Fig. 8a. Trend in chip capacity. Fig 8b. Trend in operating costs for semiconductor circuits.

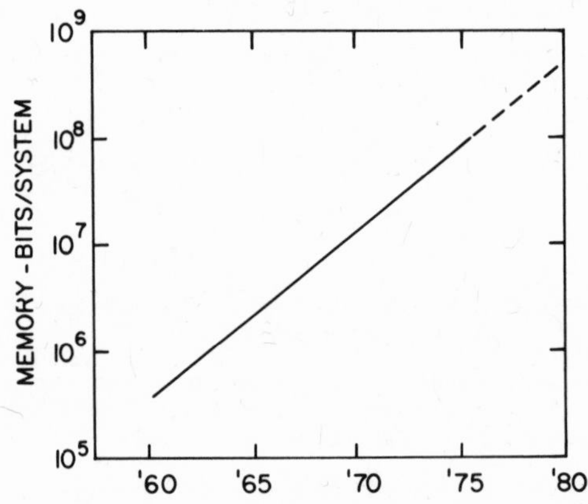

Fig. 8c. Trend in on-line memory capacity.

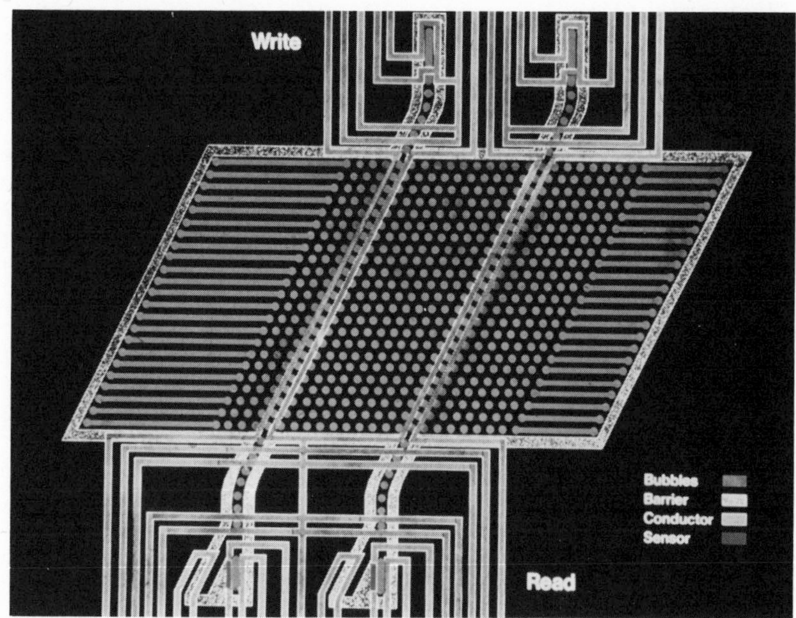

Fig. 9a. Schematic of a bubble lattice file

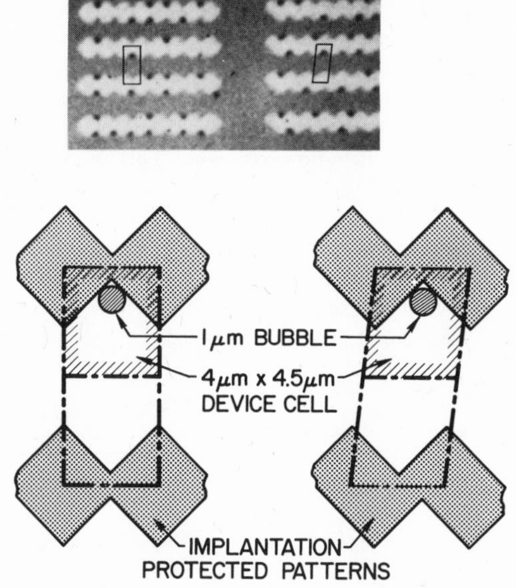

Fig. 9b. Contiguous disk bubble devices

introduced in 1966 relied on electron beam recording and CRT readout, and the Precision Instruments Unicon [17] system introduced in the late 60's relied on laser recording and readout. However, these have been replaced by magnetic read/write systems such as the Ampex Terrabit file, IBM's 3850, and similar compatible systems.

Subsequently, several video read-only systems were proposed in the early 70's. These included the RCA "SelectaVision," a holographic entry [18] which has been withdrawn in favor of a capacitance sensing device [19], the MCA Philips "DiscoVision" [20] laser-based venture, and the Teldec approach based on very high density long playing records. Of these only the last has reached the marketplace, while the others have had their market introduction postponed several times, and are now being confronted with having to overcome intial and already successful entries of VTR's, reversible magnetic tape record/playback units such as the Sony Betamax [3].

Thus, the promise of holographic [1], magneto-optic [21] and other storage based on optical devices has yet to be fulfilled. It is ironic that the anticipated advantages from the high resolution offered by optical storage have evaporated as magnetic recording, semiconductor circuits, and magnetic bubbles have achieved the density levels promised for optical approaches by using sophisticated optical and electron beam fabrication techniques [8].

If optical storage is to play a role, it must take full advantage of the parallelism offered by optical systems to achieve higher performance at a competitive price. In view of the fact that optical systems must incorporate elements such as lasers, sensor deflectors, composers, etc., which are not yet inexpensive, they would have to make an entry at the high end where their costs could be amortized over a large number of bits. Indeed, the early, though quite limited, success of the optical systems (Unicon and IBM's Mass Storage), mentioned in the foregoing, was achieved because they were aimed at the high end. The problems associated with limited field size [22], the lack of truly instantaneous page composers [23], and efficient reversible media [24] must be addressed and solved immediately, if the entrenched and the currently emerging contenders are to be overtaken before they completely monopolize the field.

IV. COMMUNICATIONS

In an automated office, one would expect relatively fast response, of the order of no more than one second, in retrieving a document from the data base and displaying it at the user's terminal. This does not pose excessive communication requirements if the document is stored in coded or machine readable form, and standard telephone lines could handle the task. However, for non-coded information, even with data compaction, the communications load becomes formidable. Thus, for binary documents with a resolution of 200 pel/cm a bandwidth of several megabits per second will be required even with data compaction of close to 10:1. With gray scale and color, a bandwidth approaching 100 MHz may be necessary if an annoying delay is to be avoided in the transfer of data from the data base to the terminal. Such bandwidths can be achieved using standard coaxial cables, but the cost can be prohibitive.

Figure 10a compares the attenuation of coaxial cables and optical fibers. Clearly for bandwidths of the order of 100 MHz, rather heavy and consequently costly coaxial cables are necessary to provide sufficiently low attenuation for distributing a signal without repeaters.

Optical fibers of very low attenuation are now becoming available, and several experimental installation based on optical fiber communications are now operational [25].

The attenuation and bandwidth of optical fibers [26] have been brought to levels which make repeaters unnecessary in most single-site establishments (Fig. 10b). Long lengths are now commercially available with attenuations of under 10 dB/km as are experimental [27] quanties of cable with $<$ 2 dB/km. At the same time cable prices have also been dropping at a fast rate (Fig. 10c) so that they are becoming competitive with coaxial cables. Prices will no doubt continue their downward spiral as usage mounts. While progress on fiber attenuation and cable cost has been rapid, progress in other areas of optical communications has kept pace. Innovation in cable conectors [28] and splicing [29] techniques are continually appearing. Light sources, especially long-lived room temperature semiconductor lasers [30], have been developed as have schemes for coupling [31] their output into and out of fibers. At the same time receivers [32] detectors and amplifiers have been developed with high sensitivity allowing low error rates even for low signal levels.

Thus, all indications point to successful entry of optical communications into the marketplace within the next decade. The target posed by existing technology in communications is moving much more slowly than in storage, and progress in optical devices has been more rapid here, leading to this optimistic assessment.

V. INPUT/OUTPUT

The important role of optical information handling devices at the input/output and terminal level of data processing systems is no longer in question. Not only are displays becoming pervasive, but optics is beginning to play a role in input devices and printers, a role which can only be expected to increase in importance with time as

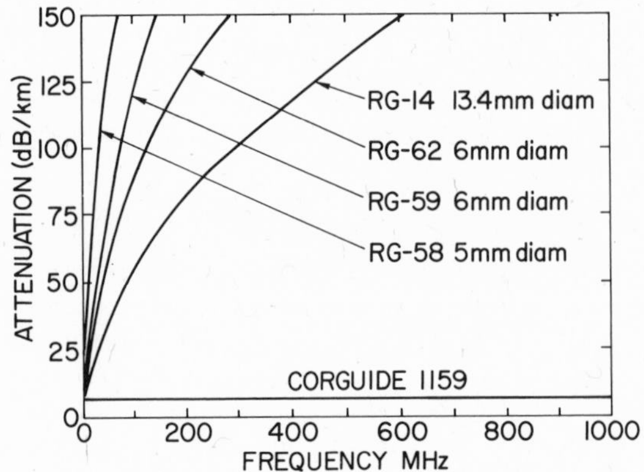

Fig. 10a. Comparison of attenuation in coaxial cables and optical fibers.

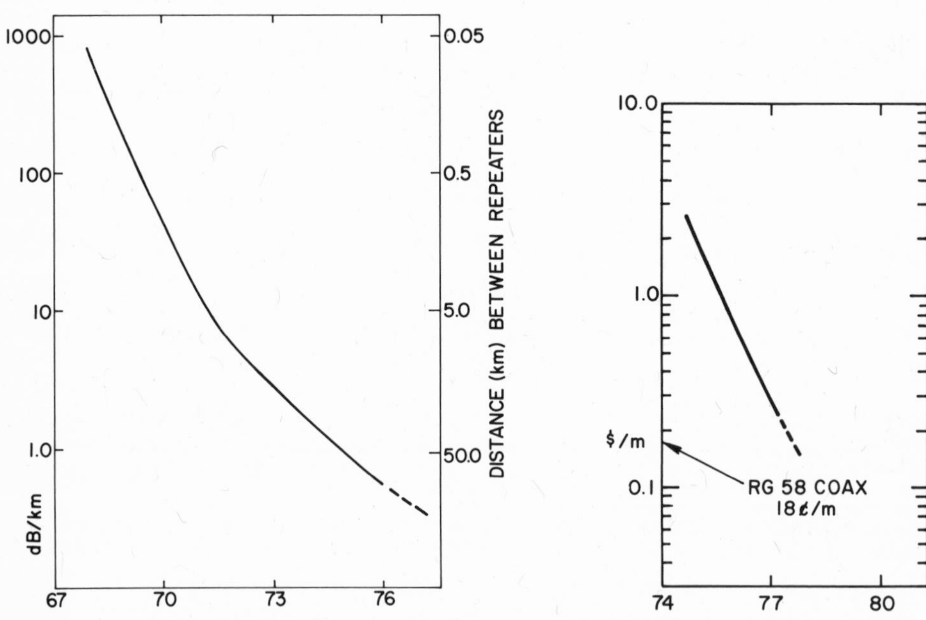

Fig. 10b. Trend in fiber attenuation.

Fig. 10c. Trend in single fiber costs.

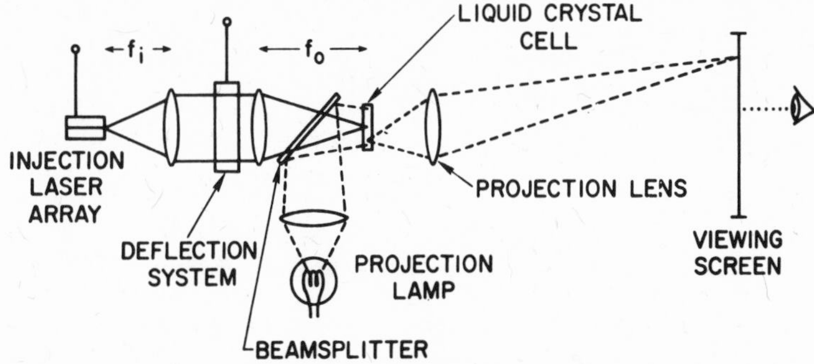

Fig. 11. Schematic of a laser-addressed liquid crystal display.

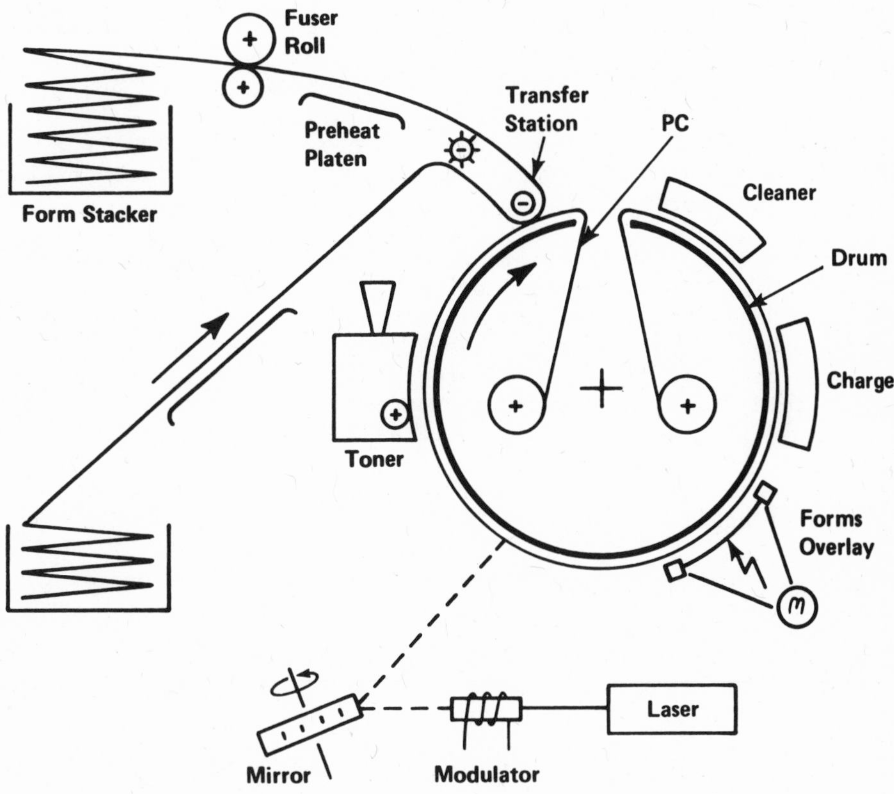

Fig. 12. Schematic of a laser-addressed electrophotographic printer.

unreliable mechanical devices are replaced by opto-electronic technologies.

In displays, while the CRT has been paramount [33] for many years, gas panels [34] have made a successful entry in the mid-range and LED's and liquid crystals [35] at the low end. At the same time new contenders such as a laser-addressed liquid crystal display [36] (Fig. 11) are bidding for entry at the high end, and electro-luminescent displays [37] in the mid-range. Today's and tomorrow's technologies are obviously optical.

In printers, while impact printers are the established technology, they appear to have reached a maturity associated with at best slow progress, making the area ripe for entries of new contenders. Optical ones have not missed this opportunity. The IBM 3800 [38], an electrophotographic device (Fig. 12), has made successful entry into the marketplace, printing up to 13,360 lines/minute (or consuming paper at almost 4 km/hr.). The Xerox 9700 is a recently announced electrophotographic printer with even higher throughput, and Fujitsu [39] has developed such a device for Kanji. Others will no doubt follow. Similarly, Xerox has been successfully marketing a laser-based Telecopier 200 facsimile system [40].

In input, not only have light pens been used on CRT's for over a decade, but scanners for document reading have been developed and are being improved in resolution and field coverage.

Today off-the-shelf arrays with 1728 elements are available from Fairchild and area arrays with 512 x 320 sensors was available from RCA while somewhat smaller arrays are available from a host of other vendors. At the same time, light beam deflectors have been developed so that acousto-optic devices can be purchased with a field coverage of several hundred spots in each of two dimensions operating at multi-MHz data rates. Furthermore, mechanical deflectors with a field coverage of more than 10^4 spots at 50 MHz data rates can be found, and more economical devices, satisfying more modest specifications are available off-the-shelf.

Clearly the success of optical devices at the I/O end of the data processing enviornment cannot be challenged.

VI. CONCLUSIONS

The role of optical information handling devices in the data processing environment was examined using the automated office as a case-in-point. It was shown that whereas they are established in Input/Output and terminal devices, and are making a strong entry into communications, optical contenders have not been too successful in the storage/memory area and will probably not play a meaningful role in the foreseeable future.

I thank Professor George Stroke for inviting my participation in the US-USSR Information Exchange Seminar on Optical Information Processing. It was this invitation that prompted the survey presented in the foregoing. I also thank many of my colleagues at IBM for many useful discussions and for some of the figures presented herein. In particular thanks are due to H. Chang, D. L. Critchlow, G. C. Feth, M. Heritage, A. L. Hoagland, E. G. Lean, R. A. Myers, E. W. Pugh, L. M. Terman, and H. N. Yu.

References

1. J. A. Rajchman, Appl. Optics, 9, (1970), 2269. See also Applied Optics, 13, (1974), 755-924, for a collection of many of the papers presented at the Topical Meeting of the OSA on Optical Storage of Digital Data. These include references to considerable prior work.

2. A. Geffon, paper #6-7, 1977 Digests of the Intermag Conference, (1977).

3. The Sony Betamax has 400 tracks/cm and about 1.6 x 10^4 flux reversals/cm of analog recording.

4. Koichi Kugimiya, et al., paper #26-8, 1977 Digests of the Intermag Conference, (1977).

5. M. Satou, et al., paper #22-8, 1977 Digests of the Intermag Conference, (1977).

6. D. L. Critchlow, Computer, 9, (1976), 31.

7. T. H. P. Chang, et al., Electronics, 5/12/77.

8. A. N. Broers, Proc. 7th Intern. Vac. Conf. & 3rd Intern. Conf. Solid Surfaces (1977).

9. H. N. Yu, et al., J. Vac. Sci. Technol., 12, (1975), 1297; and T. H. P. Chang, Proc. 8th Conf. on Solid State Devices, Tokyo (1976), 9.

10. E. Spiller, et al., Solid State Tech., April 1976, 62.

11. L. M. Terman, et al., IEEE J. Solid State Circuits, SC-11, (1976), 4.

12. See R. A. Nader, paper #21-4, S. K. Singh, et al., paper #21-5, D. J. Hayes, paper #21-6, A. H. Bobeck, et al., paper #21-7, 1977 Digests of the Intermag Conference, (1977).

13. J. L. Archer, paper #11-1, 1977 Digests of the Intermag Conference, (1977).

14. B. A. Calhoun, et al., IBM J. Res. Dev., 20, (1976), 368.

15. Y. S. Lin, et al., paper #11-6, 1977 Digests of the Intermag Conference, (1977).

16. J. D. Kuehler and H. R. Kerby, Proceedings Fall Joint Computer Conference (IEEE Computer Society, New York, 1966), 735.

17. E. E. Gray, paper #MA4, Digest of Tech. Papers, Topical Meeting on Optical Storage of Digital Data, OSA, (1973).

18. R. Bartolini, et al., App. Opt., 9, (1970), 2283.

19. "The RCA 'SelectaVision' Video Disc System," Information Display, 12, (1976), 20.

20. K. Bulthus, et al., paper #TUA3, CLEOS, Digest of Technical Papers, OSA/IEEE, (1976).

21. R. P. Hunt, IEEE Trans. Magnetics, MAG-5, (1969), 700; G. J. Fan and J. H. Greiner, J. Appl. Phys., 39, No. 2, Part II, (1968), 1216.

22. A. Vander Lugt, Appl. Opt. 12, (1973), 1675.

23. W. Stewart, et al., RCA Rev., 34, (1973) 3; G. Labrunie, et al. Appl. Opt., 13, (1974), 749.

24. "Optical Storage and Display Media," Special Issue of RCA Rev., 33, (1972), 3-310.

25. S. Shimada, paper #ThA4; M. Shimodaira, et al., paper #ThA5; I. Jacobs, et al., paper #ThB1, Optical Fiber Transmission II, OSA, (1977).

26. D. B. Keck, paper #TuD1, Optical Fiber Transmission II, OSA, (1977).

27. E. A. J. Marcatili, paper #TuC4, Optical Fiber Transmission, Technical Digest, OSA, (1975).

28. A. W. Warner, paper #5.6; A. F. Milton, et al., paper #5.7, CLEA, Digest of Technical Papers, IEEE/OSA (1977), 20.

29. C. M. Miller, paper #WA3, Optical Fiber Transmission II, OSA, (1977).

30. H. Kressel, paper #WB3; C. C. Stern, et al., paper #WB4, Optical Fiber Transmission II, OSA, (1977).

31. K. Takahashi, et al., paper #WA5, Optical Fiber Transmission II, OSA, (1977).

32. C. A. Brackett, et al., paper #ThE7, CLEOS, Digest of Technical Papers, OSA/IEEE, (1976).

33. L. K. Anderson, J. Vac. Sci. Tech., 10, (1973), 761.

34. H. J. Hoen, paper #3.1; W. E. Johnson, et al., paper #3.2, SID International Symposium, Digest of Technical Papers, (1977), 18,20 respectively.

35. J. E. Bigelow, et al., paper #6.3, SID International Symposium, Digest of Technical Papers, (1977), 62.

36. G. T. Sincerbox, paper #16.9, CLEA, Digest of Technical Papers, IEEE/OSA, (1977), 80.

37. S. Mito, paper #8.1, SID International Symposium, Digest of Technical Papers, (1977), 86.

38. M. R. Latta, et al., paper #TuD1, CLEOS, Digest of Technical Papers, OSA/IEEE (1976).

39. T. Matsuda, et al., paper #4.4, CLEA, Digest of Technical Papers, IEEE/OSA, (1977), 14.

40. P. Mrdjen, et al., paper #TuA1, CLEOS, Digest of Technical Papers, OSA/IEEE (1976).

BIOGRAPHIES OF CONTRIBUTORS

G. P. ARNAUTOV

G. P. Arnautov graduated from the Novosibirsk State University in 1963. Since then he has been working at the Institute of Automation and Electrometry (Novosibirsk, Siberian Branch of the USSR Academy of Sciences). He received his doctoral degree in technical sciences in 1974. He has more than 30 publications on information-measuring techniques and gravimetry.

R. D. BAGHLAY

Dr. Roman D. Baghlay graduated the L'vov Polytechnical Institute (the Ukraine) in 1959. Since 1959 he has been working as a researcher at the Institute of Automatics and Electrometry of the Siberian Branch of the USSR Academy of Sciences (Novosibirsk). He received his doctoral degree in 1963. He is engaged in the field of computational mathematics and mathematical physics. Dr. Baghlay has published over 40 papers.

E. S. BARREKETTE

Euval S. Barrekette received the AB degree in 1952 from Columbia College, the BS degree in Civil Engineering in 1953 from the School of Engineering, Columbia University, the MS degree in 1956 and the Ph.D. degree in 1959 in Engineering Mechanics from the Graduate Faculties, Columbia University. He was a Guggenheim Fellow from 1955-1957.

He was an Assistant Professor of Civil Engineering at Columbia University from 1959 to 1960 and had been an Adjunct Associate Professor of Civil Engineering until 1970. He joined IBM in 1960 as a Research Staff Member in the Research Division, organized the I/O department in 1972 and is now Assistant Director of the Applied Research department.

He is a member of the New York State Society of Professional Engineers, the Institute of Aeronautics and Astronautics, the American Society of Civil Engineers, the Optical Society of America, the IEEE, the Society of Engineering Science, of which he was a director from 1968-1971, Phi Beta Kappa, Tau Beta Pi, and Sigma Xi.

C. V. BYKHOVSKY

Constantine Valery Bykhovsky received his BS degree from the Institute of Fine Chemical Technology, Moscow, in 1961 and his Ph.D.

degree from the Institute of Chemical Physics, USSR Academy of Sciences, Moscow, in 1966. He published about 40 papers on the theory of chemical reaction rates, the role of metastable states in chemistry and biology, computing methods in chemical and biological reaction theory, the mechanisms of distributed memory, control and processing in cell computer nets and biological systems. Now he is with the Institute of Control Sciences, USSR Academy of Sciences, where he is the leader of the group on the study of holographic principles of control and processing in computing systems and live organisms. He is the author of the book "Cybernetics and Biological Control," Znanie, Moscow, 1977.

G. P. CHEIDO

G. P. Cheido graduated from the Tomsk Polytechnic Institute in 1962. Since 1963 he has been working at the Institute of Automation and Electrometry (Novosibirsk, Siberian Branch of the USSR Academy of Sciences). He received his doctoral degree in technical sciences in 1967. He has more than 30 publications on applied statistics and optical measuring information processing systems.

Y. V. CHUGUI

Yu. V. Chugui graduated from the Novosibirsk State University in 1968. Since then he has been working at the Institute of Automation and Electrometry (Novosibirsk, Siberian Branch of the USSR Academy of Sciences). He received his doctoral degree in technical sciences in 1976. He has more than 15 publications on holography, optical information storage and processing.

Y. N. DUBNISCHEV

Yu. N. Dubnischev graduated from the Leningrad Institute of Exact Mechanics and Optics in 1965. Since that time he has been working at the Institute of Automation and Electrometry (Novosibirsk, Siberian Branch of the USSR Academy of Sciences). He received his doctoral degree in technical sciences in 1973. He has more than 30 publications on laser optics, laser Doppler velocity gauges, and laser applications in spectroscopy.

V. M. EFIMOV

V. M. Efimov graduated from the Moscow Institute of Aviation in 1957. From 1957 to 1961 he worked in industry. Since 1961 Dr. Efimov has been working as a researcher at the Institute of Automatics and Electrometry of the Siberian Branch of the USSR Academy of Sciences in Novosibirsk. In 1965 he obtained a doctoral degree for his thesis "Investigation of Some Aspects of Time Quantization when Measuring and Control." Since 1973 Dr. Efimov has been heading the laboratory of Probable Methods of Investigation of Information Storing and Processing Systems. He has more than 40 publications on the problems of measuring and processing of random signals, optical signals are among them. He is the author of a monograph "Time Quantization in Measuring and Control Systems."

V. I. FELDBUSH

V. I. Feldbush graduated from the Novosibirsk State University in 1973. Since that time he has been working at the Institute of Automation and Electrometry (Novosibirsk, Siberian Branch of the USSR Academy of Sciences). He has more than 10 publications on space-time light modulators.

I. S. GIBIN

I. S. Gibin graduated from the Novosibirsk Institute of Electrical Engineering (the department of automatics and computing) in 1968. Since 1969 he has been working as a researcher at the Institute of Automatics and Electrometry of the Siberian Branch of the USSR Academy of Sciences (Novosibirsk). He is engaged in the development and investigation of holographic memories and optical systems of data processing mainly in the direction of studying the modes of holographic recording and reading, optimization of holographic memories, associative retrieval and spectral analysis of images. He is the author of 20 scientific publications and inventions.

A. C. GLOTOV

Alexander C. Glotov received his BS degree from the Moscow Physical Engineering Institute in 1973. He published several papers and has some patents on problems in optoelectronics. At present he is with the Institute of Control Sciences, USSR Academy of Sciences, where he is completing his Ph.D. thesis on "Semiconductor Optoelectronic Processors."

M. A. GOFMAN

Mikhail A. Gofman graduated from the Sevastopol Institute of Instrument-Making in 1971. Since that time he has been working at the Institute of Automation and Electrometry (Novosibirsk, Siberian Branch of the USSR Academy of Sciences). He has more than 10 publications on optical apertures for spectral analysis and for associative holographic memory.

S. B. GUREVICH

Prof. S. B. Gurevich graduated from the Leningrad State University, Leningrad, USSR, in 1945 and entered post-graduate courses at the same University in 1946. He received his first degree in 1948 and the second degree in 1964. Since 1959 he has been working at the Ioffe Physico-Technical Institute in Leningrad. In 1966 he became the Head of the Laboratory of Opto-electronics and Holography. He is also Vice President of Scientific Council on Holography of the USSR Academy of Sciences. He is the author of four monographs dealing with different problems of physics.

At present his fields of interest are the problems of optical information processing and holography.

M. HALIOUA

Dr. Maurice Halioua is currently Assistant Professor of Electrical Sciences at the State University of New York at Stony Brook where he has been working with Professor George W. Stroke at the Electro-Optical Sciences Laboratory since he immigrated to the United States in 1969. He also obtained both his Ph.D. degrees for his work with Professor Stroke, a Dr. Ing. in Physics in 1971 and a Dr. `es Sc. in Physics in 1974, both presented at the University of Paris in Orsay.

Before emigrating to the United States, Dr. Halioua studied at the University of Bordeaux, France, where he obtained his B.Sc. in 1962, and also at the Institute of Optics at the University of Paris hwere he obtained his Ing. Dipl. (Optics) in 1968.

Since 1972, he has also held the title of Instructor in Medical Biophysics at the Health Sciences Center at Stony Brook, in addition to his position of research associate at the Electro-Optical Sciences Laboratory.

Since 1969, he has been the co-author, with Professor Stroke, in some 20 publications in the field of image deblurring most recently of the article on the "Retrieval of Good Images from Accidentally Blurred Photographs" which appeared in Science in July 1975.

A. M. ISKOLDSKY

A. M. Iskoldsky is a physicist. Having graduated from the Leningrad and Novosibirsk State Universities and post-graduate courses at the Institute of Nuclear Physics of the Siberian Branch of the USSR Academy of Sciences he received a doctoral degree for his thesis on plasma physics in 1966. From 1963 to 1968 he was engaged in optical diagnostics of high temperature plasmas, as well as problems in high-speed electron-optical photography. Since 1968, Dr. Iskoldsky has been heading the laboratory of automation systems for the investigation of high-rate processes at the Institute of Automatics and Electrometry of the Siberian Branch of the USSR Academy of Sciences (Novosibirsk). He has more than 70 scientific publications. He is a Reader of Physics department of the Novosibirsk State University.

V. A. KHANOV

Vladimir A. Khanov graduated from the Novosibirsk State University in 1970. Since 1968 he has been working at the Institute of Automation and Electrometry (Novosibirsk, Siberian Branch of the USSR Academy of Sciences). At present he is a senior engineer. He has more than 20 publications on laser interferometers.

S. F. KIBEREV

Sergei F. Kiberev graduated from the Novosibirsk Electrotechnical Institute in 1973. Since that time he has been working at the Institute of Automation and Electrometry (Novosibirsk, Siberian Branch of the USSR Academy of Sciences). He has more than 10 publications on opto-electronic memory.

V. P. KIRYANOV

Valerii P. Kiryanov graduated from the Urals Politechnic Institute (Sverdlovsk) in 1964. Since that time he has been working at the Institute of Automation and Electrometry (Novosibirsk, Siberian Branch of the USSR Academy of Sciences). He received his doctoral degree in technical sciences in 1971. He has more than 30 publications on information-measuring techniques and laser measurements of displacements. At present he leads the group concerned with the electronics of laser measurements of displacements.

W. E. KOCK

Winston E. Kock, E.E. (1932) and M.S. (Physics, 1933), both University of Cincinnati, Ph.D. (Physics, 1934, University of Berlin) attended the Institute for Advanced Study at Princeton and was a Fellow at the Indian Institute of Science at Bangalore in 1936. He has been Director of Electronic Research, Baldwin Piano Company (where he developed the Baldwin electronic organ), Microwave Research Engineer and Director of Acoustics Research at Bell Telephone Laboratories (where he developed several microwave and acoustic lenses, directed the research on the Navy's underwater sound Jezebel-Caesar Project, and headed the group developing the picture-phone), Director of the Bendix Research Laboratories, first Director of the National Aeronautics and Space Administration's Electronics Research Center, Cambridge, Mass. and most recently Vice President and Chief Scientist of the Bendix Corporation. In 1971 he became Consultant to the Corporation and a Visiting Professor and Director of the Herman Schneider Research Laboratory at the University of Cincinnati.

Honors include the Navy's highest civilian award, the Distinguished Public Service Medal, (1964), Honorary Fellowship in the Indian Academy of Sciences (1970), an honorary D.Sc. (U. of Cincinnati, 1952), Eta Kappa Nu's Outstanding Young Electrical Engineer (1938) and the Eminent Member Award (1966), fellowship in the Acoustical Society, the Physical Society and the I.E.E.E. and membership in Tau Beta Pi, Sigma Xi and Eta Kappa Nu.

He was chairman of the Professional Group on Audio of the Institute of Radio Engineers in 1954-55 and was a member of the Governing Board of The American Institute of Physics from 1957 to 1963. He is a member of the Board of Roanwell Corporation, and Hadron, Inc., and has been Chairman of the Board of Trustees of Western College for Women, and Board member of Argonne Universities Association and the Atomic Industrial Forum. He is author of five books: SOUND WAVES AND LIGHT WAVES 1965, Doubleday), LASERS AND HOLOGRAPHY 1969, Doubleday), SEEING SOUND (1971, Wiley), RADAR, SONAR AND holography (1973, Academic Press), and ENGINEERING APPLICATIONS OF LASERS AND HOLOGRAPHY (1975, Plenum Press).

H. KOGELNIK

Herwig Kogelnik (M'61) was born in Graz, Austria, in 1932. He received the Dipl. Ing. and Doctor of Technology degrees, both from the Technische Hochschule Wien, Vienna, Austria, in 1955 and 1958, respectively, and the Ph.D. degree from Oxford University, Oxford, England, in 1960.

I. N. KOMPANETS

I. N. Kompanets graduated from Moscow Physical Engineering Institute in 1968, and received the degree of the Doctor of Physics and Mathematics in 1973. Since 1967 he has been working at P. N. Lebedev Physical Institute, USSR Academy of Sciences. At present, he is a senior researcher and the head of a research group. Dr. Kompanets is engaged in the study of opto-electronic materials, the construction, investigation of space light modulators and their applications for optical information processing.

A. M. KONDRATENKO

Anatoly M. Kondratenko graduated from the Novosibirsk State University (Faculty of Physics) in 1967, where he continued in 1967 as a staff member, and from 1969-1972 as a post-graduate student, doing his thesis in 1974 under the supervision of Prof. A. N. Skrinsky and Dr. Ya. S. Derbenev); since 1975 he has been a senior research staff member at the Institute of Nuclear Physics (Novosibirsk). His work is devoted to the theory of radiation polarization of electrons and positrons, particle spin dynamics in the storage rings and accelerators, and X-ray holography.

V. P. KORONKEVICH

In 1950 Dr. Voldemar P. Koronkevich graduated from the Leningrad Institute of Precision Mechanics and Optics. He worked at D. I. Mendeleyev State Institute of MEtrology in Leningrad. In 1956 he was given a doctor degree for his thesis "Air Dispersion in Visible Spectrum Region." In 1957 he headed the laboratory of Interference Measurements at the Siberian Institute of Metrology (Novosibirsk). Since 1968 he has been heading the laboratory of Coherent Optics of the Institute of Automatics and Electrometry of the Siberian Branch of the USSR Academy of Sciences in Novosibirsk.

He has more than 50 publications on interference measurements, high precision laser interferometers for measuring length and gravity acceleration, laser Doppler systems for determining flow velocities in fluids and gases. He is an author of the book LASER DOPPER VELOCITY METERS.

P. E. KOTLYAR

Pyotr E. Kotlyar graduated from the Novosibirsk Electrotechnical Institute in 1963. Since 1969 he has been working at the

Institute of Automation and Electrometry (Novosibirsk, Siberian Branch of the USSR Academy of Sciences). He received his doctoral degree in technical sciences in 1972. He has more than 30 publications on weak magnetic field measurements and on space-time light modulators.

A. E. KRASNOV

Andrey E. Krasnov received his BS degree from the Moscow Physical Engineering Institute in 1972. He has 7 papers and patents on the problems of holographic processors. At present he is with the Institute of Control Sciences, USSR Academy of Sciences, where he is completing his Ph.D. thesis on the logical properties of volume holograms and their application to control systems.

B. E. KRIVENKOV

Boris E. Krivenkov graduated from the Novosibirsk State University in 1974. Since 1974 he has been working at the Institute of Automation and Electrometry (Novosibirsk, Siberian Branch of the USSR Academy of Sciences). He has publications in the field of optical information processing.

S. H. LEE

Sing H. Lee is an Associate Professor of Applied Physics and Electrical Engineering at the University of California, San Diego. Prior to his current position, he was an Assistant/Associate Professor at Carnegie-Mellon University. He received his Ph.D. degree from the University of California, Berkeley and is a member of the IEEE, OSA, SPIE, PHI KAPPA PHI, SIGMA XI, TAU BETA PI and ETA KAPPA NU.

A. E. MEERSON

Aleksandr E. Meerson graduated from the Novosibirsk State University in 1970. Since that time he has been working at the Computing Center of the Siberian Branch of the USSR Academy of Sciences in Novosibirsk. He has more than 10 publications on computing mathematics and geophysics.

V. N. MOROZOV

V. N. Morozov graduated from the Moscow Physical Technical Institute in 1962 and received the degree of the Doctor of Physics and Mathematics in 1966. He has been working at the P. N. Lebedev Physical Institute, USSR Academy of Sciences, since 1962. At present he is a senior researcher and the head of a research group. Dr. V. N. Morozov is engaged in laser applications in optical information processing.

Y. E. NESTERIKHIN

Professor Yuri E. Nesterikhin graduated from M. V. Lomonosov Moscow State University in 1953. Until 1961 he was engaged in thermonuclear problems at I. V. Kurchatov Institute of Atomic Energy. From 1961 to 1967 he headed the department of plasma physics of the Institute of Nuclear Physics of the Siberian Branch of the USSR Academy of Sciences. Since 1967 he has been Director of the Institute of Automatics and Electrometry of the Siberian Branch of the USSR Academy of Sciences (Novosibirsk). In 1970 he was elected a corresponding member of the USSR Academy of Sciences.

Professor Nesterikhin has more than 90 publications on the problems of plasma physics, controlled thermonuclear synthesis, diagnostics of plasma, optical data processing holographic memories.

Professor Nesterikhin is an author of the book "Methods of High-Speed Measurements in Hydrodynamics and Plasma Physics." He is the chairman of the Council on Automation of Scientific Researches , the editor-in-chief of the journal "Avtometria" and heads the department of Automation of Physico-Technical Measurements of the Novosibirsk State University.

E. S. NEZHEVENKO

Dr. E. S. Nezhevenko graduated the Leningrad Shipbuilding Institute in 1962. He then worked as a teacher at the Polytechnical Institute (the Chair of Automatics). Since 1966 he has been working as a researcher at the Institute of Automatics and Electrometry of the Siberian Branch of the USSR Academy of Sciences (Novosibirsk). In 1974 he obtained his doctoral degree. Since 1968 he has been engaged in the problems of optical information processing. At present his fields of interest are coherent-optical methods of recognition, nonlinear processing, improvement of the image quality and investigation of optical feedback systems. He is the author of about 40 scientific publications and inventions. He is heading the group dealing with the problems of analog coherent-optical processors.

I. V. PRANGISHVILI

Ivery V. Prangishvili received his BS degree from the Georgian Polytechnical Institute in 1952, his Ph.D. from the Institute of Control Sciences in 1968 and a professorship in 1969. He has published about 60 papers on different aspects of control sciences, computer architecture and optielectronics. He is the author of three books: "Microelectronics and Iterative Structures in Logical and Computing Devices," 1967; "Iterative Microelectronic Associative Processors," 1973 and "Digital Automata with Rearrangeable Structure," 1974. At present he is with the Institute of Control Sciences, USSR Academy of Sciences as a deputy director.

W. K. PRATT

William K. Pratt received the BS degree in electrical engineering from Bradley University, Peoria, Ill., in 1959, and the MS and Ph.D. degrees in electrical engineering from the University of Southern California, Los Angeles, in 1961 and 1965, respectively.

He received Masters and Doctoral fellowships from Hughes Aircraft Company, and was employed there from 1959 to 1965. He became an Assistant Professor of Electrical Engineering at the University of Southern California in 1965, an Associate Professor in 1969, and a Full Professor in 1975. In this capacity he is presently concerned with teaching and research in the areas of image processing and laser communications. Dr. Pratt is the former Director of the University of Southern California Image Processing Institute and of the Engineering Computer Laboratory. In 1976 he was awarded a Gugenheim fellowship for research in image analysis techniques.

Dr. Pratt is a member of Sigma Tau, Omicrom Delta Kappa, Sigma Xi, and the Optical Society of America.

V. G. REMESNIK

Vladimir G. Remesnik graduated from the Novosibirsk State University in 1972. Since that time he has been working at the Institute of Automation and Electrometry (Novosibirsk, Siberian Branch of the USSR Academy of Sciences). He has more than 15 publications on solid-state physics and holography.

A. M. STCHERBATCHENKO

Anatolii M. Stcherbatchenko graduated from the Novosibirsk Electrotechnical Institute in 1962. Since 1964 he has been working at the Institute of Automation and Electrometry (Novosibirsk, Siberian Branch of the USSR Academy of Sciences). He received his doctoral degree in technical sciences in 1971. He has more than 25 publications on laser interferometer signals electronic processing. At present he leads the group on program-controlled moduli of the interference coordinate-measuring system.

A. N. SKRINSKY

Alexander N. Skrinsky graduated from Moscow State University (Faculty of Physics) in 1959. From 1955 he was on the staff of the Institute of Nuclear Physics (Novosibirsk). From 1962 he was the head of the colliding beam laboratory. He became Academician of the USSR Academy of Sciences in 1970. Main subjects of his work are devoted to developing the colliding beam method for experiments on high energy physics; theoretical and experimental studies of colliding beam behavior in storage rings; study of the quantum electrodynamics applicability at high energies; experiments on hadron production in electron positron collisions; theoretical and experimental development of methods for obtaining the polarized charge particle beams in the storage rings and their application in elementary particles physics; development of electron cooling method designed for storing antiprotons and carrying out the proton antiproton colliding beam experiments; use of synchrotron radiation of electron storage rings and, in particular, X-ray holography.

V. S. SOBOLEV

Victor S. Sobolev graduated from the Odessa Polytechnic Institute in 1952. Since 1959 he has been working at the Institute of Automation and Electrometry (Novosibirsk, Siberian Branch of the USSR Academy of Sciences). He received his doctoral degree in technical sciences in 1964. He has more than 40 publications on laser Doppler velocity guages. At present he heads the laboratory on research into gas and liquid streams using laser methods.

B. I. SPEKTOR

Boris I. Spektor graduated from the Novosibirsk State University in 1973. Since 1976 he has been working at the Institute of Automation and Electrometry (Novosibirsk, Siberian Branch of the USSR Academy of Sciences). He has more than 10 publications on optical systems for image processing.

G. W. STROKE

Dr. George W. Stroke obtained his Ph.D. in Physics from the Sorbonne in Paris in 1960, and is currently Professor of Electrical Sciences and Medical Biophysics at the Stony Brook campus of the State University of New York and Director of its electro-optical sciences laboratory. Concurrently, since 1970, he has served for three consecutive years as Visiting Professor of Medical Biophysics at Harvard University Medical School.

Dr. Stroke was previously Professor of Electrical Engineering at the University of Michigan and Head of its electro-optical sciences laboratory which he founded in 1963. Before joining the University of Michigan, Professor Stroke spent 10 years at MIT where he did research work, principally devoted to the development of the method of interferometric servo-control of grating ruling, in a collaborative effort with Dean George R. Harrison. This has earned them world fame. During his tenure at MIT, he also helped in originating a method of velocity of light measurement using microwave-cavity resonance, and participated in the Office of Naval Research Fleet Ballistic Missile (Polaris) program at the Instrumentation laboratory there.

Professor Stroke's work in coherent optics and holography originated with his work in the Radar Laboratory at the University of Michigan where he helped in initiating the work on three-dimensional "lensless photography" as a consultant in 1962-1963. He wrote the first treatise on the subject under the title, _An Introduction to Coherent Optics and Holography_ (Academic Press, 1966) which was immediately translated into Russian (MIR, 1967) and appeared in its second (enlarged) U.S. edition in 1969.

In recent years, Professor Stroke has been devoting his primary research interests increasingly to two new fields: optical and digital information processing and to the life sciences, including the development of new methods of image deblurring and communications and to the improvement of high-resolution electron microscopy. The method of holographic image deblurring which he originated in 1965 has recently permitted him, with his team at Stony Brook, to sharpen up electron micrographs of virus test specimens to a degree considered unattainable in the past, as well as to extract seemingly irretrievable information from accidentally blurred photographs, notably also of archeological specimens.

In addition to _An Introduction to Coherent Optics and Holography_, Dr. Stroke published another book (at the age of 24) as well as the 320 page "Diffraction Gratings" section of the _Handbuch der Physik_ (Springer Verlag, Vol. 29, 1967) and approximately 100 scientific papers including about 50 on holography. He is also editor of Ultrasonic Imaging and Holography--Medical, Sonar and Optical Applications (Plenum Press, 1974). A widely traveled lecturer on the subject of holography and its scientific, industrial and biomedical applications, Dr. Stroke has served in a number of United States government and other advisory capacities including, most recently as a member of the National Science Foundation Blue Ribbon Task Force on Ultrasonic Medical Diagnostics and as a consultant to the American Cancer Society. In 1971 he served as U.S. delegate to the Popov Society meeting in Moscow under the U.S. State Department Scientifc Exchange program. In 1972 he was invited by the Japan Industrial Technology Association to officially advise the Japan Ministry of International Trade and Industry on its program of large-scale development of computer pattern recognition technologies. For several years he has also been assisting the National Science Foundation in its U.S.-Japan and U.S.-Italy science cooperation programs.

Dr. Stroke was the U.S. Coordinator of the NSF-sponsored U.S.-U.S.S.R. Science Cooperation Symposium on "Optical Information Processing" held in Washington, D.C. (16-20 June, 1975) and the post-conference visits, coordinated with the U.S. Department of State, to Bell Laboratories, M.I.T. and the IBM Research Laboratories in Yorktown Heights. He is the coordinator of the Seminar Proceedings. Among his most recent scientific publications are "Retrieval of Good Images from Accidentally Blurred Photographs" (SCIENCE, July 25, 1975) and "Optical Holographic Three-Dimensional Ultrasonography (SCIENCE, September 19, 1975).

P. Y. TVERDOKHLEB

Dr. Peter Y. Tverdokhleb graduated from the Polytechnical Institute in Lvov (the Ukraine) in 1958 as an electrical engineer. Since 1958 he has been working as a researcher at the Institute of Automatics and Electrometry of the Siberian Branch of the USSR Academy of Sciences (Novosibirsk). He was engaged in the problems of investigation and development of analog-digital transducers as well as in the problems of processing the measurement results. In 1965 he was given a doctor degree.

In 1968 Dr. Tverdokhleb started investigations on developing optical (in particular holographic) information storing and processing devices. He concentrated attention on the development of space-noninvariant optical systems intended for image analysis, multichannel signal processing and associative retrieval. In this field he has about 35 scientific publications and inventions. He heads the laboratory of optical information processing.

A. A. VASILIEV

A. A. Vasiliev graduated from Moscow Institute of Electronic Engineering in 1971 and then worked at this Institute as a research worker, dealing with opto-electronic materials and liquid crystal devices. Since February 1977 he has been a research worker of P. N. Lebedev Physical Institute, USSR Academy of Science and is engaged in the field of optical information processing in real time.

M. P. ZAMPINO

Michael P. Zampino received his BS degree in 1968 with a major in economics and minors in finance and chemical engineering. He has also completed all requirements for an M.B.A. in economics at the Baruch College of the City University in New York. Prior to that, he graduated in 1962 from the famous Brooklyn Technical High School with a major in chemistry, having been born in Pittsburgh, Pa. in 1944.

His professional career has been entirely in finance and banking, with major responsibilities, in the recent years in computerized processing. He was a bank examiner with the U.S. Treasury Department from 1968 to 1970 in the Office of the Comptroller of Currency, before moving in 1970 to his present responsibilities at CITIBANK (then First National City Bank) where he is currently an Assistant Vice President in the Services Management Group (formerly the Operating Group). Among the innovations he helped to introduce to the CITIBANK (under the direction of Robert B. White, Executive Vice President and Head of the Operating Group) are the "TRACE" check-image processing system, a large random-access computerized document-image data base, used to process over 2.5 million checks at CITIBANK every day. Parts of the article on "Microimage Technology" from the present volume were also selected as a special feature in DATAMATION of October 1977, with permission from the Publisher.

V. A. ZVEREV

Vitalii A. Zverev graduated from the Gorky State University in 1950. In 1953 he received his first doctoral degree in physics and mathematics and in 1964 he received his second such degree. From 1952 to 1961 he worked at the Gorky State University, and from 1959 to 1976 at the Gorky Research Institute of Radiophysics. At present he is a vice-director of the Applied Physics Institute of the USSR Academy of Sciences (Gorky). He has more than 100 publications on radiophysics and optical methods for information processing.

AUTHOR INDEX

SUBJECT INDEX